Elizabeth

By the same author

This Land of England
The Reign of Henry VIII: Personalities and Politics
*Revolution Reassessed: Revisions in the History of Tudor Government and
Administration*
The English Court: From the Wars of the Roses to the Civil War
Rivals in Power; The Lives and Letters of the Great Tudor Dynasties
Henry VIII: A European Court in England
The Inventory of King Henry VIII
Six Wives: The Queens of Henry VIII

Elizabeth

The Struggle for the Throne

David Starkey

HARPER ● PERENNIAL

NEW YORK ● LONDON ● TORONTO ● SYDNEY

HARPER ● PERENNIAL

First published in Great Britain in 2000 by Chatto & Windus.

The first U.S. edition of this book was published in 2001
by HarperCollins Publishers.

P.S.™ is a trademark of HarperCollins Publishers.

HarperCollins books may be purchased for educational, business,
or sales promotional use. For information please write:
Special Markets Department, HarperCollins Publishers,
10 East 53rd Street, New York, NY 10022.

First Perennial edition published 2001.
Reissued in Harper Perennial 2007.

Library of Congress Cataloguing-in-Publication Data is available.

ISBN: 978-0-06-136743-4 (reissue)
ISBN-10: 0-06-136743-5 (reissue)

08 09 10 11 RRD 10 9 8 7 6 5

Contents

Illustrations

Greenwich Palace from the River Thames by Anthonis van den Wyngaerde (Ashmolean Museum, Oxford)

Henry VIII by Hans Holbein the Younger (Thyssen-Bornemisza Collection, Madrid, Spain)

Catherine of Aragon by Lucas Horenbout (National Portrait Gallery, London)

Anne Boleyn (National Portrait Gallery, London)

Jane Seymour by Hans Holbein the Younger (Kunsthistorisches Museum, Vienna, Austria/Bridgeman Art Library, London)

Catherine Parr (National Portrait Gallery, London)

Prince Edward (The Royal Collection © 2000, Her Majesty Queen Elizabeth II)

Princess Elizabeth (The Royal Collection © 2000, Her Majesty Queen Elizabeth II)

The Family of Henry VIII (The Royal Collection © 2000, Her Majesty Queen Elizabeth II)

Allegory of the Reign of Edward VI (National Portrait Gallery, London)

The Eve-of-Coronation Procession of Edward VI (Society of Antiquaries, London)

Thomas Seymour (National Portrait Gallery, London)

Mary Tudor by Hans Eworth (National Portrait Gallery, London)

Roger Ascham (Hulton Getty Picture Collection)

Anthony Denny, after Hans Holbein the Younger, engraved by W. Richardson (Private Collection/Bridgeman Art Library, London)

Thomas Parry by Hans Holbein the Younger (The Royal Collection © 2000, Her Majesty Queen Elizabeth II)

Hatfield Old Hall (© Mark Fiennes)

Thomas Wyatt the Younger (National Portrait Gallery, London)

Elizabeth's letter to Mary (Public Record Office)

The Struggle for the Throne

In 1567 Elizabeth I told a delegation of parliamentarians: 'I thank God that I am indeed endowed with such qualities that if I were turned out of the realm in my petticoat, I were able to live in any place of Christendom.'[1] Her hearers were startled by her language which, typically, was both unroyal and unparliamentary. But they knew that what she said was true.

She was then thirty-four years old. For nine of those years she had been Queen. But before that, from her birth in 1533 to her accession in 1558, she had experienced every vicissitude of fortune and every extreme of condition. She had been Princess and inheritrix of England, and bastard and disinherited; the nominated successor to the throne and an accused traitor on the verge of execution; showered with lands and houses and a prisoner in the Tower. She had known what it was like to be reduced to her shift when her father had forgotten about his three-year-old daughter in the excitement of his marriage to a new wife. She had also experienced the violent loss of her clothes when her stepfather, with her stepmother's connivance, had slashed off her dress with a knife. She had even contemplated exile and opened negotiations with the French ambassador. In and among all this she had been taught, with gentleness and great skill, the most advanced curriculum of the day. But it was her lessons in the school of life that mattered more.

Only two other English or British monarchs have had similarly dangerous and adventurous youths: Elizabeth's grandfather, Henry VII, who was a hunted exile for most of his twenty-eight years before winning the crown in hand-to-hand conflict with his rival on the battlefield; and Charles II, who was also a fugitive and exile for eleven years between his father's execution and the abolition of the monarchy in 1649 and the

Restoration in 1660. But Elizabeth's reaction was different from theirs. Henry VII's experiences bred in him a corrosive and destructive suspicion; Charles II's, an equally corrosive cynicism. Elizabeth similarly learned distrust, double-dealing and a swirling obfuscation of language in which the more she said the less her meaning was clear. That, bearing in mind the circumstances, was inevitable. But, above all, she learned a sort of humanity.

And it is this humanity—in contrast to the imperial certainties of her father or the religious extravagances of her brother and sister—which has attracted me to her and explains the writing of this book.

In this I am no different from my peers. Almost all her historians fall a little in love with Elizabeth. But most (there is no accounting for tastes) fall in love with the other Elizabeth, the bewigged and beruffed Gloriana of the later years of the reign, who, with her face caked in carmine and white lead, ruled over her court like an English Turandot, with a mixture of fear and love and incense.

This Elizabeth interests me but rather repels me. Instead the woman I have half fallen in love with is the young Elizabeth as she appears in the picture she gave to her father just before his death. Here she is pale too. But it is the natural pallor of her redhead skin. She is also serious, scholarly and self-possessed. But the painfully thin shoulders, exposed by the low, square-cut dress, suggest an aching vulnerability as well.

I still remember the first time I saw a reproduction of this picture. I glanced at it casually, admiring its marvellous painterly qualities. Then I noticed the caption and was taken aback. For I had not realized that it was Elizabeth. Many readers, I am sure, will have the same reaction when they see it for the first time on the jacket of this book: is it really her?

It is.

But it is not the monstrous mannequin of her maturity. Nor is the Elizabeth of this book the familiar queen either. Most of her biographers treat the twenty-five years of her childhood, youth and apprenticeship as a preface to her reign. Here I devote most of my book to them –

beginning at the beginning with Elizabeth as the product of the most famous and momentous divorce in history. As such, the fractured relationships of a broken family conditioned her experience and shaped her world. Too often, these relationships are understood in terms of the *a priori* assumptions of pop-psychology or proto-feminism. Instead, I revaluate them against the evidence of Elizabeth's own words and behaviour.

I then look afresh, and with new or newly interpreted evidence, at the development of Elizabeth's taste and learning; at her religiosity and personal loyalties; and – most originally – at her power and property which made her at sixteen the head of a great princely household and following. Also more or less new is my discussion of the function of court ceremony in defining the role of royal women, on the one hand, and the nature of royal religion, on the other. This, put in the abstract, sounds dull. Conveyed, as it is, through accounts of actual ceremonies as diverse as royal baptisms and funerals, coronations and maundies, it is both exotic and exciting.

Finally, I re-examine Elizabeth's unwished-for position as leader of her sister Mary's (dis)loyal opposition and try to see how, if at all, we can reconcile Elizabeth's protestations of her own loyalty with the fact that members of her household were up to their necks in every plot against the régime. The result is the rediscovery of a host of extravagant characters and madcap schemes. Particularly striking in all this is the role of women. Women schemed and plotted with the best, and the ones who took most risks were those who had already thrown over the restraints of marriage and family respectability. For them, Elizabeth, unwed and already ambivalent about the whole idea of marriage, was a model and patroness and her household a refuge from the world of men and masculine domination.

But, above all, I never forget that the years of Elizabeth's apprenticeship are a wonderful adventure story. We know they had a happy ending and that she survived and became queen. Elizabeth herself, her friends and enemies, had no such foresight. I try, as far as possible, to suspend our hindsight as well, and to tell things as they happened, with cliffhangers and narrow escapes. If the result reads like a historical thriller, I shall be well pleased.

But it is more than a thriller. For these are the years of apprenticeship of one of England's greatest rulers. The experiences I describe are what made Elizabeth human, and, as I have already suggested, humane. They are also what made her great. In comparison, the apparatus of Gloriana, on which so much ink and ingenuity have been spent, is flummery and superstructure. Here is what matters. Here are the experiences to which Elizabeth herself returned, time and time again, in each great speech, as in 1567, and at every crisis of the reign.

And here, therefore, is the intellectual and historical justification of the book. I present the argument in the last seven chapters. Four deal in detail with the events of her accession: with the choosing and establishment of her council; with the immensely subtle process by which the butterfly of a Protestant England was shaped inside the chrysalis of the dead, Catholic Marian régime; and with the legislative cliff-hanger of the Elizabethan religious settlement. But then I look at the limits that Elizabeth consciously imposed on Reform and at the promises she made in her first parliament and kept thereafter, against great difficulties, in a reign of forty-five years. These promises can be summarized in two sentences: 'I will not be like my sister Mary. And I will not rule like her either.'

But already I anticipate. The subtitle of this book is 'The Stuggle for the Throne'; the full, equally wonderful story of how she occupied it belongs to another, subsequent volume.

David Starkey
London,
March 2000

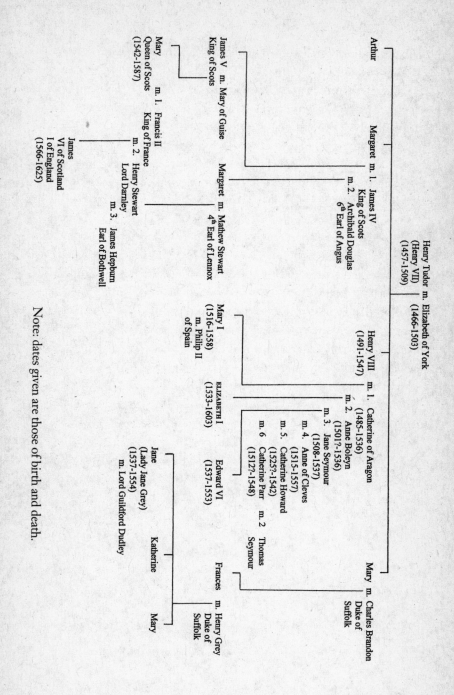

Henry Tudor m. Elizabeth of York
(Henry VII)   (1466-1503)
(1457-1509)

Arthur

Margaret m. 1. James IV
               King of Scots
          m. 2. Archibald Douglas
               6th Earl of Angus

James V m. Mary of Guise
King of Scots

Mary
Queen of Scots
(1542-1587)

      m. 1. Francis II
             King of France
      m. 2. Henry Stewart
             Lord Darnley
      m. 3. James Hepburn
             Earl of Bothwell

James
VI of Scotland
I of England
(1566-1625)

Margaret m. Mathew Stewart
               4th Earl of Lennox

Henry VIII
(1491-1547)   m. 1. Catherine of Aragon
                     (1485-1536)
               m. 2. Anne Boleyn
                     (1501?-1536)
               m. 3. Jane Seymour
                     (1508-1537)
               m. 4. Anne of Cleves
                     (1515-1557)
               m. 5. Catherine Howard
                     (1525?-1542)
               m. 6   Catherine Parr   m. 2 Thomas
                     (1512?-1548)   Seymour

Mary I
(1516-1558)
m. Philip II
of Spain

ELIZABETH I
(1533-1603)

Edward VI
(1537-1553)

Mary m. Charles Brandon
          Duke of
          Suffolk

Frances   m. Henry Grey
                Duke of
                Suffolk

Jane
(Lady Jane Grey)
(1537-1554)
m. Lord Guildford Dudley

Katherine

Mary

Note: dates given are those of birth and death.

Elizabeth

CHAPTER 1

Birth

Elizabeth, daughter of Henry VIII and Anne Boleyn, was born on Sunday, 7 September 1533 at 3 o'clock in the afternoon. It was an easy birth: mother and daughter were well and the child took after her father with his fair skin and long nose. But she had her mother's coal-black eyes.[1]

These are the ordinary, human details that might characterize the birth of any baby. But Elizabeth was royal. That meant that her entry into the world was vested with ceremony and hopes that went far beyond the ordinary. Indeed, as far as the hopes were concerned, they went far beyond what was usual even for a royal birth.

Royal births, like other royal events, great and small, from marriages and deaths to dressing and dining, were the object of an elaborate ceremonial. This was set out in the handbook of court etiquette known as *The Royal Book*. The ceremonies were already old when the Tudors came to the throne, though with his love of display, Elizabeth's grandfather, Henry VII, had added a few finishing touches. The result combined religious and courtly ceremony; it hid the pregnant queen like a mystery and it paraded the new-born infant like a pageant. For successive generations of Plantagenets, Yorkists and Tudors, this entry into the world had lent a little magic to even the briefest royal lives. In the case of Elizabeth, it formed a magnificent prologue to the superb royal performance that was to be her reign.[2]

The preparations had got underway in earnest in early August when it was decided that the birth would take place at Greenwich. This was the lovely, Thames-side palace where, forty-two years before, Elizabeth's father, Henry VIII, had been born. It was the favourite palace of his mother, Elizabeth of York, and it was to become his and his daughter's favourite too.

First, the Queen's bedchamber was prepared for her confinement. The walls and ceiling were close hung and tented with arras – that is, precious tapestry woven with gold or silver threads – and the floor thickly laid with rich carpets. The arras was left loose at a single window, so that the Queen could order a little light and air to be admitted, though this was generally felt inadvisable. Precautions were taken, too, about the design of the hangings. Figurative tapestry, with human or animal images, was ruled out. The fear was that it could trigger fantasies in the Queen's mind which might lead to the child being deformed. Instead, simple, repetitive patterns were preferred. The Queen's richly hung and canopied bed was to match or be *en suite* with the hangings, as was the pallet or day-bed which stood at its foot. And it was on the pallet, almost certainly, that the birth took place.

Carpenters and joiners had first prepared the skeleton by framing up a false ceiling in the chamber. Then the officers of the wardrobe had moved in to nail up and arrange the tapestry, carpets and hangings. At the last minute, gold and silver plate had been brought from the Jewel House. There were cups and bowls to stand on the cupboard and crucifixes, candlesticks and images for the altar. The result was a cross between a chapel and a luxuriously padded cell.[3]

By the third week of August all was ready, and on the 26th there took place the ceremony of the Queen's 'taking her chamber'.[4] First, she went in procession to the Chapel Royal and heard mass. The company then returned to the Queen's great chamber, which was the outermost room of her suite. There, standing under the cloth of estate or canopy that was the mark of her rank, she took wine and spices with the assembled company. Her lord chamberlain now called on everyone present to pray that 'God would give her the good hour', that is, a safe delivery. Another procession formed and accompanied the Queen to the door of her bedchamber. At the threshold, the males of the court took their leave of her and only her women entered.

Her confinement had now begun. The Victorians used the word as a euphemism, but the etiquette of the English court confined a pregnant queen indeed in a sort of purdah. Thenceforward, until the birth and her 'churching' thirty days after, she dwelt in an exclusively female world, attended solely by women.

These ceremonies were ambivalent. They emphasized that childbirth was a purely female mystery. And they paid the tribute of the dominant male world to that mystery. But they did so on strict conditions: the queen, literally, had to deliver. They also underscored how inconceivable, how monstrous even, was the notion of an unmarried and childless queen. For a queen was a breeding machine, or, as the Spanish ambassador put it only a little more elegantly, 'the entire future turns on the accouchement of the queen'.[5] Elizabeth's career was to mount a magnificent challenge to this received wisdom; her mother's, on the other hand, was to be an awful example of its truth.

But at least on this occasion, Anne Boleyn did deliver, going into labour less than a fortnight after having taken to her chamber. Immediately, work was started to prepare the very different stage-set that was needed for the christening. This was as public as the confinement was private. It was to take place, not in the Chapel Royal, which lay at the east end of the river façade, but in the Church of the Observant Friars, which was situated at some distance from the main palace, to the north-west. Once again Elizabeth was following in her father's footsteps, as this was also where he had been christened. The chosen route, which led from the great hall to the west door of the church, was turned into an outdoor corridor. Holes were dug for posts on which were mounted frames and rails and the whole was hung with tapestry. In view of the distance, hundreds of pieces must have been used. Inside the church was built a large octagonal stage, three steps high, with railings and a reinforced section in the middle to carry the weight of the great silver font, which was brought specially from Canterbury, as it had been for every royal christening since that of Edward, Prince of Wales, the ill-starred son of Henry VI and Margaret of Anjou. Time was short, and the men worked on Sunday and the Feast of the Nativity (getting double pay for so doing). In just three days and with only hours to spare, they had finished. It remained only to conceal what they had done with rich fabrics. The woodwork of the stage was covered with red and blue cloth; the church and the west doors were hung with cloth of gold; even the font was lined with soft linen lest the metal abrade the tender skin of the royal child.[6]

The following day was the 10th. Early in the morning the christening procession assembled. The heralds carried their tabards. Attendants and serving men bore unlighted torches. Lords and ladies carried the equipment needed for the ceremony: a gold cellar of salt, for the exorcism of the child; great silver gilt basins in which the godparents could wash off traces of the holy oil with which the child was anointed; a chrisom-cloth, to be bound over the crown of the baby's head after she had been anointed with chrisom; and a taper, to be lit after the baptism was completed. Elizabeth herself was carried by a duchess, and her long train by three peers and peeresses of the royal blood. Four barons carried a canopy over her. She was christened by the Bishop of London, and the Archbishop of Canterbury was her godfather. Three times the baby was plunged in the waters of the font. Then the lighted taper was thrust in her hand. At this moment, all the torches were lit; the heralds put on their tabards and the trumpets rang out in honour of Elizabeth, Princess of England and of France.[7]

But still, despite the preparations and the clockwork perfection of the ceremony, all was not quite right. Some hated the fact that she had been born at all. One of the officiating clergy was asked whether the baby had been baptized in hot or cold water. 'Hot,' he replied, 'but not hot enough.' And everybody, including her own father and mother, agreed that she was the wrong gender. What made it worse was that all the best astrologers and doctors had predicted a boy. Henry VIII had spent the last anxious days of his wife's confinement pondering whether to call his forthcoming son Edward or Henry and organizing the celebratory tournament. The Queen herself had the letter announcing the birth written in advance. Confidently, it gave thanks to God for sending her 'good speed, in the deliverance and bringing forth of a prince'.[8]

In the event, the King had to choose another name (he opted for his mother's). The joust was cancelled. And the letter was hurriedly altered by the addition of an 's', so that God was thanked for the birth of a 'princes'.

Nobody was deceived. It was not for this, for a daughter, that Henry VIII had risked his kingdom by divorcing his first wife, and imperilled his immortal soul by breaking with Rome. It was for a son.

CHAPTER 2

Family

One of the many ironies in the life of Elizabeth is that she, who was to be the Virgin Queen, had a father who married six times. There are two reasons for a man to marry so often. Either he does not take marriage seriously enough or, alternatively, he takes it far too seriously and expects too much from it. Henry VIII, though we think of him as a woman-devouring Bluebeard, in fact belonged to the second category, and spent his life in two quests: for a son and for a happy marriage, of which a son was an integral part. The repercussions of his quests go far to explaining Elizabeth's eventual attitude, both to marriage and offspring.

Henry's attitude was rooted in his own experience of family. For he had a remarkably happy childhood. As a second son, he was brought up with his sisters. The burdens and responsibilities were thrust on his elder brother, Prince Arthur, instead. Arthur was brought up alone, at a distance from the rest of the family, and was subjected to a punishing programme of scholarly and political training to fit him for the throne. As less was expected of Henry, fewer demands were made on him, and the dominant influence on his upbringing was his mother, Elizabeth of York, who was intelligent, beautiful and, in a royal family of stridently assertive personalities, a reconciler and healer.[1]

But in 1502 Arthur suddenly died. The following year Elizabeth of York died in childbirth, desperately trying to have another son. The baby, a girl as it happened, died too. Henry, aged twelve, was now both motherless and heir to the throne. On the single thread of his life hung the whole future of the Tudor dynasty. The pattern of his life changed accordingly. He was summoned to court, where his widowed father watched over him obsessively.

Henry's father, Henry VII, was a very different personality from his mother. He had been hardened by years of exile. Now he was ageing rapidly and, with the death of his wife, he had lost the main stabilizing influence in his life. Predictably, tensions between father and son simmered, and Henry greeted his death in 1509 with scarcely disguised jubilation: he was not yet eighteen; he was king and he had the whole world at his feet.

Strikingly, his first act (after overthrowing his father's two principal ministers and disavowing their policies) was to get married. His bride was Catherine of Aragon, daughter of the 'Catholic Kings' of Spain, Ferdinand and Isabella.

By their marriage, Ferdinand and Isabella had united their two hitherto warring kingdoms of Castile and Aragon. It was a parallel achievement, though on a much greater scale, to that of Henry's parents. For the marriage of Henry Tudor, the heir of Lancaster, and Elizabeth of York, had been the first, vital step in the ending of the Wars of the Roses. So it was natural that the two new, successful régimes should enter into an alliance that was both diplomatic and dynastic. But Catherine's husband-to-be was not Henry but his elder brother, Arthur.

After protracted negotiations between the two sets of parents, Catherine had embarked for England in the autumn of 1501. Symbolically, she sailed in the teeth of tremendous storms.[2] The marriage of Arthur and Catherine took place in November in Old St Paul's in London. A huge, elevated walkway was built from the west doors to the high altar, with a platform on which the wedding itself took place. Both Arthur and Catherine wore white, and, before hearing Mass in the sanctuary, turned to wave to the people. The crowd of thousands went wild. But it was Henry, aged only ten, who stole the show. He escorted the bride to the ceremony, and, at the ball afterwards, threw off his gown to dance more freely in his doublet and hose. His parents applauded indulgently.[3]

Soon after the wedding, the groom returned with his bride to Ludlow to resume the government of his principality of Wales. The following spring was unusually cold and wet, even for the Welsh Marches. Arthur became ill, worsened rapidly and on 2 April died. His widow, numb with shock and sick herself, was carried back to England in slow stages in one of the Queen's litters, heavily trimmed in mourning black. But grief was not allowed to get in the way of business. The alliance between England and Spain was bigger than the life of a prince. And, in any case, there was now another Prince of Wales, Henry himself.

So it was quickly decided that Catherine should remarry with Henry. The couple were betrothed; the permission to marry, known as a dispensation, which was required from the Pope in view of Catherine's relationship with Henry as his brother's widow, was obtained; there was even a proxy marriage. But marriage proper was conditional on Henry's giving his consent once he became of the canonical age. Whilst his father lived, this consent was never given. Instead, there was perpetual bickering about Catherine's dowry, which left Catherine, still in England, in an undignified limbo. In theory she was Princess of Wales. But she was a princess without a prince, as Henry and Catherine were not allowed to live together. Worse, she was a princess without an income as her father and father-in-law made her a pawn in their disagreements about money. And worst of all, though unknown to her, she was a bride without a groom as, acting under his father's orders, Henry had secretly repudiated the marriage.

Catherine's position, like so much else in England, was transformed by Henry's accession: the difficulties over money, which had seemed so insurmountable while his father lived, were brushed aside in a headlong rush to the altar. Was Henry in love with Catherine? Or was he in love with the idea of being married? Did Catherine, with her gentleness combined with strength of character, remind Henry of his mother? Or was it that being married would affirm his full status as an adult – something which was very necessary in view of the tiresome fact that his council persisted in treating him like a minor, instead of the grown man of seventeen years and ten months that he knew himself to be. Or was it, most unromantically of all, that an alliance with Spain would come in useful in the war against France that he was determined to launch?

Probably, in view of the mixture of human motives, a bit of each was involved. But Henry certainly thought he was in love with Catherine. And Catherine knew that she was in love with Henry. It was her duty as a wife and daughter of Spain. It was also, bearing in mind the youthful Henry's charm, stunning good looks and boundless vigour, a pleasure. In the coming years, that love was to be sorely tested – on both sides. Catherine's survived; Henry's did not.

At first, however, everything seemed perfect. They were married privately on 11 June 1509 and, a fortnight later, on the 24th, were crowned together. It was Midsummer's Day. And indeed the young royal couple – he just eighteen, she twenty-three – seemed another Oberon and Titania, a fairy king and queen. However, it was not Fairyland but Camelot that provided the model for their court, as it plunged into a round of revelling and jousting in which the lead was soon taken by the King in person. Henry's starring role in the tournament split opinion. Youth cheered him on; age worried about the defiance of convention and the risk to the succession.

But the succession, too, seemed in hand. Within a month or two of the wedding Catherine apparently became pregnant, as on 26 February 1510 a warrant was issued to the royal storehouse and purchasing agency known as the Great Wardrobe. It ordered the keeper to supply materials for the Queen's taking to her chamber and also the royal nursery – 'God willing'. The minutest details were specified. The great cradle of estate was to be re-covered in red cloth of gold and the royal arms repainted to included Catherine's. Beds were to be provided for the nurse and the rockers (whose job, of course, was to rock the cradle), and hangings for the chamber of the Lady Mistress. The cradle was probably the one in which Henry himself had lain, and the Lady Mistress was certainly Mrs Elizabeth Denton, who had brought up Henry and his younger siblings.[4]

There are two versions of what happened next: the official version put out by Catherine herself; and the real story given in confidential despatches by her confessor and the Spanish ambassador. According to Catherine, she miscarried of a girl in May and Henry was wonderfully supportive. The truth was much more tangled. Catherine actually miscarried in January. But the fact was kept secret. Moreover, as she remained swollen, Catherine and her doctors persuaded themselves that somehow she was still pregnant. On the assumption that she was coming to term, she took to her chamber, with all the attendant ceremonies, whereupon the swelling disappeared. The King was furious; the council openly questioned whether she was capable of conceiving and the Spanish ambassador wondered how anybody could think they were pregnant while they were still menstruating.[5]

It was a tricky situation. Fortunately for Catherine, at this point she did become pregnant and was delivered of a boy on 1 January 1511. It was the best New Year's gift his parents could have received. The baby was christened Henry and his father mounted the most splendid tournament of his reign in which he rode as Catherine's 'Sir Loyal Heart'. Days later little Henry was dead.[6] Catherine's weary story of miscarriages and infant deaths resumed, punctuated by the birth of a single child who lived. It was a girl, Mary, who was born on 18 February 1516.

Relations between elder and younger sisters are often difficult – particularly when there is an age gap of seventeen years, as there was to be between Mary and her half-sister Elizabeth. But fate also cast them as opposites in appearance and character and opponents in religion and politics. Even at the time, Mary's birth raised as many problems as it solved.

Catherine of course regarded Mary, her only surviving child, with a simple, uncomplicated love. She also took charge of her upbringing. Religious and learned herself, she made sure that her daughter was given teachers and priests who inculcated her with the same qualities.[7] For her father, on the other hand, it was her role as his heiress that mattered.

This raised profound questions. England had no formal bar to female succession, such as existed in France. And all the rival lines that had fought for the throne in the Wars of the Roses based their claims on descent in the female line. But no woman had actually sat on the throne in her own right. The nearest example was the Empress Mathilda, daughter and heiress of Henry I, in the twelfth century. But it was not a happy precedent as Mathilda's claims led to decades of civil war and were made good only in her son, Henry II, the founder of the house of Plantagenet.[8]

How could Henry VIII avoid a similar fate for his daughter? For a decade after the birth, it has been argued, English foreign policy had a single principal goal: to find for Mary a husband who was sufficiently powerful to make sure that she succeeded unchallenged to the English throne. The quest, in France and Spain, proved abortive.[9] Instead, unprecedented measures were taken to try to solve the problem at home. In 1526 Mary was proclaimed Princess of Wales and sent off to rule the principality. It was the first (and, so far, the last) time that a woman had been given the title in her own right. And it was intended as a prelude to the equally unprecedented event of the accession of Mary as Queen Regnant. A few months previously, to provide a spare as well as an heir, the King also advanced his bastard son, the Duke of Richmond, to extraordinary honours in the north of England.[10]

Neither expedient, however, was really satisfactory. There were obvious difficulties about the succession of a bastard. Mary herself seemed the better bet. But then in 1527 her legitimacy was questioned as well. The doubter was one of the French envoys, in London to negotiate, yet again, about a possible marriage between Mary and a French prince. But, he asked, was Mary all that valuable a catch? Did not the fact that her father had married his brother's widow raise doubts about the validity of the marriage and therefore doubts about Mary's status as well? Similar anxieties had surfaced before, but they had been suppressed. Now, suddenly, they struck an answering chord in Henry VIII's mind. In Leviticus, God had threatened childlessness as the punishment for any that should transgress the law and marry his brother's widow. Henry interpreted his failure to have a son as the equivalent to childlessness. Therefore by extension, he reasoned, both he and his marriage were accursed.[11]

Why had doubts arisen so urgently, almost twenty years after the event? The continued intractability of the succession was certainly one reason. So, less honourably, was Catherine's altered appearance. Years of abortive pregnancies had taken their toll and her once pretty face and figure had become dowdy and shapelessly fat.[12] Finally, and least honourably of all, Henry had fallen violently in love with another woman – Anne Boleyn.

Anne Boleyn was as different as possible from Catherine, or at least the woman that Catherine had become. In family, she was of course at a disadvantage – though not as great a disadvantage as snobbish commentators, now as well as then, suggest. Instead of Catherine's unimpeachable royalty, Anne's ancestors in the male line were a characteristically English mixture of country gentlemen and great merchants; while in the female line she descended from the great Irish noble house of Butler and the greater English one of Howard. In all other respects, however, apart perhaps from the beauty of holiness, Anne outscored Catherine hands down. She was petite and slim, with pert breasts (which Henry worshipped as her 'pretty duckies') and deep, dark eyes and lustrous auburn hair. By sixteenth-century standards, her colouring was too dark and idiosyncratic for her to be a beauty. But what she lacked in conventional good looks she more than made up for in charm and character. For she had been schooled in the court of France. There she had acquired a mastery of the French language, style, dress-sense, and a certain *je ne sais quoi* in her dealings with men.[13]

    This almost clinical detachment marked her out at the English court, where there were lots of handsome young women who were only too willing to throw themselves at Henry's feet – or wherever else he required. One of these was Anne's elder sister, Mary, who had been Henry's mistress for a time. Mary had yielded quickly and the King had lost interest almost as quickly. Anne resolved not to make the same mistake. She made plain that although she would give Henry her heart she would not give him her body until they were married. It was the first time since the death of his father that anybody had dared to say 'no' to Henry. He found it puzzling at first, then fascinating. Finally, changing that 'no' into a 'yes' became a grand and all-consuming passion.[14]
We can trace the King's changing moods and Anne's management of them in the extraordinary series of love-letters he wrote to her. They have

finished up, by some quirk of history, in the Vatican Library. And they are extraordinary, not only for their contents, but also for the fact that they were written at all. Henry, though he was the most bookish English king since the so-called twelfth-century renaissance, hated the physical business of writing. And a glance at his heavy, awkward handwriting shows why. So there can be no more striking testament to his devotion to Anne than his willingness to fill so many pages with sighs of passion, scraps of court news and obscure intimate riddles and private letter-games that once spelled out their love but are now largely indecipherable.[15]

These letters are a uniquely intimate record of a royal love affair. But as that love was royal it had also to be played out on the largest, most public stage of Church and State. And circumstances combined to inflate it into a mixture of high comedy, profound tragedy and great epic. It was not normally thus. For usually the Pope, to whom such matters were referred, was smoothly understanding of the marital problems of monarchs. Moreover, Henry had been a conspicuously loyal son of the Church and had been rewarded, as recently as 1521, with the title of Defender of the Faith. Naturally, he felt that the Pope owed him a favour. But the Pope, Clement VII, faced a more pressing obligation. For in 1527, the year that Henry's affair with Anne had begun, the troops of Charles V sacked Rome, captured the Pope and established a Habsburg hegemony over Italy. And, unfortunately for Henry, Charles V was Catherine of Aragon's nephew.

Charles V – the lantern-jawed young emperor – had come to dominate Europe in a way that neither of his more showy contemporaries, Francis I of France and Henry VIII of England himself, could hope to do. Once again, marriage was the key. Charles's grandfather was Maximilian of Habsburg. He had high-sounding titles: he was Archduke of Austria by inheritance and Holy Roman Emperor by election. But in reality he was a second-division power. He took the first step to family greatness by his marriage to Mary of Burgundy, the heiress to the Netherlands. They had a son, Philip, surnamed the Fair. Between the Burgundian lands and France lay a long-standing enmity. Also hostile to France was the newly consolidated Spain of Ferdinand and Isabella. So it was natural for Ferdinand and Maximilian to ally, and to cement their alliance with a marriage between Joanna, Catherine of Aragon's elder sister, and Philip the Fair. England, also traditionally hostile to France, joined in the family party. But all calculations were upset when Juan, the only son and heir of Ferdinand and Isabella, died. This left Joanna as heiress of Spain. Philip the Fair made good her claims to Castile but then died too. Joanna, always neurotic, became mad, but her son Charles was sane enough. He was also now heir to the greatest inheritance in Europe. At first, its parts were ruled in Charles's name and their interests by his wily grandfathers, Ferdinand and Maximilian. But, with their deaths, the rights to Spain, the Netherlands and Austria all passed to Charles V. The only problem was making his rights good. Burgundy saw him as a native son and so welcomed him with open arms. But Spain, distant and xenophobic, where he had never stepped foot, was another matter. Ironically, it was his uncle by marriage, Henry VIII, who supplied both the money and the diplomatic leverage that smoothed his passage to Spain. Many times Henry must have cursed that decision, since Charles's power was to be the most effective challenge to his ambitions. Henry's daughter Elizabeth must have cursed harder (and she could curse with the best), since Charles's son Philip was to threaten both her life and her throne.

It was with the crisis in Henry's marital affairs that the Spanish pigeons first came home to roost. For Charles V made it clear that he would never allow the Pope to annul Henry's marriage to his aunt Catherine and permit him to marry his new love, Anne Boleyn. Henry tried every expedient. Rome, for its part, deployed all the delaying tactics for which it was famous. Rome hoped that Henry would lose interest in Anne. Henry hoped that Charles's grip on Italy would be broken by the French. Both trusted that something, anything would turn up. Nothing did. Charles's power endured, and so did Henry's love. Indeed, it was made all the stronger by the obstacles placed in its path: marrying Anne became a matter of satisfying his kingly power and pride as much as his sexual lusts.

Finally, Henry cut the Gordian knot. Frustrated of a settlement in

Europe, he decided to resolve things at home. The theory was dealt with by a commission of clerics and academics, led and managed by Henry himself. This commission discovered legal and historical precedents which showed (to its own satisfaction) that Henry himself was rightly head of the Church and that the Pope's claim to this position was a mere usurpation – albeit a thousand-year-old one. The practical details were left to parliament, which enacted a series of statutes that first restricted and then ended papal power and constituted instead Henry as Supreme Head on Earth of the Church of England.[16]

One element remained to be put in place. The divorce crisis had also turned England's foreign relations upside down. France, the historic foe, was confirmed as the new ally; the Habsburgs, the Tudors' traditional friends, became the new enemy. But would the friendship of Francis I of France extend to countenancing the promotion of Anne Boleyn from royal mistress to Queen Consort? In October 1532, Henry VIII and Anne Boleyn (newly created Marquess of Pembroke for the occasion) met Francis at Boulogne and were given a lavish reception. Queen Claude, as both a good Catholic and good wife, flatly refused to turn up to honour the usurping mistress. But the endorsement of the King of France and his sister, Margaret of Angoulême, was enough for Henry and Anne.

Confident now that the way was open for her to become Henry's wife, Anne's 'no' finally turned into a 'yes' and, perhaps on the voyage back to England, she allowed Henry to sleep with her for the first time. By December she knew she was pregnant. Events, so long in suspended motion, now moved at breakneck speed. On 25 January 1533, Henry and Anne were secretly married. This left the King temporarily bigamous. But on 23 May, the newly appointed Archbishop of Canterbury, Thomas Cranmer, who was a client of the Boleyns, used the new-found autonomy of the English Church to declare that the marriage of Henry and Catherine was null and void. On 1 June Anne was crowned Queen in a magnificently staged spectacle. And on 7 September, Elizabeth was born. The mountains had laboured and, it seemed, produced a mouse.[17]

CHAPTER 3

Infancy and Mother's Death

Despite the disappointment with Elizabeth's gender, her parents put a brave face on the matter. She was immediately proclaimed Princess, that is (in the absence of the hoped-for brother), heiress apparent to the throne. And she was moved into the specially prepared nursery suite at Greenwich. A chamber had been adapted for 'the lady princess' by building a wainscot screen. And a purpose-made joined or folding table had also been built for her nurse to roll her clothes on. At the head of her little establishment was her Lady Mistress, Margaret, Lady Bryan, the widow of Sir Thomas Bryan and the sister of the late Lord Berners. Lady Bryan had fulfilled the same role for Henry VIII's first daughter, Mary. Now she transferred her affections and skills with equal enthusiasm to the upbringing of Mary's successor and (so Mary thought) usurper, Elizabeth. A grateful Henry VIII recognized Margaret Bryan as a baroness in her own right.[1]

Then, in December, came the decision to set up an independent house and household for the little princess. The house chosen was Hatfield, situated in pleasant and airy countryside some twenty miles to the north of London, where a little court was established. This was normal procedure for a royal infant, but the peculiar circumstances of Elizabeth's family life were to give it a special importance. Years later, Elizabeth made the point herself when she quoted St Gregory to the effect that 'we are more bound to them that bringeth us up well than to our parents, for our parents do that which is natural for them – that is bringeth us into this world – but our bringers up are a cause to make us live well in it.' And indeed the members of her household, not of her family, were to be the principal influences on her as a child and young woman.[2]

Not, at first at least, that her parents were at all deficient in 'natural' affection. Indeed, her mother plainly adored her. Anne Boleyn took pleasure in giving Elizabeth the prettiest clothes and equipment: caps, which were made to measure, in purple and white satin, each covered in a caul or net of gold; needle ribbon to roll her hair; gold and silver trimmings for her 'little bed' or cradle, and fringes for her 'great bed', in which she lay in infant state when she received important visitors. In the spring of 1534, Anne came all the way to Hatfield to see Elizabeth and make sure that she was settled into her new life.[3]

At the end of March Elizabeth was moved to the rebuilt 'Prince's Side' at Eltham, which had been her father's principal boyhood home. It lay only five miles from Greenwich, and a fortnight later both the King and Queen paid a sort of official visit to their daughter. They found her, according to one of the courtiers who accompanied them, 'a goodly child as hath been seen, and her Grace is much in the King's favour, as a goodly child should be, God save her'. In October 1535, when she was two years old, there was a formal parental consultation on her progress, the upshot of which was that the King, 'with the assent of the Queen's grace, hath fully determined the weaning of my lady princess'. Lady Bryan was instructed accordingly.[4]

But behind the bright little princess there was another figure, who followed her like a gloomy and reluctant shadow. For there were two victims of the annulment of Henry VIII's first marriage. The first was Catherine herself, repudiated after twenty-four years of devoted married life. And the second was her daughter, Mary, who, at the vulnerable age of seventeen, found herself declared illegitimate. These personal blows were accompanied by demotions in status. Catherine was reduced from Queen, as Henry VIII's wife, to mere Princess Dowager, as Arthur's widow; while Mary was demoted from Princess and inheritrix of England to the Lady Mary, the King's bastard daughter. Both mother and daughter steadfastly refused to recognize either the loss of their formal titles or the transfer of these titles to their successors, Queen Anne and Princess Elizabeth. This was not mere standing on dignity but a desperate attempt to protect the self-identity that was so threatened by the actions of their husband and father. Catherine's resistance was serenely regal, even saintly; Mary's, understandably, had more than a touch of teenage hysteria. Against the two women the King's council wielded every weapon of humiliation and calculated insult. Worse, from Catherine's point of view, was the enforced absence of her daughter, whom she was not allowed to see even when she was dying. For Mary, however, the council devised an even sharper torment by requiring her to be perpetually in the presence of her successor Elizabeth.

The decision was taken at the same meeting which had resolved to set up Elizabeth in her own household. Mary's independent household, it was decided, should be dissolved. Instead, she and the rump of her servants were to be sent to join Elizabeth at Hatfield. And, to add insult to injury, charge of the joint household was given to Anne Boleyn's aunt and uncle, Sir John and Lady Shelton. The messenger nominated to convey the news to Mary was Thomas Howard, Duke of Norfolk. He was both England's premier peer and Henry VIII's leading general. And, when he confronted the outraged girl, he needed all his prestige and military firmness. Mary resisted to the point at which physical force was about to be used to move her. Then she condescended to enter the litter. (It was just as well that Mary's own sense of dignity spared her violence, for Norfolk's compunction would not have held him back: later, he was to order his servants to pinion his own wife.)

When Mary arrived at Hatfield, she set herself to be as difficult as possible. And, since she was a Tudor, she succeeded. She spent days in her chamber, weeping uncontrollably. She fell ill. When she convalesced, her peculiar dietary requirements set the household in turmoil and greatly increased its costs. And there was one thing she would not do. She would not acknowledge Elizabeth as Princess. She was, she explained, prepared to call her sister, just as she referred to the King's bastard son, the Duke of Richmond, as her brother. But she would never address her as Princess. And, despite Lady Shelton's threats (rather unwillingly administered), her father's studied neglect, the loss of her jewels and the refusal to serve her food in her chamber which forced her to dine in the hall, she stuck to her guns. Elizabeth was a highly precocious child, so it is just possible that some of her earliest memories were of Mary, red-eyed and resentful, stamping her foot as she refused to curtsy to her baby half-sister.[5]

We know these details of Mary's treatment thanks to the despatches of Eustace Chapuys, Charles V's ambassador in England. Charles V was furious to learn that his power, which had been able to block Henry VIII's divorce in Rome, did not extend to protecting his aunt and his cousin in England. So he let his ambassador off the leash. Chapuys proved an enthusiastic mastiff. He schemed and he plotted and he cajoled. Mary and Catherine were reinforced in their resistance to Henry's orders and the Habsburg embassy extended an open welcome to any dissident members of the English court. To judge from Chapuys, there seem to have been a lot of them – and in the highest places in the council and the privy chamber, which served as the King's personal staff. Some were undoubtedly sincere and were to pay with their lives; others were simply reinsuring themselves against a possible turn of the tide. It was a disturbing foretaste of the role of the Spanish embassy as a centre of espionage and intrigue under Elizabeth herself.

Chapuys's despatches also reveal something else that was to dog Elizabeth for the rest of her life. We think of her as treading the primrose path of popularity. But she also aroused a venomous personal hatred. It began in her cradle. For Chapuys, Elizabeth was 'the little bastard' and her mother, Anne Boleyn, 'the great whore'. Many others, at home and abroad, used the same language or worse. One reason for the loathing was an outraged sense of personal and dynastic loyalty. Chapuys and others were devoted servants of the Habsburgs and of Catherine and her daughter. In their view, Anne Boleyn had usurped a place that was rightly another's and so, though she was too young to know it, had her daughter. And they were determined that both mother and child should pay the price.

But there was an ideological aspect to the strength of feeling as well. Anne Boleyn has gone down in history as one of the great *femmes fatales*. It is not a role that we associate with fervent religious commitment. But Anne and her brother George, Lord Rochford, were enthusiastic sponsors of the new developments in religion. Both drew their inspiration from reformed circles in France. Rochford imported advanced religious works from there, one of which he translated and dedicated to his sister. Anne Boleyn had large, lavishly illustrated copies of the Psalms and other religious texts in French prominently displayed in her chamber, and encouraged her ladies to read aloud from them. The result was a deliberate contrast to the household of Catherine of Aragon. Catherine's household was famous for its orthodox, Catholic piety; Anne's court, she was determined to show, would be equally pious, but with a radical, Reforming edge.

There has been much dispute as to the extent of this radicalism. Predictably, Chapuys denounced Anne and her brother as out-and-out heretics and Lutherans. Most modern historians have preferred to label them instead as 'Evangelicals'. Lutherans mounted a challenge to the fundamentals of Catholic belief on the sacraments. 'Evangelicals', on the other hand, differed in approach, not doctrine. They emphasized the importance of preaching the word of God rather than relying on the ceremonies of the Church. Recently, a re-examination of Rochford's imported French books has suggested that Chapuys was more nearly right than wrong.[6]

The commitment of the Boleyns to Reform was partly temperamental. But it was also a logical consequence of their position. Catherine's own Catholic piety, added to the refusal of Rome to annul her marriage, meant that the defence of the Aragonese marriage became increasingly identified with fervent Catholic orthodoxy. And Catholic orthodoxy, for its part, flung its weight behind the defence of the Queen and her daughter when John Fisher, the most conspicuously devout of the bishops, put himself forward as Catherine's principal champion. None of this left the other side with much choice: the Boleyns had to back Reform; and the Reformers, on the whole, backed the Boleyn marriage. There were some fence-sitters, most prominently Henry VIII himself. He was proud of his orthodoxy and was profoundly unhappy at the identification of the 'right' attitude to his marriage with the 'wrong' beliefs in religion. But even he, to begin with, could not afford to be too fussy about his supporters.

So the lines of future political and religious division were emerging. They did not yet constitute factions, much less parties. Instead, there were nuclei – a family connection here, a network of like-minded friends and associates there – round which larger structures might form. And from the beginning, Elizabeth was at the centre of events. In the fullness of time, she would become a leader; even in the cradle she was a symbol.

These were the large issues raised by Anne Boleyn's apparent triumph. But, in view of human pettiness, small things counted as much or more. And what counted against Anne was her strident self-assertiveness. Instead of setting herself to win hearts and minds, she trod on toes – the more sensitive and powerful, the better. On one occasion, according to Chapuys, she drove the Duke of Norfolk from the room with a choice of words you would not use to a dog. This trait irritated her friends, outraged her enemies, and finally began to grate on Henry himself. Sparky independence was one thing in a mistress, quite another in a wife. So when she dared to remonstrate with him about some woman who had caught his eye, he told her to be silent, as her betters had been before her.[7]

All these tensions came together in the first months of 1536. On Friday, 7 January, Catherine of Aragon, who had been suffering from increasing ill health, died at Kimbolton. The news reached London on the Saturday. The King's first reaction was relief: 'God be praised,' he exclaimed, 'we are free from all suspicion of war!' By the Sunday, his mood had changed to heartfelt (and heartless) rejoicing. He entered the Chapel Royal clad all in yellow and Elizabeth, who was visiting the court for the Christmas festivities, was carried in to Mass, too, with the sound of trumpets. After dinner, Henry took his little daughter in his arms, 'like one transported with joy', and showed her off to the assembled company.[8]

Anne Boleyn rejoiced as loudly as her husband. But, in fact, Catherine's life had been one of the chief props of her own position: while his first wife lived, Henry could not be seen to be getting rid of his second. Anne Boleyn's other defence was the fact that she was again pregnant. But this defence was soon removed as well, for on 29 January, the day of Catherine's burial at Peterborough Abbey, Anne miscarried of a fourteen-week-old foetus. It was just possible to identify it as male.

Anne was reported to be terrified that she was about to repeat the experience of her predecessor. The rumours had already started that she could not bear another child to term. The King was in 'great distress'. And he had fallen for another woman, Jane Seymour.[9]

Anne's premonitions proved correct. And it seems that she sealed her fate by the method she chose to try to win Henry back: she flaunted her sexuality. At the beginning of their relationship, this tactic had driven him into her arms with jealous lust. Now, it provided the evidence that was used to bring her down on charges of multiple adultery and incest with her brother, a favourite musician and three Gentlemen of the King's Privy Chamber.[10]

Anne's terror was that, like the adulterous Guinevere in the *Morte d'Arthur*, she would be sentenced to be burned at the stake, and with no Lancelot to rescue her. Instead, Henry, as he boasted in her death warrant, was merciful. She was to be beheaded and not burned. Moreover, she who had been so French in life was executed in the French manner: kneeling upright with a sword, rather than prostrate on the block with an axe. 'I heard say the executioner was very good,' she remarked to the lieutenant of the Tower, 'and I have a little neck.' She was correct on both counts, and her head came off with a single stroke.[11]

CHAPTER 4

Childhood and Education

Anne Boleyn's death was a terrible blow for Elizabeth, and her father's role in it more terrible still. But how deep the wound went we do not know as Elizabeth never wrote or spoke a single word about it: her father's name was to be constantly on her lips, her mother's and her mother's death, never.

Was this silence the result of a repression of a trauma too hurtful for the conscious mind to acknowledge? Most writers have thought so. But, despite the pathetic story, told much later, of the accused Anne Boleyn holding out their daughter to Henry VIII in mute appeal, Elizabeth was almost certainly not at Greenwich at the time of her mother's arrest, but twenty miles away, at Hunsdon in Hertfordshire. Mary, we know, had been moved there from Eltham at the end of January and, as she was at this time a mere appendage of her younger sister's household, we can assume that Elizabeth had gone there too after spending the Christmas holidays at court. And at Hunsdon, airy and rural on its hill between three river valleys, the horror of the events in the capital was inevitably muted.[1]

There was some impact, of course, as the shower of lovely clothes which Anne Boleyn had lavished on her daughter suddenly dried up. This was the sort of thing which an alert three-year-old would notice and question. It also had a practical impact as the child quickly outgrew her now static wardrobe. Within a few weeks, she had literally nothing to wear. We can imagine her bitter, childish humiliation; we can also speculate that it was to avoid its repetition that the adult queen was to fill her wardrobes to overflowing with hundreds of dresses.

At this point, Lady Bryan wrote a lengthy letter to Thomas Cromwell, the royal Secretary. She made the point, very forcibly, about the child's shortage of clothes; she also asked for guidance about her status and how she was to be treated. This has been taken to mean that Elizabeth was in deep disfavour in the aftermath of her mother's death. Actually, it shows no more than that she had been temporarily forgotten, which, with the excitement of the King's new marriage to Jane Seymour, was hardly surprising. Such neglect was heartless. But it was not serious.

The rest of Lady Bryan's letter dealt with nannyish concerns – Elizabeth's problems with teething (which meant that Lady Bryan allowed her more of her own way than she thought really proper), and her general perfection apart from that – and the latest episode in Lady Bryan's quarrel with the male head of the household, Sir John Shelton. The issue this time was where Elizabeth should eat. Shelton wanted her to 'dine and sup every day at the board of estate'. Lady Bryan, on the other hand, wanted her to continue to eat in her chamber, apart from the occasional feast day. For Shelton, it was a question of economy; for Lady Bryan, a matter of making sure that the child ate and drank what was suitable for her age.

Lady Bryan, who, having brought up two Tudors successfully, was necessarily a formidable character, got Cromwell on-side and got her way.[2]

The quarrel is an interesting sidelight on the tensions within the joint household of the two royal sisters. More importantly, it shows that there was no real question of Elizabeth's losing her status as the King's daughter; the only issue was how soon she would assume it by dining in state.

The other, much more serious, dispute in the sisterly establishment was the one-sided one between Mary and Elizabeth. This, too, was appeased in the wake of Anne Boleyn's fall. Mary, who was actually tender-hearted to excess when issues of principle were not involved, quickly made her peace with Anne Boleyn's ghost and prayed that 'that woman' might be forgiven. Mary was also forced to end her quarrel with her father by a mixture of brutal applications of the stick and lavish promises of the carrot. By June, she was browbeaten into submission. One of the first fruits of a warmer relationship with her father was Mary's discovery that Elizabeth, far from being a monstrous little bastard, was actually a 'toward' child, who would be a joy to Henry. Thereafter, the two sisters lived amiably under the same roof, exchanging presents and doing favours to each other's servants. And, at times, they became really close.[3]

Also important in the all-round reconciliation was Henry VIII's new wife, Jane Seymour, who went out of her way to show favour to Mary, without forgetting Elizabeth. But Jane's great achievement was in giving birth to the longed-for son, Edward. As both male and unquestionably legitimate, the boy displaced both of his sisters from the succession and made the dispute for precedence between them otiose. The new pattern of family relations was symbolized by his christening: Mary acted as his godmother while Elizabeth (who had to be carried herself) took the second most important part and bore the chrisom-cloth. The pattern shifted again with Jane Seymour's sudden death a fortnight later from puerperal fever. Her son was sent to be brought up in his sisters' household and they became genuinely fond of him: Mary, so much the elder, was motherly in a distant kind of way; Elizabeth was much closer as a sister to a clever little brother.[4]

Edward's birth also had a more direct impact on Elizabeth's upbringing. Lady Bryan's position as Lady Mistress of the royal nursery meant that she was *ex officio* in charge of each royal baby on its arrival in the world. So, with Edward's birth, her attentions were immediately transferred from Elizabeth to the new baby. The loss of Lady Bryan, who had acted as her surrogate mother, could have been another heavy blow to Elizabeth. Fortunately, Cromwell replaced her as Elizabeth's principal gentlewoman with Catherine Champernon. It was an inspired appointment.[5]

Kate, as she was usually known, was the daughter of Sir Philip Champernon of Bere and Modbury in Devon. Sir Philip engaged in learned discussions about his family history with the antiquary, John Leland. It also looks as though he gave his children the benefits of the new interest in classical scholarship. One daughter, Joan, turned out to be something of a blue-stocking, while Kate, soon after her appointment, took Elizabeth's education in hand herself. Later, Elizabeth paid tribute to her for her 'great labour and pain in bringing of me up in learning and honesty'. Roger Ascham, the greatest educationalist of the century and briefly Elizabeth's tutor as well, also sang Kate's praises but counselled, delicately, against her forcing the pace with Elizabeth too much. 'The free [sharp] edge is soon turned, if it be not handled thereafter,' he remarked sententiously. But Ascham, with all his experience, under-estimated the girl: Elizabeth's wits were tough as well as sharp and, far from being blunted, acquired an extra edge from Kate's vigorous honing.[6]

The results were soon known at court. In December 1539, Thomas Wriothesley, shortly to be appointed royal Secretary, had business at Hatfield with Mary. He also paid a courtesy call on Elizabeth who, as usual, was keeping a joint household with her sister. Probably he began by patronizing the little girl. If so, he was taken aback by the reaction. Though she was only six years old, she had, he wrote, spoken to him with as much assurance as a woman of forty. 'If she be no more educated than she now appeareth to me,' he continued, 'she will prove of no less honour and womanhood, than shall beseem her father's daughter.'[7]

Wriothesley, almost certainly, had paid the same compliment to Elizabeth herself. Instantly his earlier patronizing tone would have been forgiven. For to be her father's daughter was her proudest boast. It was also true. Back in 1499, when he was eight, Henry had confronted the famous scholar Erasmus. Erasmus was at least as difficult to impress as Wriothesley. But impressed he was. Even twenty years later he recalled the child's poise, precocious learning and (since he was a boy) wicked teasing.[8]

The precociousness she inherited from her father meant that Elizabeth quickly exhausted Kate's own stocks of learning. Fortunately help was to hand. In 1544 her brother Edward entered his sixth year and, as he explains in his *Chronicle* (he was precocious too), his education was put on a formal footing with the appointment of male tutors. The most important was John Cheke of St John's College, Cambridge. One of the key voices recommending Cheke was Anthony Denny, then the King's favourite body servant. Denny was Cheke's fellow Johnian; he was also, as he had married Joan Champernon in 1538, Kate's brother-in-law. As is the way of such things, one contact led to another. Cheke's star pupil at St John's was Roger Ascham and Ascham in turn had a favourite student in William Grindal. Soon Grindal turned up at the court of the royal children with a letter of introduction from Ascham which recommended his appointment as Elizabeth's tutor. With such backing, the deal was as good as done and he took up the post immediately. Ascham himself became both a frequent visitor and correspondent. He was especially close to Kate and to the man she married about this time, John Ashley.[9]

Elizabeth now found herself at the heart of a remarkable circle, in which friendship, scholarship and, it soon became clear, religion were all interwoven. Grindal was her main teacher, and took her through an increasingly difficult programme of Latin and Greek. Ascham himself taught both Elizabeth and Edward the italic hand of which he was a master. It is a beautiful but laborious script and there were many broken pens and frayed tempers *en route* to Elizabeth's own mastery which is displayed most flamboyantly in her signature with its swirling, cascading flourishes. Elizabeth's broken silver pen, Ascham wrote at the end of one of his letters to Kate, should be sent to him for repair. In return, he sent her another pen, an Italian book and a book of prayers. Elizabeth also called on Edward's teachers for particular tasks, so that Jean Belmain, for example, who was Edward's French tutor, helped to perfect her already excellent knowledge of the language. Belmain, despite his nationality, fitted perfectly in this little world and became genuinely close to her.[10]

And it is easy to see its attractions. Most of the time the royal children were in Hertfordshire. It was an upland county with dense forests and a scattering of royal houses in the clearings: Ashridge, Hunsdon, Hatfield and Hertford Castle. When the children were bored with books, there were gardens for walking and parks for hunting. Here Ascham, despite his learning, would have taken the lead. Author of a text book on archery, he was also an amateur of the cockpit.[11] This was a taste which remained with Elizabeth for life: she loved baiting both birds and animals, the bigger the better. For the rest, the royal children and their servants made their own entertainment: they had minstrels, players and fools. And, in their household, there was the human comedy of so many lives under one roof: people who got married, had children and needed godparents for them. The court and the continuing saga of Henry VIII's marriages must have seemed very far away.

There is no evidence that Elizabeth met her father's fourth wife, Anne of Cleves, during her six months of marriage to the King. This might have spared her the knowledge that her father had extricated himself from this detested match by declaring himself impotent on his wedding night. Elizabeth did meet the fifth wife, Catherine Howard. She travelled in her barge and received some presents of jewellery from her. But the jewels were noted to be of small value and there is no suggestion of intimacy.[12] This, too, was a lucky escape. Henry, like a modern middle-aged executive, thought he was rediscovering his youth by marrying his sexy young queen. Instead, he was only preparing the ground for his own humiliation. When Catherine Howard's inevitable adultery was discovered, he wept bitter, self-pitying tears.

With Henry's sixth and final wife, Catherine Parr, it was a different matter. She went out of her way to establish good relations with all the King's children. And she and Elizabeth seem to have become particularly close, corresponding and exchanging presents. But still it is important to get the relationship right. Catherine did not, as is usually claimed, reunite the royal family. Instead, as we have seen, Henry's children had been living together amicably for several years. Nor did Catherine educate Elizabeth. Elizabeth, ten years old at the time of the wedding, was already formidably learned. Instead, it was Catherine, a bit of a late developer in these rarefied circles, who was taking Latin grammar lessons and being instructed in the italic hand. A year or two later, Prince Edward, only nine years old but already an accomplished prig, wrote to his stepmother on her new-found achievements with exquisite patronage. His teacher would not believe, he said, that her last letter was really in her own hand till he saw her signature, which (he graciously acknowledged) was written equally well in the same style.[13]

Nevertheless, there *was* a change in Elizabeth's position in these final years of Henry VIII's reign. But it came about primarily as a result of Elizabeth's new relationship with her father, not with her stepmother, who was never more than a moon to Henry's sun.

CHAPTER 5

Rehabilitation

Henry VIII had seen little of his second daughter since her mother's execution. But he had been kept in touch with her progress by letters and reports. And these, like Wriothesley's in 1539, were overwhelmingly favourable.

Then, in 1542, father and daughter met in person. Henry, still a disconsolate widower after the disastrous failure of his fifth marriage, was trying to drown his sorrows in work. But in September, in the course of a visit to inspect the fortifications on the Essex coast, he decided on a little family diversion and summoned both his daughters to dine with him at the ancient royal hunting lodge of Pyrgo Park to the north of Romford. It was almost certainly the first time that the King and his two daughters had eaten together, and the occasion was regarded as sufficiently important for the Lord Chamberlain's department to draw up a formal 'ordinance' to deal with the arrangements.[1]

All went well. Henry clearly liked what he saw and decided to begin the formal process of Elizabeth's rehabilitation. This was to prove much less fraught than Mary's equivalent reconciliation with her father in 1536. Mary had nailed herself to the cross of her mother's memory; Elizabeth did not make the same mistake. Instead, she was very much her father's daughter and he came to reciprocate by taking a warm, fatherly pride in her (whatever his residual feelings about Anne Boleyn might have been). Within six months, in April 1543, when the King floated the idea of marrying Elizabeth to the son of the Scottish Regent, the Earl of Arran, he praised her as being 'endowed with virtues and qualities agreeable to her estate' and dropped heavy hints that he was already considering her restoration to a place in the succession.[2]

Soon, however, Henry was concerned with his own marriage, his sixth. In June, the King's daughters were invited to court to inspect the King's future wife, Catherine Parr, and in July they were guests of honour at the wedding. Mary stayed on at court, but within the month, Elizabeth, who was, after all, still only a child of ten, had gone back to the schoolroom with Edward. But foreign courts had caught the scent of the revolution in the King's family arrangements and in December the Lady Regent of the Netherlands pointedly asked the English agent how the royal family were and whether they all 'continued still in one household'.[3]

In 1544 the revolution became official. In parliament, held in the spring, Mary and Elizabeth, in that order, were formally restored to a place in the succession after Edward. Curiously, they were not legitimated at the same time. But then Henry always liked to have his cake and eat it. The new arrangements were signalled, as usual, in court ceremony. On 26 June Henry ate with all three of his children in a grand dinner at Whitehall and the dinner was followed by a 'void'. A void was a large reception in which wine and sweetmeats were served. Everyone stood, which reduced the formalities of etiquette and made it a sociable occasion. This, in other words, was Henry introducing his children and successors to the court. The point was driven home by a notable absentee. The current, childless, Queen was not present.[4]

And it is this occasion, almost certainly, which is commemorated in the great painting, now at Hampton Court, known as *The Family of Henry VIII*. It is set at Whitehall, portions of whose buildings and gardens are visible in the background. The royal children stand on the steps of the throne. The King sits, as he had probably done in reality, because the pain in his leg made standing for long periods (the usual procedure at a void) intolerable. And, in place of the absent Queen Catherine, Jane Seymour, the dead mother of Henry's heir, is shown sitting next to the King. Elizabeth is in the humblest place, on her father's left, and her hands are modestly crossed in front of her. But there was triumph in her heart.[5]

Elizabeth's rehabilitation had come at just the right moment. She was now in her early teens and at her most impressionable. This meant that the humiliations and neglect of childhood – such as they were – were forgotten, and the memory of her mother's death buried further still.

Instead, she had a place in the succession, at court and, increasingly, in her father's affection. She rejoiced in them all, especially the last. Which is why her memory of her father, formed in these few years of the mid-1540s, was so benign: for her, he was not a wife-murdering monster, but a loving parent, a formidable ruler and model to which she aspired. Fortunately for her country, she would emulate only the form and not the substance: like her father, she would bite men's heads off; unlike him, she would rarely cut them off.

Elizabeth's rehabilitation also coincided with a major shift in the direction of royal policy. This, too, was to have momentous consequences for the impressionable little girl who, in reality as well as in the Hampton Court painting, now stood so near to her awe-inspiring father. For the last twenty years, Henry's concerns had been domestic. He had been faced with the complications of his family life – his marriages, divorces and the problem of the succession – and with religion. But by 1544 these problems seemed solved. He had settled the succession; he had arrived at a broad-based religious settlement that commanded the assent of most of his subjects – apart from the fanatics on both sides; and he had a wife in whose loyalty and good sense he trusted as much as he once trusted Catherine of Aragon's. Settled at home, the old Henry felt able to return to the preoccupations of his youth. That meant foreign war, in particular war against England's traditional enemy, the French.

For months, the King and his councillors had been plotting and organizing. There had been another diplomatic revolution. England had broken with the French, whose support had been necessary to the first steps of the Reformation, and had flung itself once more into the arms of the Habsburgs. It was just like the old days. Like the old days, too, was Henry's enthusiastic work in overseeing preparations for war: the raising of troops, the gathering of munitions and victuals, the provision of transport by land and sea, the nomination of commanders. The detailed work, of course, was left to members of the council, but there is no doubt that it was the King who was in charge, both of the broad picture and (sometimes with devastating effect on the erring underling in question) on points of detail as well. It was the old soldier fighting once more the battles of his youth. But the battles were to be for real, on the field of northern France, and not on the carpet of the Privy Chamber, with his impressive collection of model ships and guns.

By late June, the largest and most powerful amphibious force that England had assembled since the height of the Hundred Years War under Edward III was ready to set sail. And, as in the days of Edward III and Henry's boyhood hero, Henry V, Henry had decided to take command in person. At the dinner at Whitehall on 26 June, the King bade his formal farewells to his children; on 6 July he completed the arrangements for the regency government during his absence and on the 11th he departed for France. He sailed from Westminster to the mouth of the Thames; then rode overland to Dover and arrived at Calais on the night tide on the 14th. On the 25th he departed in state to lay siege to Boulogne. Boulogne was France's principal Channel port and massively fortified. But on 18 September, after a set-piece siege that was commemorated in a series of great wall-paintings at Cowdray House, Henry entered the city in triumph. He resolved to add it to the English dominions in France and set in train a massive (and massively expensive) programme of refortification to protect it from French attempts at recapture. On 30 September, he returned to England, a conqueror.[6]

Modern historians have been dismissive of Henry's laurels. Boulogne was insignificant. Nor could even a much more substantial conquest have justified the cost, which left England exhausted and half-bankrupt. And there was scandalous confusion among the English forces, in particular after the King's return. There is some truth in these criticisms. But most contemporaries did not see it that way. For them, the conquest of Boulogne had inflicted a staggering blow to French prestige. And French humiliation was completed when, the following year, a revenge attack was launched on the Isle of Wight. It was a duel by proxy between the two kings: Francis I directed the assault from Le Havre; Henry organized the defence from Portsmouth. And Henry won. Despite the loss of the *Mary Rose* in manoeuvres in the Sound, the French were repulsed with heavy losses. Henry had proved that he could take a French city at will;

the French, in contrast, were unable to occupy even a foot of English soil.

On 31 July, just as the Boulogne campaign got underway, Elizabeth had ended her first letter to Catherine Parr with a prayer to 'the Lord God to send [Henry] best success in gaining victory over his enemies so that your highness, and I, together with you may rejoice the sooner at his happy return'. Elizabeth's prayers were granted. Henry was victorious; his return was happy and Elizabeth was there to join in when the King was reunited with his family at Leeds Castle. Elizabeth was also present when, two years later in 1546, the Admiral of France, the most powerful servant of the second most powerful monarch in Christendom, came to London to sue for peace with her father.[7]

The departure of a father to war and his triumphant homecoming is one of the most powerful experiences of childhood. It coloured thousands of young lives in the First and Second World Wars. There is no reason to suppose its effects were any less profound on Elizabeth. These events confirmed the heroic stature of her father in her eyes. And they determined her sense of England's proper place in the world and of the proper role of the monarch at the head of his armies. Later, in her own reign, we can still sense her frustration that events excluded England from its rightful place in the sun and that her gender prevented her from commanding her troops in person. We can only regret it, too: she would have been a formidable Boadicea.

CHAPTER 6

Stepmother: Catherine Parr

In 1544, this year of war abroad and loving reconciliation with her father at home, Elizabeth stepped into the limelight of history which she never left. *The Family of Henry VIII*, painted in 1544/5, is her first surviving portrait; and in July 1544 she wrote her first surviving letter. The picture tells a straightforward story. But the letter, like much of her subsequent correspondence, is difficult to interpret.

It is addressed to her stepmother, Queen Catherine Parr. Like most of the letters of royal children, it is only secondarily a vehicle of communication. Its primary purpose, instead, was to show off the latest of Elizabeth's scholarly attainments. This was a mastery of Italian, which she may have already been learning from a native speaker, Battista Castiglione. So the letter is written in an elaborate, courtly Italian, which Catherine, almost certainly, could have understood only with the help of a translator. Modern commentators have fared even less well, for they have difficulties not only with the language Elizabeth deployed but with the conventions that both the contemporary writer and the addressee took for granted.[1]

Elizabeth began by deploring the unkindness of fortune and the 'continuous whirl of human affairs',[2] which had deprived her of the Queen's presence for a whole year. Worse, they had just missed each other again. This 'exile' would be intolerable if she did not expect to enjoy the Queen's company again soon. Meanwhile, Elizabeth was grateful for the Queen's kindness in mentioning her every time she wrote to Henry, for she had not 'dared' to write to the King herself.

The problem, of course, is in the coincidence of 'exile' and 'dared'. Historians of a sentimental disposition have put the letter's two key words together to paint a picture of Elizabeth as a wretched child, who was not only cut off from her family but in such deep disgrace that she was forbidden even to write to her father. The picture would indeed be pitiable if it were true. But it is not. Instead, as we have seen, Elizabeth had just been reconciled to Henry and restored to the succession. And far from being 'exiled' from her family she had last dined with her father and her two siblings only a month before the date of the letter. The Queen, however, had not been present. And it is to this event that Elizabeth is referring when she regrets having recently missed seeing Catherine. But she attributes the fact, not to disfavour, but to the 'continuous whirl of human affairs' – that is, to the busy and divergent schedules of a king's wife and a king's daughter.[3]

Elizabeth's not 'daring' to write to Henry, on the other hand, is explained by etiquette. There was an etiquette about both speaking and writing to the King. The rule essentially was (as indeed it still is) that you did not speak unless the monarch first spoke to you. In face-to-face conversation, this was easily handled. Letters were rather more difficult. Fellow-monarchs alone were allowed the privilege of unsolicited correspondence. Ministers, mistresses, wives and ambassadors could also initiate business as required. Everybody else, however, had to approach the King through one of these privileged intermediaries. The rule applied even to royal children. This is why most of Mary's letters were addressed to her father's minister, Cromwell, rather than to her father himself. But Mary, it might be objected, was negotiating a stony path back to favour at the time. But the same is true for that apple of his father's eye, Prince Edward. None of Edward's early letters displaying his boyhood accomplishments were addressed to the father who was undoubtedly so proud of them; instead, the Prince invariably wrote to the Queen. Elizabeth, in not 'daring' to write to her father directly, was simply following the same rule of royal good manners.

So the Italian letter, on which so much ink has been spent, did not signal even a storm in a tea-cup. And in any case Elizabeth was soon taking tea, or its sixteenth-century equivalent, with Queen Catherine in person. When Elizabeth wrote the letter of 31 July she was staying at St

James's; meanwhile the Queen and the King's other two children were at Hampton Court.

This location had been chosen by Henry himself as part of the arrangements he had made for his family and his kingdom during his absence in France. Edward, he decided, should be brought from his usual residence in one of the Hertfordshire manors assigned for his use and lodged instead at Hampton Court. This was his birthplace; the palace also contained an extensive and recently rebuilt Prince's Lodging next to the Queen's Apartments. Henry also decided that this was an appropriate moment to move to the next stage in Edward's upbringing. So the women who had hitherto brought him up were discharged from their posts and in their place a household of male officials and tutors were appointed to guide him towards manhood. This is the moment, vividly recalled by Edward at the start of his *Chronicle*, when his infancy ended and his serious education began.[4]

As so often in royal lives, one generation found itself treading in the footprints of its predecessors: over half a century before, in 1492, when Henry himself was a one-year-old baby, his father, Henry VII, had launched the invasion of France that was more or less obligatory for any self-respecting late-medieval English king. His goal had been Boulogne (though he was bought off by a pension before the siege became serious) and, before his departure, he had also made a settlement of the royal family. This envisaged that Henry (then the second son) would be brought up with his sisters under his mother's general supervision. Henry had duly moved in with his mother to his birthplace at Greenwich.[5]

Informally, much the same happened in 1544. Queen Catherine joined Edward at Hampton Court almost immediately and Mary followed soon after. Initially, however, Elizabeth was left behind at St James's. It is unclear why. But fear of the plague, which was rampant that summer and had perhaps affected a member of her household, is a more likely explanation than misbehaviour. At all events, when she wrote on 31 July, Elizabeth expected that her 'exile' from the Queen's presence (or should it be quarantine?) would soon be over. And so it proved. She joined the rest of the royal family at Hampton Court in August and remained at the Queen's court throughout the summer and autumn.[6]

As usual, the court went on a progress in the early autumn, though, because of the King's absence and the plague, its extent was limited to a southern half-circuit of the capital from west to east. At the beginning of September it was at Woking and remained there till the end of the month. Then, by easy stages and with a halt at Nonsuch for dinner and an overnight stay at Beddington, it moved east and north to Eltham and then south via Otford to meet the King, triumphantly back from the wars, at Leeds Castle on 3 October.

Throughout this period, Elizabeth was in attendance on the Queen and we catch glimpses of her in Catherine's accounts. In September, while staying at Woking, Elizabeth instructed her gentleman usher, Mr Cornwallis, to ride with her gift to the christening of Mr Cotton's child at Penn in Buckinghamshire; later she was at Beddington and Eltham and gentlemen were paid for their attendance on her. Mary seems to have followed the same itinerary, though at some point Edward returned to Hertfordshire.[7]

These months were Elizabeth's first extended stay at court since her earliest infancy. But the court that she now experienced, fully and for the first time, was a court without a king. Instead, it was presided over by a queen. Catherine, admittedly, was Queen Regent (as she formally signed herself) rather than Queen Regnant. Nevertheless, Henry had left her with the full exercise of royal authority during his absence. The pomp was hers: so it was the Queen who sat in state in the presence chamber; the Queen who was served at table bareheaded and on bended knee, and the Queen in whose name polite notes of thanks were sent to deserving commanders. But the powers were hers as well: legal process and the formal witnessing of acts took place in her name and she was given commissions to issue warrants for the expenditure of money and to raise and despatch troops. And her letters to Henry show that these were no mere formalities. They note the sending of money, munitions and troops. To her husband, she used the proper tone of humility of a royal wife. But when she wrote to Henry's council she adopted the full royal style. 'Right trusty and right well-beloved cousins,' her letters began, 'we greet you well.' They ended 'given under our signet' and they were signed at the top 'Catherine the Queen'.[8]

It was an impressive transformation for the daughter of a northern knight who had been born, far away and with no fanfare at all, in Kendal Castle in 1512. But, when all was said and done, Catherine was a figurehead – for her husband abroad and for the council which he had left attendant on her at home.

The reshaping of the King's council was one of the most important achievements of Henry's reign and a principal legacy to his successors – and to Elizabeth in particular. In 1540, following the fall of Thomas Cromwell, Henry had reconstructed his council. It was given a fixed membership; an official hierarchy, based on the holding of certain ranking offices; a secretariat and an official record; and formal powers to summon individuals before it by legal process. It was, in short, a new council and, naturally enough, it acquired a new name. First in common speech and then in official documents, it became known as the 'Privy Council'. It is a name that echoes through the history of Elizabethan England.

But the most important change in Henry VIII's new council was one of practice. Hitherto, meetings of the earlier, more loosely defined council had been erratic and concentrated in the 'term-time'. The terms, which corresponded roughly to the modern academic terms (though they were even shorter), were the periods, from the late autumn to the early summer, when the central law courts in Westminster were open. All other state business, including sessions of parliament when it met, tended to concentrate in these periods too. But the council set up in 1540 was expected to meet more or less daily, in vacations as well as in terms. This represented a major break with established practice, and councillors remonstrated at the unwonted intrusion into their free time. But Henry was implacable and used his own letters to hold to the grindstone of business the noses of his councillors – from his brother-in-law and old jousting companion, the Duke of Suffolk, to the Lord Chancellor (who sorely missed his country garden and country pursuits with his wife).

Another difference was one of location. The new council was a court council, which met in a room in the King's Privy Lodgings (whence the name 'privy council'). But, despite its physical closeness to the King, it never met in his presence. Instead, it communicated with him by intermediaries, of whom the most important was the King's Secretary.

Also, because it was a court council, the new council followed the King on royal progresses. In such circumstances, the council tended to split into two. One group of councillors was left to mind the shop in London; while another attended the King on his travels. But it was the group with the King who conveyed orders to the group in London. Henry's departure to the front in 1544 had a similar effect: part of the council accompanied him to France, while another group remained at home in attendance on the Queen Regent.[9]

The war of 1544 presented the newly established Privy Council with a major challenge. The war was hugely expensive and ran wildly over budget. It was fought on two fronts as the Scots, as usual, took the opportunity to try to stab England in the back. And counsels were divided – literally – by the absence of the King abroad. As is inevitable in war, there were many disasters. But they tended to be military rather than administrative. The new council had proved its worth.

Now, it would be absurd to think of the eleven-year-old Elizabeth knowing, much less understanding, the finer points of the reorganization of the Privy Council. But she was certainly familiar with the broad outline of events in 1544. The Italian letter shows that, even before she went to live at court, she knew about her stepmother's letters to her father in France and that she fully grasped what was at stake in the war. When, soon after, she went to court, she found herself that much nearer the centre of events. Indeed, her rank gave her the best vantage point of all in the Queen's chamber. There she would have seen the state accorded Catherine as Regent, and she might have been present when councillors submitted formal business.

The sight of some of the most powerful men in the realm bowing low before a woman was unusual in Tudor England. I like to think it made a deep impression on Elizabeth. As, indeed, did the whole pattern of her experiences in those summer and autumn months of 1544. Elizabeth saw a court fronted by a woman and managed by a council. Between them, they successfully ran a country racked by war, plague and religious division. It could have served as a precedent for her own government as Queen. True, it was awe of the absent Henry which held dissenters in check. But then, all Henry's children, including Elizabeth herself, were similarly to invoke his name at moments of crisis. And is there really all that much difference, she might later have reflected, between a distant king, who sends his orders from beyond the sea, and a dead one, who rules from beyond the grave?

CHAPTER 7

Reformed Religion

When Elizabeth spent most of the latter half of 1544 with her stepmother, Catherine Parr, she encountered not simply the regal waxwork of the Queen Regent, but a real, complex and attractive woman. Catherine's marriage to Henry was her third. But she was only thirty-two and still clearly sexually attractive. Her qualities of mind and character were considerable, too. Stories of her learning are, as we have seen, exaggerated; so too is her healing role in the royal family. Nevertheless, her effect on the young Elizabeth was very great. She was a role model for female rule and she was, above all, a powerful example of a particular sort of piety. It would be wrong to say that Elizabeth simply took her religion from Catherine; other persons and subsequent events exerted too strong an influence for that. But these times with her stepmother laid foundations which never shifted.

An interest in religion was as usual among great ladies of the early sixteenth century as, say, an interest in clothes. And many, like Catherine Parr herself, combined the two. There was, of course, nothing new in a fashion for religiosity. Religion was one of the few areas where late-medieval women were allowed a high degree of autonomy and initiative – even while they were married but especially in widowhood. Many took advantage of the opportunity offered for self-expression. Famously devout ladies included three generations of the royal family: Henry's great-grandmother, Cecily, Duchess of York; his grandmother, Lady Margaret Beaufort; and his first wife, Catherine of Aragon.[1]

But the Reformation introduced a complicating element: when did an interest in the 'hotter' sort of religion turn into the patronage of dissent, even heresy? Anne Boleyn stepped into this minefield and it blew up in her face. Catherine Parr was to play the same dangerous game – though, as she was 'the queen who survived', rather more successfully.

Of the facts there is no doubt, as both friends and enemies agreed on them. Catherine read widely in the scriptures and received formal instruction in them. She held religious 'conferences' in her apartments, especially during Lent. Then, each afternoon, one of her chaplains would preach a sermon in her privy chamber, in which he would attack 'abuses', that is, traditional religious practices in the Church. The audience would include ladies and gentlemen of Queen Catherine's household, many of whom shared her religious enthusiasms, and anyone else who was interested. She wrote prayers and meditations herself and published two volumes, one in 1545, and the other, more radical, collection in 1547, after Henry's death.[2]

Catherine, in short, was running a Tudor Open University course in religion at Henry's court. She even tried to recruit Henry himself, by engaging him in vigorous religious disputation. Henry, who was more used to regarding himself as a professor than a student in these matters, was not amused.

Elizabeth, on the other hand, was certainly a receptive student and we can imagine her listening, intent and white-faced, to the lectures in the Queen's privy chamber, or standing by Catherine as she wrote or read. In religion, at least, Elizabeth was the pupil and Catherine the tutor.

But what of the curriculum? In other words, just how radical were Catherine's beliefs? Her writings and letters, which are remarkably consistent, show that her faith revolved round four tenets: three positive, and one negative. First, man must submit himself absolutely to the will of God. Second, man was wholly sinful. His sinful nature meant that he could never fulfil the commandments which God had laid down as necessary for salvation. By God's justice, therefore, he was condemned to damnation. Instead, third, he could be saved ('justified') only by his faith, not by his actions or 'works'. All this led, fourth, to the negative conviction that most of the traditional apparatus of Catholic worship – the saints, images, pilgrimages and the liturgy itself – were, at best, a mere distraction and, at worst, a devilish conspiracy by the Pope as

Antichrist to deflect the soul from its quest for the saving, lively faith that could come only from the word of God in the scriptures.[3]

Now much of this – in particular points one, two and four – was common ground. It was with the third that Catherine moved onto dangerous territory. For 'justification by faith', as it was known, was the principal article of faith of the great German reformer Martin Luther. Henry had considered the doctrine in his negotiations with the German Lutheran princes, with whom he seemed to have much in common, but had rejected it decisively. In espousing it, therefore, Catherine was challenging Henry directly.

She survived because, as she put it, she disputed with Henry as a loving and, it must be said, a very clever wife.

We can see her technique in one of the letters she wrote to Henry during his absence in Boulogne. Her love for him, she explained, made her submit absolutely to his will. She said this in no expectation of reward; nor did she deserve any. Instead, she explained daringly: 'I make like account with your majesty, as I do with God.' She knew she was a great 'debtor' or sinner towards God, because she was unable to recompense His benefits (by following His commandments). And as a sinner she would die, 'But yet I hope in His gracious acceptation of my good will.' That is, she trusted that God would accept her faith, and save her in spite of her deficiencies. Then came the punch-line. 'And even such confidence I have in your majesty's gentleness.' She hoped and knew – she was saying – that Henry would receive her into the salvation of his arms even though she had done nothing either to deserve or recompense his love.[4]

This is masterly. As he read the letter, Henry would have realized immediately that Catherine was explaining her relationship to him in terms of the doctrine of justification by faith. She is the unworthy soul; he is the bountiful God who will receive her faithful love in place of the duty she owes but can never pay. Put in these gratifying terms, and by a pretty, yielding woman, Henry found even the doctrine of justification by faith acceptable.

The modern reader, on the other hand, might find that Catherine's figure of speech comes near to blasphemy, and might also be disconcerted by its mingling of sacred and profane love. But then Catherine was a sexual creature, too. Within a very few years she would find herself a slave to earthly passions – and indulge them with earthy gusto. It was to be another lesson for Elizabeth to learn, this time painfully.

At the time, however, with a husband like Henry, it was easier for Catherine to worship his kingship than his body. Indeed, king-worship amounted to a fifth article of Catherine's faith. Henry was God's chosen instrument for the extirpation of papal abuses in England. He was favoured by the God of battles so that those who served him were victorious in the field (whereas those who served a reprobate, like Francis I, were defeated). He was destined, she was sure, 'after this life to enjoy the kingdom of God's elect'.[5]

For Catherine, therefore, the King of England and the King of Kings came near to fusing into a single entity. Catherine was not alone in this view. The illustrations in Henry's own Psalter, or Book of Psalms, showed him as both David and (by analogy) as Christ; while on the title page of the Great Bible he impersonates God the Father himself. Elizabeth was a devout little girl and she would have read the Bible every day of her life. Always, as she opened the book, there was that image. It showed her father sitting in majesty. She had seen him similarly enthroned and had stood beside him. It would hardly be surprising if she too came to fuse her earthly and her heavenly father.[6]

It is not clear where Catherine obtained her faith. Her second husband's family, the Nevilles, were religiously conservative and were, moreover, connected by marriage with families such as the Throckmortons, who were leading opponents of the Reformation. And the officiating bishop at her wedding with Henry was Stephen Gardiner, Bishop of Winchester, who was the most outspoken and vehement episcopal conservative. All this makes it unlikely that Catherine was already religiously radical when she married Henry. And it makes it nigh on impossible that, as some traditional accounts have it, she was manoeuvred into the marriage by the Reformed faction at court to advance the Reformation in the country by advocating it in Henry's bed.[7]

Rather, I suspect, it was the other way round and that it was her marriage to Henry that led to her interest in Reform. We know, because she said so herself, that she regarded Henry's offer of marriage as a sign of divine 'calling' or vocation. The question was what was she called to. Her answer seems to have been that, like the Biblical Queen Esther, it was her mission to save God's people, that is, the Reformed, from persecution by the King's evil councillors. She went to the task with a will.[8]

This means that in religion as in scholarship, Catherine was a freshman, who was learning on the job. Hence her need for scripture lessons alongside her instruction in Latin grammar and italic handwriting. She borrowed her humanist teachers from among her stepchildren's tutors; she got, almost certainly, her principal religious instructor from among her husband's councillors. During her regency, as we have seen, a group of privy councillors was appointed to attend on her at court. Chief of these attendant councillors was Thomas Cranmer, Archbishop of Canterbury. He was Elizabeth's godfather; he now seems to have acted as Catherine's religious guide.

Seen in this light, the religious picture at court during Henry's absence on the Boulogne campaign becomes suddenly intense and more dramatic. The Queen Regent was undergoing the first exquisite pangs of conversion. One moment, she was lamenting the recent past, in which her soul had been enslaved to Catholic superstition; the next, she was rejoicing that she had been given knowledge of the pure word of God. Meanwhile Cranmer, who was both councillor and confessor to the Queen, was also experiencing a religious crisis, in which his own faith, usually so slow and cautious in its evolution, was making one of its periodic leaps into the light. Finally, the Queen's ladies, echoing the Queen's enthusiasms, turned the court into a sounding board for the preacher's pulpit. It was like a revivalist meeting. In the middle was Elizabeth – young, intelligent, eager to please and, as always, supremely sensitive to atmosphere.[9]

Naturally, Elizabeth became a religious enthusiast too. As she explained later, when she was a young woman, 'I studied nothing else but divinity'; only when she became Queen did she apply herself to what we would now call political science. As usual, she exaggerated for effect – but not by much.[10]

Naturally, too, her stepmother was the first beneficiary. She had sowed the seed (or at least had watered the seed that Catherine Champernon had planted). And she received its first fruits. It took the form of Elizabeth's 1545 New Year's gift to the Queen.

Shortly after Henry's return to England from France and his reunion with his wife and daughters at Leeds Castle and Otford, Elizabeth had rejoined Edward's household at Ashridge in Hertfordshire. And it was there in late November or December that she started work on her present for Catherine. Hitherto she had given the usual expensive trinkets. This time it would be more personal: it would be by her and about her. And, above all, it would be about their joint faith. The new religion centred on the Book and books. It was intellectual and demanded learning or at least literacy. Elizabeth's present for Catherine reflected all this: it was a book and it advertised both her learning and (perhaps) her feminine accomplishments.[11]

Elizabeth chose the subject of her book with great care. It was her own translation into English of a French religious poem, *Le miroir de l'âme pécheresse*, by Margaret of Angoulême. Margaret, who had become Queen of Navarre by marriage, was the favourite sister of King Francis I. She used this favour to act as the leading patroness of Reform at the French court. She had also, as we have seen, acted as hostess to Elizabeth's mother when the still-unwed Henry and Anne Boleyn visited Francis I at Boulogne in 1532. So the choice of author sent one message to Catherine, who now found herself emulating Margaret's role at the English court. But the content of the poem sent another, more important one. Its theme, as Elizabeth summarized it in her prefatory letter addressed to Catherine, was the inadequacy of the human soul:

how she (beholding and contemplating what she is [in the mirror or glass of the title]) doth perceive how of herself and of her own strength

> she can do nothing that good is or prevaileth for her own salvation,
> unless it be through the grace of God, whose mother, daughter, sister
> and wife by the Scriptures she proveth herself to be. Trusting also that,
> through His incomprehensible love, grace and mercy, she (being called
> from sin to repentance) doth faithfully hope to be saved.[12]

Now two things are remarkable about this summary, which is wholly
faithful to the poem. The first is its quality as précis. It manages to reduce
the complexities of the doctrine of justification by faith to two clear,
simple sentences. This would have been no mean achievement for an
adult; for an eleven-year-old girl it is astonishing. If it were her own
work, and there is every sign that it was, it shows that her boast about
devoting herself to divinity was no idle one. But just as striking is the
question of gender. The loving, supplicating soul is female; the
beneficent God is, of course, male.

This should set bells ringing, for it is a precise echo of Catherine's
letter to Henry the previous July. That had dealt with the doctrine of
justification, and had explained it in terms of the love of the
soul/Catherine for God/Henry. So the letter uses the same trope as
Margaret of Angoulême's poem. And since there is no trace of originality
in Catherine's thought or writing, it is almost certain that the poem was
the source for the letter. I would go further: not only was Catherine
reading and studying the *Miroir* herself during the summer of 1544; she
was also reading it with Elizabeth.

And she was reading it with a purpose. Her mission was no less than
the conversion of the King's younger daughter. The poem was the
perfect choice of vehicle. Its author was a princess born and a queen by
marriage. She employed the language of courtly love. And she showed,
both in her life and writing, that even royalty could be Christian.
Elizabeth, as her translation demonstrates, was instantly responsive: she
became that summer and autumn, as she was to remain after her own
fashion, a good daughter of the (Reformed) Church.

But that made her more, not less, her father's daughter. Here again
the poem spoke directly to Elizabeth's condition, as, with a different

emphasis, it had spoken to Catherine. For Catherine, its message had been the wifehood of the soul. For Elizabeth, naturally enough, it was the daughterhood. God, the poem said, was kind to daughters as (at least recently) Henry had been to her. And he was a merciful judge, who did not execute adulterous wives. Or at least, as Henry had done with her own mother, He merely sentenced them to a merciful death.

It is all, apparently, deliciously ironical. And psycho-feminist historians have seized on the ironies (as well as some of the slips of Elizabeth's pen) to read into her translation of the *Miroir* a message of suppressed hatred of her father for his treatment of both her and her mother.[13] But there is no need for such subtleties with Elizabeth. When she wrote that she loved God, she meant it. And when she wrote that she loved her father, she meant it, too. For she felt, understandably, that they had both given her cause.

Elizabeth's translation of the *Miroir* is, therefore, one of the foundation-texts for her biography. But it is also a vivid insight into the moment. It shows, reassuringly, that this infant phenomenon, who was capable of summarizing the sort of idea (justification by faith) which leaves today's undergraduates blank, was equally an eleven-year-old child, who had left everything to the last minute. She began writing out her presentation copy neatly, and with good intentions. But, as time ran out, her handwriting declined and the number of mistakes increased. Shamefacedly, the prefatory letter confessed her errors to Catherine and begged her to keep them to herself. The date of the letter says it all: it was only finished on New Year's Eve. This would scarcely have given the messenger time to ride the twenty odd miles to London, through a winter's night and morning, to present it to the Queen at Greenwich. Probably, indeed, it was late. 'There is nothing,' Elizabeth bewailed, 'done as it should be.'[14]

CHAPTER 8

Royal Father

Elizabeth's anxieties on New Year's Eve 1544, as she bade her messenger make post-haste to court, were short-lived. The last two years of her father's reign were to be the happiest and most secure of Elizabeth's life. In the schoolroom she was the favourite pupil of talented teachers and gained in confidence accordingly. At court, on her occasional visits there, her position and status were also of the most honourable. She was My Lady Elizabeth's Grace, the King's daughter. And she was a leading member of the Queen's formal establishment of ladies-in-waiting. This made her the third ranking lady of the court: first came her stepmother, Queen Catherine, next her elder half-sister Mary, and then Elizabeth herself. Most importantly of all perhaps, in view of the internationalism of monarchy, her status was recognized abroad. The exception, of course, was at the Habsburg courts. These remained implacably hostile to her as the 'whore's daughter' and, consequently, persisted in treating her as a bastard. Elsewhere, however, it was taken for granted that her restoration to the succession meant that she was legitimate in the only sense that mattered: her potential claim to England, guaranteed by parliamentary statute, was worth the stain on her escutcheon. So from Scotland to Denmark she was regarded as a desirable royal bride and negotiated for accordingly. Much ambassadorial ink was spilled on the negotiations. But none came to anything. It was a pattern that was to repeat itself *ad infinitum*.[1]

Later, of course, it was Elizabeth herself who spun out such negotiations into ever-finer threads of obfuscation. But, while her father lived, she probably knew little of them.

Equally, we know little of Elizabeth's movements for most of 1545. But in November both she and her sister Mary paid a pre-Christmas visit to court. Maybe they looked dowdy after their sojourn in the country. At any rate, the King instructed the Great Wardrobe to supply precious materials for dresses in the latest fashion for his daughters and their ladies. He was similarly generous on each of Elizabeth's subsequent visits to court. Those early anxieties over clothes were over. And it was her ever-bountiful father who had banished them.[2]

Elizabeth also tested Henry's bounty in a more direct fashion on this November visit. Royal letters were issued, ordering the appointment of John Huddleston, BA, to a fellowship at the King's Hall, Cambridge, on the next vacancy. The King's Hall, soon to be absorbed into Henry VIII's vast deathbed foundation of Trinity College, was medieval England's graduate school for royal administrators and places there were valued accordingly as offering a fast track to a comfortable, often powerful, bureaucratic post. This plum was secured for Huddleston 'at the suit of the Lady Elizabeth'. Nothing seems to be known about the beneficiary. But he would have been brought to Elizabeth's attention by her teachers, who formed a major 'family' within the Cambridge mafia. More important is the fact that this was Elizabeth's own gesture: not only was she secure in her father's favour; she had shown that she could secure it for others as well. The royal favour was the fountain-head of power in Tudor England. When she was the source herself, she was to control its flow with a care that bordered on stinginess. Patronage—despite this early instance of her tapping Henry VIII's generosity—was one area where Elizabeth was not her father's offspring, but the prudent Henry VII's granddaughter.[3]

By the end of November, Elizabeth had left the court to return to one of her usual nursery homes at Hertford Castle. Clearly, the great court celebrations of Christmas, New Year and Epiphany were still felt to be too taxing for the now twelve-year-old girl – though her own household would have made merry in time-honoured, but more modest, fashion. But as her servants busied themselves with seasonal preparations, Elizabeth had more serious business on her mind. Last year she had translated, written out and perhaps bound a book as a New Year's gift for her stepmother. This year, she would make one for her father. And there would be no getting things wrong this time – despite the fact that the project was a much more demanding one.

Elizabeth's gift for Catherine had involved the comparatively simple exercise of translating from French into English. And her prefatory letter was in English too. This was probably as much as Elizabeth could manage at the time; it was also at the right level for Catherine herself, who, after all, was still only a tyro in scholarship. This year it was different. Elizabeth herself, thanks to the devoted labours of William Grindal, had come on by leaps and bounds; while her father, the intended recipient, was the best educated crowned head in Europe – or at least fancied himself as such. So the translation was to be *from* English, and it was to be, not into one, but three other languages: Latin, French and Italian. The prefatory epistle was composed in Latin, too. Even the handwriting was improved. In place of the rather unformed script of the previous year, she wrote in a muscular, regular italic that was the equal of her best subsequent efforts. For page after page the hand flows on, without break or blemish.

Only the text was from the same world as previously. Then, Elizabeth had taken a reforming religious treatise by the Queen of Navarre; now she did a triple translation of a similar work by the Queen of England. This was Catherine Parr's own *Prayers and Meditations*, which had been finished and published earlier in 1545. The choice was a gracious one, contriving to pay simultaneous tribute to her stepmother and her father alike. It shows Elizabeth's own continuing interest in religion. It also shows Henry's new openness, in these final years of his reign, to mild religious experimentation – at least when women were the mediators. What is most revealing of all, however, is the prefatory letter. I have already argued that the years after 1542 had seen a comprehensive *rapprochement* between Henry and Elizabeth, which wiped out permanently and utterly whatever resentments she may have harboured about her own earlier treatment or that of her mother. This letter is the final proof.

Its phrasing (behind which it is not unfair to detect a helping hand from Grindal's sophisticated Latinity) is sonorous. But it does seem sincere. Moreover, instead of the usual rag-bag of compliments of such compositions, it focuses narrowly on two things: Elizabeth's feelings about her father, King Henry, and her regard for the sort of theocratic kingship which Henry had created.

First, then, her feelings about Henry. She hails him as 'matchless and most benevolent father'. And she proclaims her determination that, as his daughter, she would be 'not only an imitator of his virtues but indeed an inheritor of them'. Henry probably chuckled indulgently at the youthful arrogance of the words. But Elizabeth, I think, wrote them in cold earnestness. In any case, they were prophetic.

Equally revealing, about both the present and the future, is what Elizabeth has to say about her second theme of kingship. Since philosophers, she observes sententiously, 'regard [a king] as a god on earth', nothing should be more acceptable to him than theology – 'the study', and here her pen takes flight, 'which lifts us to heaven and renders us heavenly while on earth and divine while yet in the flesh, and which, though we were in the toils of endless and infinite troubles, even then restores us to our happiness and felicity'. This last point, about the consolations of theology, reads like an afterthought. But events were soon to test both her faith and her scholarship. Neither was found wanting.[4]

Nor did her view of the divinity of kings or of the appropriateness of divinity for kings change much in the future. As Queen, she was to be as committed to the royal supremacy over the Church as Henry. This was because her learning taught her that the supremacy was legitimate in both history and divinity. But there was a stronger and more emotional reason as well. She valued it, above all, because she had inherited it from her father. The Church of England was his creation. As his daughter – his heir as well as his imitator, in her own phrase – she was determined to preserve as much of it as possible as closely as possible. This accounts for her fierce, possessive pride in her inheritance: it was her Church – not her parliament's, still less her clergy's. And if she wished to help herself or her favourites to its property or regulate its ceremonial, that was her affair as well.

In other words, there is very little in the adult sovereign which is not to be found in this letter of the twelve-year-old princess.

It is, in short, an astonishing achievement. Her New Year's gift for Catherine had been impressive; her gift for her father was, as she intended, prodigious. She also managed to finish it in time. Her prefatory letter was dated at Hertford on 30 December. This year, her messenger could have ambled and still delivered the present in time.[5]

CHAPTER 9

Father's Death

W hile Elizabeth was working on her New Year present for her father, Henry himself was occupied with the concluding business of the parliamentary session. The session was a busy one and the two Houses worked right through to the beginning of the holidays. In the nick of time, however, the legislative programme was completed and on Christmas Eve 1545 parliament met for the formal business of the prorogation, or breaking-up for the holidays. The King sat enthroned, with the Lords Spiritual on his right hand and the Lords Temporal on his left, all in their parliamentary robes of scarlet trimmed with white fur of miniver. The commons stood at the bar, and their speaker gave an oration. Normally, the Lord Chancellor replied on the King's behalf. This time, however, Henry spoke for himself. Perhaps he had a premonition that this was – as indeed it proved to be – his last parliament. Slowly, he heaved his great bulk to his feet and unfolded, as he put it, 'the secrets of my heart'.[1]

The speech from the throne was a genre that Elizabeth later made her own. Henry's last speech shows that in this too she was following in her father's footsteps. She could not have been present in person on that Christmas Eve as she was far away at Hertford Castle. But Henry's words were noted by one of the MPs present, Edward Hall. He was putting the finishing touches to his chronicle history of England, which began in 1399, with the fall of Richard II, and ended in his own times, with 'the triumphant reign of Henry VIII'. Two years later, the chronicle was published, with the speech given in full. Elizabeth must have devoured it. Its ideas became hers, and its manner and phrasing she made her own. When Henry had spoken, the atmosphere was emotional and many present had wept, or 'water[ed] their plants', as one reporter quaintly put it. As she read, Elizabeth may have wept too, for the memory of a father who by then was dead. But, above all, she was stunned, as his hearers had been, by the force of his words.

Henry implied that he was speaking *extempore*, or off-the-cuff. It may have been so. But nevertheless his speech followed all the rules of rhetorical composition that formed so important a part of both his and Elizabeth's education. First came the *explanation* of why he had chosen to speak; next the *deprecation* of the speaker's praise of his talents; and then the *gratulation*, or thanks, for parliament's generosity in voting him a subsidy, or tax, and endowing the Crown with the lands and property of the chantries, colleges and hospitals. In both cases Henry promised to use their generosity, not for his private benefit, but for the public good. It was at this point that the speech shifted emotional gear. 'No prince in the world,' he assured them, 'more favoureth his subjects than I do you, nor no subjects or commons more love and obey their sovereign lord than I perceive you do me, for whose defence my treasure shall not be hidden, nor, if necessity require, my person shall not be unadventured.' This has the authentic Elizabethan ring. Or rather, when Elizabeth struck this note herself, she was a true Henrician.

Henry now warmed to his theme. Despite the mutual bond between himself and his subjects, division and discord walked abroad in the kingdom. The reason, he averred, was a lack of charity. 'Charity,' he said in the new English translation of the scriptures, 'is gentle, charity is not envious, charity is not proud.' But where was charity when the clergy were divided among themselves, with 'the one call[ing] the other heretic and Anabaptist, and he call[ing] him again Papist, hypocrite and Pharisee'? And where was love when the laity railed against the clergy?

Her father's ideal of union and concord, of a Church broad enough to accommodate everybody of good will, was to be the key to Elizabeth's own settlement of the Church. And none of the mutual name-calling of radicals and conservatives, Papists and Puritans, laymen and churchmen, would ever deflect her from it.

But great speeches also depend for their impact on telling little phrases, in which the speaker familiarly seizes his listeners by the arm.

Elizabeth knew this, and supplied them in plenty. So did Henry. There was homely language: the Bible in English, which he had given his subjects for their private edification, was instead, he protested, 'disputed, rimed, sung and jangled in every alehouse and tavern'. There was even neologism, as Henry invented new words because the old ones were not vivid enough to characterize the folly of religious disputation. 'Some,' he said, 'be too stiff in their old *Mumpsimus*, other be too busy and curious in their new *Sumpsimus*.' Henry's old teacher, John Skelton, the jangling, macaronic rhymester, would have been proud of him. So too was his daughter, imitator, heir and pupil, Elizabeth.[2]

The parliament ended and the Christmas and New Year festivities over, the cycle of the royal year began its usual revolution. Removals, at monthly or bi-monthly intervals, from one palace to another, were punctuated by celebrations of the great feasts of the Church which were, simultaneously, major ceremonies at court. No one could have known that the cycle was turning for the last time in the present reign, least of all Elizabeth.

Instead, she would have been enjoying her new-found importance. For, at thirteen, she was fast approaching the sixteenth-century threshold of adulthood for women. Her education of course continued away from court. But her visits to court were becoming more frequent and, when she was there, her role in ceremonies more prominent. She was probably at court at Whitsuntide, in July, for the reception of the special embassy from France and again at Christmas.[3]

With these more frequent visits, she would have noticed the rapid deterioration in her father's health. He was now scarcely able to walk. Mechanical means were needed to mount him on his horse; he was dragged along the endless galleries and enfilades of his palaces in a special wheeled chair or 'tram', and he was winched upstairs in a lift. The cycle was now a vicious one: the less exercise he took, the fatter he got and the worse his problems became. There were good days, of course, when he was something like his old self. But there were more bad ones, when the pain and infection from his ulcerated leg left him speechless, breathless and black in the face.[4] His mood became black, too, and he enforced his plea for charity in his 1545 speech from the throne by striking indifferently at religious conservatives and radicals alike.

Christmas was especially trying. As the court maintained a public face of rejoicing, behind the scenes – in the King's bedchamber and council chamber and in hurried whispers in the galleries – was unfolding a complex political crisis that would shape the future of both England and Elizabeth herself.

On 12 December, two of Elizabeth's closest relations on her mother's side were sent to the Tower on charges of treason. Thomas Howard, Duke of Norfolk, was England's premier peer and his son Henry, Earl of Surrey, its brightest intelligence. But neither rank nor brilliance availed. The father, with cold cunning, prostrated himself before the King's wrath, and, with fortune on his side, survived. The son defended himself proudly, and was destroyed. The formal charge against him was that he had appropriated one of the royal coats of arms. The real one was his determination that his father should be protector or regent in the minority which, everyone now knew, would follow the King's imminent death.[5]

Henry, however, still had enough life in him to direct personally the investigation of the charges against Surrey and to give effect to his own wishes for the structure of the minority government. These were set out in his will, which was signed and witnessed, so the text claimed, on 30 December. In fact, it was 'signed', not with the King's own hand but, like almost all documents at this period, with a stamp of the King's signature. Under Elizabeth, this fact was to call the whole validity of the will into question; at the time, it passed without remark. Instead, it was its contents that mattered. The will provided, in notably conservative terms, for the repose of the King's soul in the next world and the security of the succession in this. It ratified and amplified the terms of the act of succession of 1544 and set out the powers and personnel of a council of regency. This excluded both the Howards and their ally, Bishop Gardiner, the leading conservative churchman. Henry, it seemed, had seen the future and it was Reformed.[6]

On 18 January Surrey was beheaded on the scaffold at Tower Hill.

His death, at the age of only thirty, left both an indelible memory and an enduring example. Poet and soldier; now puffed up with pride, now prostrate with depression; the inventor of the sonnet and blank verse, the two key forms of English poetry, yet vain, solipsistic and self-destructive – he was a prototype for the favourites and would-be favourites of Elizabeth's own court. Henry VIII, before he destroyed him, had doted on him as a second son. Elizabeth, more than once, was to repeat the same pattern with other handsome men. But her love was to be not parental but sexual – or at least that's what it pretended to be.

The execution of Surrey's father, the Duke of Norfolk, was due to take place on the morning of 29 January. But he was spared by another, greater death than his own. Late in the evening of the 28th, Sir Anthony Denny, the Chief Gentleman of the Privy Chamber, summoned up courage to tell the King that he was dying. Henry ordered that Archbishop Cranmer should be sent for. 'But first,' he said, 'I will sleep a little.' By the time Cranmer arrived, Henry was speechless. Instead, he clutched the Archbishop's hand when Cranmer asked him of his faith and assured him of his salvation. He was still holding it, with the grip that had held a thousand lances, when, towards two o'clock in the morning, he died.[7]

'The King is dead; long live the King!' The formula in which the heralds announced a royal death expressed the theory of the continuity of monarchy. The practice on this occasion was different. Instead of being proclaimed from the rooftops, Henry's death was kept a profound secret for three days. Parliament, which had reconvened on 14 January, remained in session and the ordinary public rituals of the court continued uninterrupted in the name of the dead King.

Behind this smoke-screen of normality (and behind the closed doors of the privy lodging) a pre-arranged plan was set in motion to seize the reality of power in the new reign. The plotters were led by three men: Sir Anthony Denny, the Chief Gentleman of the Privy Chamber; Sir William Paget, the King's Secretary, and Edward Seymour, the Earl of Hertford. Each had a different, but complementary, contribution to make. Denny's office gave him control of access to the private apartments. This control was used to keep the fact of the King's death secret. Paget was the chief political intelligence among the plotters; he was also, as Secretary, in charge of the administrative staff of the Privy Council. Finally, Hertford was the public face of the plot. As the victor over the Scots at the Battle of Pinkie, he was by far the most successful military commander of the last years of Henry VIII's reign. He was also, as Prince Edward's senior uncle on his mother's side, the closest living relative of the new young King. This combination of military prestige and proximity of blood outweighed his relatively modest origins as the son of a mere knight and made him a plausible candidate as regent for his nephew.[8]

The fall of the Howards (which the triumvirate had certainly helped to bring about) removed the principal obstacle to their plans; the Council of Regency appointed by Henry's will provided the means. Now it remained only to set the machinery in motion.

The first few hours after Henry's death were spent in tying up loose ends: opponents were squared and spoils divvied-up. Next Hertford rode to Hertford Castle to take control of the person of the boy who was now King. But even now Edward was kept in the dark about his father's death. Instead, he was told that he was being taken to London for his creation as Prince of Wales. On the way back to London, the party halted at Enfield. Elizabeth had taken up residence there soon after the ending of that last, lugubrious Christmas at court. Hertford and his party brought brother and sister together and, only then, broke the news of their father's death.[9]

It was a surprisingly considerate gesture. It also underscored how close the two were known to be. Recently, probably, they had exchanged portraits. In the background of Edward's is Hunsdon House. This was one of the Hertfordshire houses where Edward and Elizabeth had spent most of the last decade studying, playing and learning their future roles under the same roof. Now, suddenly, their father's death made that future their present. Edward was King. But what was Elizabeth?

Edward's *Chronicle* describes the bare facts of the meeting. Later historians have embroidered it to represent the two children throwing themselves weeping into each other's arms. Perhaps they did. Certainly Elizabeth had, did she but know it, much to weep about. The security and happiness of her father's last years were over. She was to know no real peace until, fifteen years later, she wore his crown.[10]

CHAPTER 10

Brother: King Edward VI

How quickly men forget.

Henry VIII had bestridden England like a colossus and, on 31 January, when his death was finally announced to a dumb-struck parliament, everyone, led by the Lord Chancellor, wept copious tears. Were they crying for Henry? Or for fear of their own, uncertain futures, now that the royal rock on which England had been rebuilt was gone?[1]

Within a few weeks, however, Henry's own son and heir, Edward, was writing to Elizabeth that their grief for their father should be moderated, as his soul was certainly in heaven.[2]

Edward, of course, had plenty to divert his mind. He was now at the centre of events as the politics and pageantry of the ending of one reign and the beginning of a new reign – *his* reign – got underway.

The action, rather like one of the contemporary pageants or dumb shows which were mounted to celebrate great events, unfolded at several levels. Below were the solemnities of the dead King's funeral; in the centre, the splendours of the new King's coronation; while above, and only intermittently visible, though giving form and meaning to the whole, were the political manoeuvres which shaped the new régime. The three scenes were played out simultaneously, and to an astonishingly tight timetable: from Henry's death to his son's coronation took a bare three weeks.

On 31 January, Edward, accompanied as always by his uncle, Edward Seymour, Earl of Hertford, entered London and took up residence in the Tower. The same day, the Council of Regency, having first sworn to maintain Henry's will, then immediately broke it, in spirit if not in letter, by nominating Hertford as Protector or Regent. On 2 February, Henry's body, after being disembowelled, cauterized, embalmed and chested in lead, was carried to lie in state in the Chapel Royal at Whitehall. On the 16th he was buried at Windsor and on Sunday, 20 February, Edward VI was crowned at Westminster Abbey.[3]

The burial and the coronation were separated by only four days. But they were worlds apart. The funeral was conducted by Bishop Gardiner of Winchester. He had been excluded from the regency council because of his religious conservatism. But he was loyal to the last to his old master's memory and it remained to him to have the last word also. At the funeral, he preached a sermon on the text from Revelations: 'Blessed are the dead which die in the Lord.' A better choice, he must soon have bitterly reflected, would have been Jesus's own words in St Matthew's Gospel: 'Let the dead bury their dead.'[4]

For soon after Henry had been buried with all the pomp of the old Church, his son was crowned in the most radical ceremony in the thousand-year history of the English coronation. The King's youthfulness was used as an excuse to shorten the ritual and eliminate some of its most offensively 'superstitious' features. But what the coronation lacked in tradition it more than made up for in verbal drama. This was supplied by Gardiner's opponent, Archbishop Cranmer. Cranmer had already concurred in the council's decision that, with the death of the King, the spiritual powers of the bishops had died also and had to be received again at the hands of the new King as Supreme Head of the Church. Now he made the coronation, with its unparalleled opportunities for publicity, into a parable of the limitless powers of the royal supremacy.

Traditionally, it was the archbishop who administered the coronation oath and anointed and crowned the king. Cranmer did all three (though he shared – again a unique touch – the actual imposition of the crown with the Lord Protector). But then Cranmer did what no archbishop had done before: speaking directly to the King and congregation, he explained that the anointing was not necessary. Instead, the King was God's anointed, not by the application of oil, but by virtue of the quasi-divine powers bestowed on him from on high. As for the oath, the King was honour-bound to observe it. But, should he break it, no earthly man – neither the Archbishop, nor any other of the King's subjects, nor any external power – had the right to hold him to account. Instead, elected by God, the King was accountable only to God.

The royal supremacy, in short, meant an autocracy, in which the king-cum-supreme-head would rule absolutely over the souls as well as the bodies of his subjects. Indeed, he would rule *especially* over their souls. As far as the temporal powers of the kings of England were concerned, the traditional role of parliament put real limits on royal authority. Not so with the new powers of the Supremacy. The papal monarchy over the Church, which Henry VIII had made his own, was an absolute one. So too was that other model for Henry's kingship, the monarchy of the Old Testament kings.

Cranmer's coronation address invoked this precedent too. 'Your predecessor Josiah', he explained to Edward, had renewed the covenant between Jehovah and the Jews. He had overthrown the idols of rival gods, desecrated and destroyed their altars and slain their priests. He had also succeeded at the age of eight. 'Your majesty', Cranmer told the new young King, '[is] a second Josiah, who reformed the church of God in his days.'

Edward, there is every reason to believe, took the Archbishop at his word and saw his kingship directly and literally as Cranmer had painted it: he would be God's instrument and the reform of the Church was his God-ordained task. Josiah, according to the Book of Kings, ruled for thirty-one years in Jerusalem. Edward was to be allowed only a fraction of that in England. But, as each passing month brought his majority nearer, he grasped more firmly at the reality of the absolute power which Cranmer had held before him, so tantalizingly, at his coronation.[5]

Elizabeth, as her letter to her father in December 1545 makes clear, fully endorsed the view that kings were gods on earth. But she was not present to witness the coronation, with its epiphany of her little brother into a priest-king. Nor was she present at her father's funeral.

Indeed, of Henry's immediate family, only his widow, Queen Catherine, attended the interment, with its interminable masses of the

Trinity, of Our Lady, and of requiem. She was accorded royal honours. But circumstances contrived to emphasize how transitory had been her part in Henry's life. She watched the ceremony from high up, through the richly carved windows of the Queen's Closet at St George's Chapel. This had been built for Henry's first queen, Catherine of Aragon. And it was next to his third queen, Jane Seymour, that Henry was buried.[6]

Henry's burial place was a gesture full of meaning. The world, for the foreseeable future, belonged to the Seymours: to Jane's son, Edward VI, and to his Seymour uncles, Edward (the Protector) and Thomas (the Lord Admiral).

Elizabeth got on well with her half-brother and was sympathetic to the religious radicalism of the new régime. But her relations with Edward's maternal family were more problematic. She was the daughter of Anne Boleyn, the woman whom Jane Seymour had rivalled and supplanted, and between the Boleyns and the Seymours there was no love lost. There was also, it soon became clear, no love lost between the Seymour brothers either.

These eddying currents provided at once opportunities and dangers for Elizabeth. She was to experience both in full measure.

CHAPTER 11

Stepfather: Thomas Seymour

Henry had realized that the world would be a difficult place for his daughters after his death and he had done his best to look after their interests in his will. This confirmed the family settlement of 1544. Mary was given the priority in the succession. Otherwise the two sisters were treated identically. Each was given a dowry of £10,000 on her marriage 'to any outward potentate' and an income of £3,000 a year before her marriage. If, however, either married without the consent of the council, she was to lose her place in the succession and be penalized with a reduced financial provision.[1]

An income of £3,000 placed them both in the highest ranks of the aristocracy. Mary assumed her position immediately.[2] Elizabeth, on the other hand, was still too young to live alone. Instead, she was sent to stay with her stepmother, Catherine Parr.

Queen Catherine had been well provided for, too, by Henry's will. For her 'great love, obedience, chasteness of life and wisdom', he left her £3,000 in plate, jewels and household stuff and £1,000 in money, as well as her dower in lands. This enabled her to continue to live in regal style in one or other of her many houses. Her favourite was Chelsea, and it was there that Elizabeth joined her soon after Henry's death.[3]

The arrangement seemed a sensible one for all parties. Catherine and Elizabeth had lived together before, during Henry's absence in Boulogne. They shared the same interests in learning, where Elizabeth was the leader, and in religion, where Catherine took the role of tutor. But, in the event, the move turned out to be a misfortune for both women. For it brought Elizabeth into contact – too close contact – with the leading malcontent of the new reign: Thomas Seymour, the King's younger uncle.

The division of spoils among the new ruling élite had taken place in the first, frantic weeks of the reign. It was hammered out at the council board in private and announced in a great court ceremony on 17 February. The date was strategically chosen: it was the day after the funeral of Henry, whose supposed intentions were invoked to justify the lavish distribution of offices, titles and land, and it was three days before Edward's coronation, at which the newly promoted would parade in their finery. Edward Seymour, the Lord Protector, was made Duke of Somerset; John Dudley was made Earl of Warwick and Lord Great Chamberlain, and Thomas Seymour was made Lord Admiral and Lord Seymour of Sudeley.[4]

Thomas Seymour's share – a barony, a great office of state, which gave him augmented rank in both the council and the Lords, and a substantial land grant – was not a mean one. But then Thomas Seymour had no mean opinion of himself. He was not unintelligent. But his heart ruled his head and his ambition ruled his heart. And his ambition was insatiable. He was physically impressive, too. Tall, well-built and with a dashing beard and auburn hair, he was irresistible to women: *dévôte*, blue-stocking or *politique*, they gladly gave up religion, learning and prudence at his beck and call. In short, if Surrey were the intellectual ancestor of the Elizabethan favourite, Seymour was its physical prototype. Inevitably, Elizabeth fell for him. She did not quite sacrifice her prudence. But her prudishness and protestations of virtue were exposed as a sham.[5]

Not that Thomas left it to Elizabeth to make the first move. For he was a man with a grievance about the fact that his elder brother, Edward, monopolized all real power. Thomas felt instead that the early fifteenth-century precedent of Henry VI's minority should have been followed, when authority had been divided between the king's two uncles. Now, in 1547, his scheme was that Somerset should rule the kingdom as Protector while *he* took charge of the King's person as his Governor. In view of the intensely personal nature of Tudor monarchy, the proposed division of powers was hopelessly unrealistic. But realism was not Seymour's strong suit.[6]

Unable to bring his existing relationship with the royal family into play, Seymour decided to forge another. His first idea seems to have been to marry either Mary or Elizabeth. This was a non-starter. Henry's will had required his daughters, 'on his blessing', to be ruled by the advice of his council, especially in the matter of marriage. And the consent of the council would never have been given for a match with Seymour. Frustrated of Henry's daughters, Seymour turned instead to the next best thing: the King's widow. Catherine was rich; she was also willing. She had been in love with Seymour before Henry's proposal of marriage called her to higher things. Now, after taking three husbands for duty, she was at last able to take one for pleasure and she rushed at the chance. The two were married secretly in mid-April.[7]

Catherine's conduct was understandable. Even so, it hardly measured up to the qualities of 'great love, obedience, chasteness of life and wisdom' which her late husband had seen in her. Worse was to come. Publicly, Catherine continued to enact the part of the great and pious lady by publishing the second and more radical volume of her religious writings. Privately, it was another matter as she found herself aiding and abetting Seymour in the most sordid episode in Elizabeth's life.

Seymour was now Elizabeth's stepfather. He was also, since she lived under Catherine's roof, effectively her guardian. In both capacities it was his duty to protect and nurture her. Instead, at the least, he abused his trust; he may even have sexually abused her. For Elizabeth was not only a king's daughter, with a fortune and a claim to the throne. She had also blossomed into a pretty teenage girl.

We can see what Seymour saw thanks to Elizabeth's second surviving portrait. It dates from the last year of Henry's reign or the first two or three of Edward's, when Elizabeth was in her early teens. The artist is unknown, but he was a master, who responded to his subject with a delicate sensitivity. Elizabeth stands. Her body is slightly turned to the right. But her eyes confront the observer directly and hold him with a steady stare. The set of the jaw is firm, too. Her auburn hair, parted in the middle, is her father's, as is her pale complexion, her delicate mouth and her long, slightly arched nose that would turn into the imperious eagle's beak of the older woman. But the eyes, coal-black and profound, are her mother's, Anne Boleyn's. She holds a book, with gold mounts at the corners, in front of her. The gesture displays her hands, with their long, slender, beringed fingers, in which she took such pride. It also suggests that she has been interrupted in reading. And it was clearly serious study: a slip of paper marks one place in the book; her left index finger another. To her right is another, much larger book, open on a velvet-draped lectern. She is ready, you feel, to receive Grindal or Ascham and begin her exercises in Latin or Greek or theology.

But she is no blue-stocking. She wears the latest French hood. This reveals the face, rather than concealing it like the native English gable head-dress. Her dress, with its long, wide sleeves, is in crimson damask and it is open at the front to show a magnificent underskirt, richly worked in gold embroidery. Her under-sleeves are of the same fabric. Her hood, necklace, dress and girdle are trimmed with lustrous pearls. The effect is rich, fashionable, yet elegantly restrained. There is also a hint, just a hint, of her breasts under the tightly stretched fabric of the bodice.

Finally, there is the background. Tawny curtains are pulled back on either side, leaving her head and milky-white shoulders outlined against the dark void between them. A doorway seems to lie beyond. To the left is the dull glint of a rich counterpane. She is standing in her bedroom in front of her bed.[8]

Not too much should be read into this. Peace and quiet were difficult to find in early Tudor households and bedrooms were made to serve many purposes. Her father's illustrated Book of Psalms, for instance, shows him, too, reading in his bedchamber, though he had studies, libraries and closets in plenty.

But, come the night, the books would be put away; the splendid dress with its underskirts and trimmings would be removed; the bed curtains would be drawn; and Elizabeth, wearing only a shift, would settle to sleep. The dark eyes would close and the firm jawline relax a little.

This was Seymour's opportunity. Early in the morning, he would come into her bedchamber, of which he had pocketed the key. If she were up but not dressed, he bade her a hearty good morrow and 'struck her on the back or buttocks familiarly'. If she were still in bed, he pulled back the curtains to 'make as though he could come at her'. Elizabeth would then seek refuge by retreating to the furthest corner of the bed. On one occasion, when the household was staying in his own London house, he entered clad only in his night-shirt and gown, or, as the contemporary account put it, 'bare-legged'. The modern equivalent of 'bare-legged' is 'without trousers' and the innuendo is the same.[9]

Seymour did not go so far. But he did pretty well everything else. He snatched kisses from Elizabeth; 'played' with her maids and stole embraces from her under his wife's very nose. But most extraordinary are the incidents where Seymour's antics were actually assisted by his wife, the pious, learned and Protestant Catherine Parr.

Early on two mornings, for instance, Catherine joined her husband in his visit to Elizabeth's bedchamber, where they both tickled the girl in bed. Later, in the garden, Catherine held Elizabeth while Seymour cut her dress into a hundred pieces. Reproved for her behaviour by Kate Ashley, Elizabeth said simply that 'she could not strive with all', with both her step-parents.[10]

What are we to make of all this? When an indignant Kate Ashley went on to challenge Seymour for behaviour that risked wrecking Elizabeth's reputation, his reply was as bold as brass. With an oath, he swore: 'I will tell my Lord Protector how I am slandered; and I will not leave off, for I mean no evil.' This is the 'reproof valiant'. Some modern writers go further and offer the 'lie circumstantial'. Perhaps, one has speculated, the dress-cutting incident was in fulfilment of some wager.[11]

It is interesting to consider how these excuses would play in front of a modern panel of social workers and paediatricians, all sensitized to the faintest hint of child abuse.

It is also interesting to wonder how Catherine herself reconciled her behaviour with her conscience. She had become pregnant soon after the marriage. Maybe it was the effects of this pregnancy – her first at the age of thirty-six – which unbalanced her judgement. Maybe she was trying to keep Seymour's love by going along with his infatuation for his pert and pretty stepdaughter. Maybe, though it is hard to see how, she was trying to stop things going further.

By May 1548, however, she had decided that things had gone too far. So Elizabeth was sent off to stay with Sir Anthony Denny and his wife at Cheshunt. Denny was one of the leading figures of the new régime; while his wife Joan, née Champernon, was Kate Ashley's sister. It was an ideal arrangement: Denny's position offered protection while his wife proved a kindred spirit, sharing both her sister's learning and her commitment to the new religion.

Before Elizabeth left her stepmother's roof she had an interview with Catherine. Catherine warned her of the dangers to her reputation posed by her conduct, while Elizabeth, on her own admission, 'answered little'. This was uncharacteristic: Elizabeth did not normally take reproof lying down. Her silence suggests her complicity, at least, in what had taken place. Elizabeth had tasted forbidden fruit and found she enjoyed the taste. Catherine, for her part, as Kate Ashley told Elizabeth's cofferer, Thomas Parry, was jealous at an obviously mutual attraction.[12]

Absence, however, quickly made the heart of both stepmother and daughter grow fonder. Catherine had retreated to Seymour's principal seat at Sudeley in Gloucestershire for the birth of her child. By July she was missing Elizabeth and desired her with her till 'I were weary of that country'. Elizabeth replied in equally affectionate terms, joining her wishes with those of the Dennys and Kate Ashley for Catherine's 'most lucky deliverance'.[13]

The signs were mixed. Catherine was frequently sick in the early stages of pregnancy. But later the child kicked so vigorously that both mother and father were convinced that it was a boy. In the event, the delivery on 30 August went well, though the child turned out to be a girl. Three days later, however, Catherine fell ill of puerperal fever. In pain and perhaps delirious, her old jealousies and fears returned and she accused her husband publicly of wanting and perhaps hastening her death. Seymour turned on the charm to comfort her and silence the witnesses. His efforts seem to have worked. On 5 September she made her will. This was nuncupative, or oral, her last wishes being taken down at her dictation and witnessed. The phrasing duly reflects the irregularities of speech: she left 'all' to Seymour 'wishing them to be a thousand times more than they were or been'.[14]

On 7 September she died and was buried at Sudeley in what has been called the first Protestant royal funeral. So strange, indeed, were the proceedings that they needed a sort of explanatory commentary from Miles Coverdale, the translator of the Bible, Catherine's almoner and later Bishop of Exeter. Echoing Cranmer's role at Edward's coronation, Coverdale explained away what was left of the traditional funeral ritual: the offerings of money were made only for the poor, not for the soul of the deceased, he insisted, while the candles were there only to honour her worldly status, 'and for none other intent nor purpose'. Coverdale's equivocations are an interesting anticipation of the debate on ceremonies which was to be one of the main points of contention between Elizabeth and the vast majority of her clergy.[15]

Seymour seems to have been shocked or at least surprised by his wife's death and his first thought was to break up the great household that had attended on her. The household had been required by her status as Queen Dowager, but it was also a sign of his own ambitions. And these soon reasserted themselves. So the decision to dissolve the household was reversed too. Instead, Seymour decided, it should attend on Lady Jane Grey, Henry VIII's great-niece and granddaughter of the King's sister Mary, Queen Dowager of France and Duchess of Suffolk. Jane had lived with Catherine Parr; now she was brought back to stay with Seymour on the vague understanding that he would marry her to her cousin, King Edward.[16]

Meanwhile, Seymour had revived the project for his own marriage to Elizabeth. Now that a marriage rather than a liaison was in the offing, Kate Ashley reversed her earlier disapproval and became an enthusiastic advocate for Seymour. Not that Elizabeth needed much persuasion. Those early-morning slaps and tickles had been so far from outraging her maidenly modesty that, whenever Seymour's name was mentioned in her presence, she smiled with pleasure. She particularly enjoyed hearing him praised.[17]

For Seymour the prospective marriage was as much about business as pleasure. He had earnest conversations with Thomas Parry, Elizabeth's chief accounting officer, about the state of her finances. 'How many servants she kept?' 'What houses she had, and what lands?' Where were they? Were they leased out or in hand? And had the property due to her under the terms of her father's will been finally made over to her by the formal grant known as letters patent? Then they discussed the savings that could be made by pooling and rationalizing his resources and hers.[18]

It is easy to mock this as the behaviour of a fortune-hunter. But such enquiries were integral to most Tudor upper-class marriages, in which the personal was interwoven with the sort of accountancy we associate with a company take-over or merger. And for good reason, since, proportionately, the assets involved were on a similar scale.

Inevitably word of all this leaked out. It was said that the reason Seymour kept Catherine's ladies together was for them to attend on Elizabeth after her marriage to him. And Seymour and one of his senior colleagues on the Privy Council, John Russell, Earl of Bedford, had a stand-up row on the subject.[19]

Elizabeth kept her head. Her liking for Seymour was obvious. But when Parry asked her point-blank: 'If the council would like it' – that is, if the legal conditions for her marriage established by her father's will were fulfilled – 'whether . . . she would marry with him?', she replied with studied vagueness: 'When that comes to pass, I will do as God shall put in my mind.' It was, prophetically, the sort of reply that she was always to give to questions about her suitors. In the future, such 'answers answerless' were part of a diplomatic game; in 1548, they were a matter of life and death – Seymour's death and, as it turned out, Elizabeth's life.[20]

As if his marital schemes were not enough, Seymour was also intriguing directly with the King against his brother Somerset. Somerset kept the King on short rations; this made Edward an easy target for Seymour's gifts of money and little luxuries. Seymour was reported even to be planning to capture the probably willing Edward and use his possession of the young King's person to force a change of government. Somerset had no choice. Seymour was arrested, interrogated and condemned. And Elizabeth herself narrowly escaped the fate of her admirer.

Soon after Seymour's arrest on 16 January 1549, Kate Ashley and Thomas Parry were also detained. When, a few months previously, Kate had told Parry of Elizabeth's feelings for Seymour and that she had seen them embracing, Parry had sworn himself to secrecy. 'He would rather be pulled with horses than he would disclose it.' In the event, he proved less resolute and it seems to have taken only a couple of nights in a cold cell in the Tower for both Elizabeth's governess and her cofferer to sing like canaries.[21]

All this was reported to Elizabeth in lip-smacking detail by Sir Robert Tyrwhitt, who had been sent to Hatfield by the council to interrogate her. Both Tyrwhitt and his wife had been in Catherine Parr's service: he as her Master of the Horse; she as one of the ladies of her privy chamber. In this capacity she had noted, with evident indignation, Seymour's treatment of his wife and his behaviour to her on her deathbed. With such a vantage point, Tyrwhitt himself was fully aware of the rumours about Elizabeth and Seymour. He must have thought it an easy matter to wring confirmation out of Elizabeth. But he misjudged her badly.[22]

When he told her of the arrests of Kate and Parry, she 'was marvellously abashed, and did weep very tenderly a long time'. But she brushed aside her tears long enough to ask if they had confessed. Tyrwhitt showed her their signatures on their depositions; assured her that they had told everything and advised her to do the same. Elizabeth was not so easily outwitted. She called Parry a 'false wretch' but otherwise said nothing. Tyrwhitt now tried threats, reminding her that 'she was but a subject'. That failed too. Finally, he had to admit 'that he was not able to get anything from her but by gentle persuasion'. 'For I do assure your grace,' he continued in his report to Somerset, 'she hath a good wit, and nothing is to be gotten from her but by great policy.'[23]

And even with his 'great policy', all that he got from Elizabeth was a confirmation of the tales told by Kate and Parry. She was willing to admit to gossip about Seymour's interest in marrying her and to the discussions about her estates. But throughout she insisted absolutely that both she and her servants were of one mind that she should 'never . . . marry, neither in England nor out of England, without the consent of the king's majesty, your grace [Somerset], and the council's'. Elizabeth, clearly, knew the clauses of her father's will by heart. Henry VIII, with his pernickety attention to verbal detail, would have made a good lawyer; his daughter, in this as in so many ways, was a chip off the old block.[24]

Tyrwhitt, for his part, was frustrated. He was sure that Elizabeth and the pair in the Tower were singing from the same hymn-sheet (he says 'they all sing one song') because 'a secret pact had been made between [them], never to confess to anything to the [in]crimination of the other'. But in the face of their mutual agreement on the key point he could prove nothing. Sensing her antagonist's defeat, Elizabeth began to move to the offensive. In her letter to Somerset of 28 January, she boldly required the Protector to counter the rumour that she was 'in the Tower, and with child by my Lord Admiral'. These were 'shameful slanders', which the council should officially repudiate.[25]

At the beginning of February her composure was temporarily shaken when she learned that Kate, under intense questioning, had revealed details of Seymour's old romps with her, including the dress-cutting episode. The stories damaged her dignity. But they did nothing, as her forensic mind quickly spotted, to imperil her head. So she soon resumed her high tone.

Somerset himself reproved her 'that I seem to stand in mine own wit in being so well assured of mine own self'. Elizabeth made a token excuse. But her modest mode was never very convincing and she immediately went on to repeat her demand for a proclamation officially repudiating the slanders circulating about her. Somerset had agreed to take action against any tale-bearers that she could name. She thanked him. But, as she shrewdly pointed out, for her personally to turn in offenders would be 'but a breeding of an evil name of me that I am glad to punish them and so get the evil will of the people, which thing I would be loathe to have'.[26]

The tone of high dignity changed once more to girlish rage and impotence when she was informed that Lady Tyrwhitt would replace the disgraced Kate as her Lady Mistress. First she stormed: 'Mrs Ashley was still *her* mistress, and that she had not so demeaned herself as that the council should now need to put any more mistresses upon her.' Then came the tears: 'she wept all that night and loured all the next day'. Tyrwhitt, sensing that controlling Elizabeth was a problem that no mere man could deal with, drily observed that, 'if I should say my fantasy', it was 'more meet that she should have two [lady mistresses] than one'.[27]

By 7 March, however, her confidence was fully restored and she boldly wrote to the Protector, requesting that the council 'be good' to both Kate Ashley, because of her care in bringing Elizabeth up, and to her husband, because he was Elizabeth's kinsman.[28]

A fortnight later, on 20 March, Seymour was beheaded. The execution was unpopular and Hugh Latimer, the star preacher of the new régime, was wheeled out to try to quell the murmurs by blackening Seymour's memory. Latimer's indictment was comprehensive: Seymour was a debaucher of women, an atheist or at least a disbeliever in the immortality of the soul, and, of course, a traitor. 'He was, I heard say, a covetous man: . . . I would there were no more in England. He was, I heard say, a seditious man, a contemner of common prayer. I would there were no more in England. Well he is gone. I would he had left none behind him.'[29]

Kicking a man when he is down, still more when he is dead, is something that even the worst of us revolt from. Only the very righteous, indeed, like Latimer, seem able to speak ill of the dead with such conviction.

Elizabeth, in this respect as in most others, was made of mortal clay. Her loyalty was to Seymour. Her corresponding contempt for the Latimer we can only guess. But her reaction probably marked the first step in the road which led this 'Protestant princess' to despise her clergy almost as vehemently as Seymour is supposed to have 'contemned' common prayer.

CHAPTER 12

Adulthood

Most of us have a moment when we can say we grew up. For Elizabeth, that moment was the Seymour affair. Before, there was much that was girlish in her behaviour; after, almost nothing. It was also an initiation, and a brutal one, into the world of adult sexuality. Almost all the men that she subsequently loved, or pretended to love, resembled Seymour. And all the affairs ended in the same way, in frustration and, in the case of the last, again in death. Elizabeth's motto was '*semper eadem*' – 'always the same'. Nowhere was this truer than in her sexuality.

We can approach the question in two languages: our psychology or their theology. We can put Elizabeth on the couch or in the confessional.

The psychological dossier might go like this. Elizabeth had been abused by Seymour. Like many abused children, she had fallen in love with her abuser. But this love was associated with feelings of guilt and self-loathing which prevented its fulfilment. In the fullness of time, the circle turned, and Elizabeth, also like many abused children, became a sort of abuser in turn. Denied fulfilment herself, she took pleasure in denying it to others. She played with her suitors like an angler with a fish; and she enforced a reluctant celibacy on her favourites and maids.

Elizabeth, of course, would have understood none of this. If she thought about her experiences at all she would have approached them in the language of theology. She had been taught a stern and unbending faith. Desire, she knew, was inseparable from sin and sin led inevitably to punishment. Her relationship with Seymour and its denouement on the scaffold was a textbook illustration of the truth of this teaching. It would be hard to think of a more effective antidote to desire, or of something more liable to quench its first flames.

So the two languages, though very different, deliver a similar verdict. This is scarcely surprising, as the psychologist is today's priest and his 'science' a modern religion. But the fact that their conclusions are similar does not make them correct. Both are dangerously, mechanistically, determinist. And Elizabeth was the most wilful, the most mercurial of women.

She was also, within a highly restricted circle, the most loyal. Here again the Seymour affair has a message and a much less equivocal one. It was quickly grasped by Tyrwhitt. 'Her love for Ashley', he noted, 'is to be wondered at.' So much so indeed that 'she will no more accuse Ashley than herself'. Nor would Elizabeth betray Parry. Parry was, as Holbein's unsparing portrait suggests, a fat, self-satisfied Welshman. He was also, as Tyrwhitt had no difficulty in demonstrating, an embezzler who was abusing his office to cheat Elizabeth. None of this mattered to her. Parry was hers, and that was enough.[1]

This loyalty to her servants was to be another enduring characteristic. It goes back, I suppose, to the insecurities of her relationships with her parents. With her mother, her relations were early and strong. With her father, they were late and strong. In between, there was a gap that represented most of her childhood. Into the void had stepped her servants and officers, people such as Kate Ashley. They were her fixed points in a changeable world and on them her security depended. In her letter to Somerset pleading for Kate, she had distinguished between the duty we owe to our parents and to those who bring us up and she had placed the latter higher than the former. She meant exactly what she said and her conduct shows that her words were translated into action.

Elizabeth's loyalty to old servants is one of the few sentimental streaks in her character. Utterly unsentimental, however – though it relied heavily on the manipulation of sentiment – was her attitude to popular opinion. Throughout the affair, Elizabeth showed herself acutely sensitive to her public reputation. She worried about the effect of the lurid rumours of her having given birth to Seymour's bastard. But she was equally concerned about the effect of a too-crude attempt to suppress the stories. Already, in her mid-teens, Elizabeth was a mistress of propaganda. Propaganda has been defined as 'that branch of the art of lying which consists in very nearly deceiving your friends without quite deceiving your enemies'. In the fullness of time, Elizabeth's people were to be more than deceived; they were to be besotted. The Seymour affair and its aftermath show her honing her talents.

In short, Elizabeth's behaviour in the Seymour crisis is almost beyond praise. She was decisive yet properly calculating. She played her hand to the full, yet never overplayed it. It is an astonishing performance for someone only in their sixteenth year. But historians have made her actions seem even more remarkable than they were by emphasizing that everything she did was alone and unaided.

Now, in a formal sense, Elizabeth *was* alone. She had left Denny's tutelage late in 1548 and set up her household (140 strong, according to Parry) at Hatfield. But that does not mean that she was bereft of advice. Soon after Seymour's arrest, Denny himself and William Paulet, Lord St John, arrived unexpectedly at Hatfield. First they dined; then they arrested Kate Ashley and Parry. After dinner we know that Kate and Parry were able to agree on tactics to cope with their forthcoming ordeal.[2] It seems inconceivable that Paulet and Denny had not also discussed strategy with Elizabeth herself. Denny was one of the smoothest operators of the age; Paulet, one of its great survivors. He served four Tudors – Henry VIII, Edward VI, Mary I and Elizabeth herself – continuously from the days of Cardinal Wolsey to his death, still in harness as Lord Treasurer, in 1572. By then he was senescent, crotchety, corrupt, proto-Catholic and hopelessly incompetent. But Elizabeth had a soft spot for him. If there was a single thread in his career, it was the continuity of his experience as a royal councillor. He, therefore, is the most likely source of the advice to Elizabeth to insist under interrogation that she and her servants had been prepared to entertain Seymour's advances only subject to the agreement of the council. It was this advice that saved Elizabeth. If my guess about Paulet's role as its author is right, it is easy to see why he joined the select group of her servants, whom only death would part from her.[3]

CHAPTER 13

Hatfield: Further Education

The years of the Seymour affair, 1548 and 1549, were not all bad for Elizabeth. She emerged from the affair itself toughened and tempered. She also – when the ugliness of court politics did not obtrude – enjoyed a remarkably happy personal life.

Elizabeth's relations with her half-brother and -sister remained good, even though Edward and Mary were increasingly at loggerheads with each other. And Elizabeth's intellectual development advanced in leaps and bounds. Above all, she was domestically secure in a household that was staffed with old servants who had become old friends. And even when a new face had appeared – as, for instance, when Kate Champernon had married John Ashley (or Astley) in the mid-1540s – it tended to be a sympathetic one. Ashley was Elizabeth's kinsman; he also shared fully in the intellectual and religious tastes of her household. A little later, he recalled his experiences in her service. He remembered:

> our friendly fellowship together at Cheshunt, Chelsea and here at Hatfield, her grace's house; our pleasant studies in reading together Aristotle's *Rhetoric*, Cicero and Livy; our free talk mingled always with honest mirth; our trim conferences of the present world, and too true judgments of the troublesome time that followed.[1]

The tone of Ashley's remarks will be familiar to anyone who has attended a college reunion and reminisced about their memories of the interesting tutorials, the bosom friendships and the long hours setting the world to rights as the fire burned low and the lights dimmed. Ashley's memories were rose-tinted, as such memories usually are. But they touched an essential truth. When Erasmus had described Henry VIII's court as being like a university, it was the old flatterer laying it on with a trowel. Elizabeth's household, however, with its central mission to educate its young mistress and its close connections with St John's, Cambridge, was *genuinely* a university extension college. Indeed, her household was more advanced than its sister-foundations by the Cam and the Isis. Unlike any university college till the twentieth century, it was co-educational. And it was run by the greatest and most influential teacher England produced before the age of Arnold at Rugby and Jowett at Balliol. This was Roger Ascham, who himself combined the careers of don and courtier. Ashley's recollections are addressed to Ascham; while Ascham, for his part, regarded Ashley's friendship 'amongst my chief gains gotten in the court'.[2]

And Ascham is our principal witness to Elizabeth's life and development in these formative years. William Grindal, Elizabeth's former tutor and Ascham's protégé, had died prematurely in early 1548 and Ascham succeeded him immediately. Elizabeth gained intellectually, for Grindal, despite all his qualities, did not have the outsize personality of the great teacher. And we gain too. For Ascham was not only the greatest teacher of the age; he was also one of its most notable gossips. We get the flavour of the man from his wide and chatty correspondence and his *Schoolmaster*, published in 1570. Ostensibly, the *Schoolmaster* is a book on educational method; actually, it is Ascham's memoirs. And Elizabeth, as Ascham's star pupil, figures heavily in both the letters and the book. Between them, they offer a testimonial to her achievements; they also show how the early Elizabethan style was formed. And this is crucial. For with Elizabeth style was everything.

First, Ascham describes Elizabeth's scholarly routine. The beginning of the day, he writes, 'was always devoted by her to the New Testament in Greek, after which she read select orations of Isocrates, and the tragedies of Sophocles, which I judged best adapted to supply her tongue with the purest diction, her mind with the most excellent precepts, and her exalted station with a defence against the utmost power of fortune'. For her religious instruction, 'she drew first from the fountains of Scripture', and subsequently elaborated her knowledge by the study of

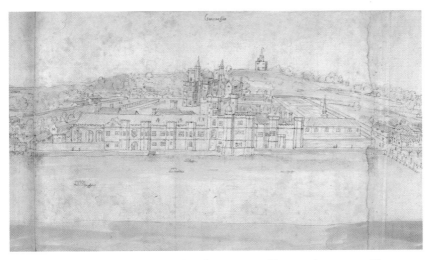

Elizabeth was born at Greenwich Palace in 1533. She was given a magnificent christening in the Friars' Church – the long building to the right with the central spire.

Henry VIII.
In order to marry Elizabeth's mother, Anne Boleyn, he broke with Rome and made himself Head of the Church of England. These decisions set English history on a new course and shaped the lives of his three children. Elizabeth, nearest in character to Henry VIII, was the most successful in exploiting his legacy.

(*Left*) Catherine of Aragon, Henry's first wife, fat of face and figure after fifteen years of miscarriages and still-births had failed to produce a son. (*Right*) This portrait of Anne Boleyn captures something of the sexuality which so attracted Henry VIII. Elizabeth never mentioned her mother after Anne's execution for adultery and incest.

(*Left*) Jane Seymour, Henry's third Queen, represented as the submissive little wife in contrast to Anne Boleyn's vampish style. She died soon after giving birth to Henry's longed-for son, Prince Edward. (*Right*) Catherine Parr, Henry's sixth and last wife. In 1544 Elizabeth spent a formative few months in her household, experiencing at first-hand both female rule and the Queen's religious enthusiasm.

Prince Edward, painted for his father in the last months of Henry VIII's life. In the background is Hunsdon House, one of the Hertfordshire manors where Edward was brought up with his half-sisters. His relations with Elizabeth, only four years older, became close, while her education benefited from the presence of Edward's tutors.

Probably done for her father at the same time, this painting shows Elizabeth as the model Renaissance princess. A finger and a book-mark keep her place in the book she is holding; the Bible is open on a lectern. But there is also a hint, in the dark eyes and the bed with its curtains open behind her, of the sexuality she inherited from Anne Boleyn.

Henry VIII and his family (detail).
This painting commemorates Henry's decision to reinstate both his daughters, Elizabeth on his left and Mary on his right, in the succession after his son, Edward, who stands at his knee. Jane Seymour, by then dead but Edward's mother, is shown next to Henry, rather than his then wife, Catherine Parr, who was childless.
From this point, Elizabeth's relationship with her father became close and she idolized him.

The official version of the reign of Edward VI: above left, Henry VIII,
on his death-bed, hands over power to his son who sits enthroned (centre) with his
older uncle, Edward Seymour, Duke of Somerset, standing beside him as Lord
Protector. Right is the Council, with Protestant 'goodies' (above) and (below) the
Catholic 'baddies', who witness with horror the destruction of Popery.

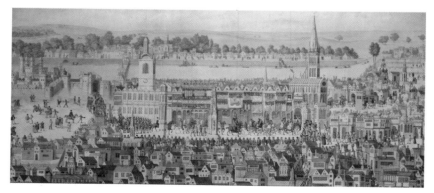

The Eve-of-Coronation Procession of Edward VI.
There were the usual pageants, dealing with Edward's imperial crown, his parents,
his education and his embracing of true religion. Elizabeth listened to the speeches
explaining the pageants at her coronation with rapt attention; the nine-year-old
Edward was more interested in the antics of a Spanish acrobat at St Paul's.

Thomas Seymour, Lord Seymour of Sudeley, was Edward VI's younger uncle. He wanted to share in the power of his elder brother, Lord Protector Somerset. To this end, he married Henry VIII's widow, Catherine Parr. The goings-on when he tried to make love to Elizabeth in her stepmother's house became public knowledge. Seymour was condemned for conspiracy and Elizabeth, aged only 16, had her first brush with the law.

Mary Tudor, born in 1516, remained devoted to her mother's memory and her religion. She became the focus of Catholic opposition to Edward VI's increasingly Protestant policy.

Roger Ascham. Gossip, teacher, sportsman and Elizabeth's most influential tutor. He was a leading light in the golden generation at St John's College, Cambridge, in the 1530s.

Sir Anthony Denny, as Chief Gentleman of the Privy Chamber, was Henry VIII's chief of staff. After the Seymour affair, he gave Elizabeth a sympathetic refuge in his house at Cheshunt.

Thomas Parrie.

Sir Thomas Parry was a leading member of the 'Welsh Mafia' in Tudor times. Described by the Spanish ambassador as 'a fat Welshman', he was Elizabeth's principal business manager during her years as princess. After her accession, he became a leading councillor but died in 1560.

Hatfield Old Palace in Hertfordshire. Elizabeth acquired the house – built by Cardinal Morton – in 1549, and it became her principal residence until her accession.

'St Cyprian, the *Common-places* of Melancthon, and similar works, which convey pure doctrine in elegant language'. In Latin, she read with Ascham 'almost the whole of Cicero, and a great part of Livy: from those two authors her knowledge of the Latin language has been almost exclusively derived'. Somewhere, time was also found for modern languages and for her training in oral proficiency. This was notable too: 'French and Italian she speaks like English; Latin with fluency, propriety and judgment' and Greek passably. Her handwriting, in either the Greek or Latin alphabets, was supremely elegant, and she was an accomplished musician – though, according to Ascham, she did 'not greatly delight' in it.[3]

It is a catalogue of formidable achievement. We should, of course, discount a little for Ascham's partiality. But only a little, as his compliments are by no means indiscriminate. Take, for instance, his comments on Elizabeth's oral Greek, which he notes she spoke 'frequently, willingly and moderately well'. At first sight this sounds like praise, albeit fairly faint. But a practised writer and reader of student reports will sense that what Ascham is really saying is that 'Elizabeth's confidence outruns her ability'. This sort of coolness is reassuring; it means that we can take his praise elsewhere more seriously.

So Ascham's reliability is not the problem. Instead, curiously, our difficulty lies with Elizabeth's very achievement. This is so substantial, and in fields so alien, that many writers are merely dazzled by it. But, fortunately, Ascham, garrulous and egotistical as always, offers plenty of clues as to what he thought the effect – as opposed to the content – of his teaching should be. He singles out three things. It should inculcate moral precepts or maxims; it should fortify the mind against adversity; and it should offer a model of style. Even theology, where we might think content was all-important, was not free from this tyranny of style, and Ascham explains that St Cyprian among the ancients and Melancthon among the moderns had been chosen for particular study because they 'convey pure doctrine in elegant language'.

Much of this was common to the general pedagogy of the age. And there are clear signs that Elizabeth absorbed it. She does, genuinely, seem to have found her learning a comfort and, even in old age, translated Boethius's *Consolation of Philosophy*. Less happily, her later letters and speeches are filled with saws, phrases and fables culled from her reading. Bearing in mind how much and how widely she had read, her repertoire is rather small and conventional: the wisdom of Solomon, the presumption and punishment of Prometheus, the perils of Scylla and the lure of the Sirens make tediously frequent appearances. But then, her education did not favour originality. For her, as for Ascham, repetition held no disgrace: if a thing had been said once supremely well, why ever say it differently?[4]

On the other hand, the tortuous prose of Elizabeth's maturity, of which the mosaic of allusions is a part, does present a difficulty. For it is contrary to Ascham's stated preference for a plain, elegant, unforced style. Authors who follow this model are praised; those who deviate from it are pilloried. So Edward Hall, the chronicler, is denounced for his 'indenture English', his 'strange, inkhorn terms' and his repetitiousness, 'as though he had been, not writing the story of England, but varying a sentence in Hitchin school'. Sallust, another complex writer, is similarly denigrated, as one in whose 'writing is more art than nature, and more labour than art'. These are shrewd criticisms, sharply expressed. One wonders what he would have made of one of his former pupil's more elaborate compositions. Instead, wisely, Ascham's *Schoolmaster* has nothing whatever to say about Elizabeth's style. Praise is lavished on her other accomplishments. But of her writing, there is not a word.[5]

It was very different while Elizabeth was still Ascham's pupil. Then his letters claim her as one of his own. He aimed to inculcate his style, partly by his choice of authors who exemplified it, and partly by his way of teaching them. This was by double translation. His pupils had to translate a passage of (say) Latin into written English. Then, sometime later, when the immediate memory had faded, they had to retranslate it into Latin, trying, as nearly as possible, to reproduce the original word for word. The method, Ascham claimed, not only taught students grammar and vocabulary; it also forced them to enter into the mind of Caesar or Cicero and, as it were, relive their style from the inside.[6]

By these means, Ascham told his correspondent, Elizabeth came to prefer a

> style that grows out of the subject; chaste because it is suitable, and beautiful because it is clear . . . Her ears are so well practised in discriminating all these things and her judgment is so good, that in all Greek, Latin and English compositions there is nothing so loose on the one hand and so concise on the other which she does not immediately attend to, and either reject with disgust or receive with pleasure as the case may be.

Clearly, there was a change in Elizabeth's tastes *en route*. One of Elizabeth's shrewder biographers is Mandell Creighton. He was a Victorian bishop as well as a historian and with a cool clarity he saw through Elizabeth's performances: 'her love of simplicity soon passed away. Indeed, it was never real, and Ascham's mention of it shows that Elizabeth was acting a part.'

Cool indeed. But, I fancy, like many Victorian judgements, it is also too harsh. For Elizabeth's preference for simplicity at this time extended to many areas other than her literary tastes. Her letters were direct and straightforward. She also, again according to Ascham, chose plainness in dress: 'with respect to personal decoration,' he writes, 'she greatly prefers simple elegance to show and splendour, so despising the outward adorning of plaiting the hair and wearing of gold.'[7]

Even more enthusiastic testimony for Elizabeth's chaste style comes from John Aylmer. Aylmer, later Bishop of London, was a friend of Ascham and the beloved tutor of Lady Jane Grey. This means that he knew Elizabeth when both she and Jane were living in the Seymour household. Armed with this eyewitness knowledge, he is emphatic. 'I know it to be true,' he writes, that Elizabeth, for the whole of Edward's reign, never wore the rich jewels and clothes left her by her father. Instead, she offered a 'more virtuous example' than the writings of Saints Peter and Paul, her 'maidenly apparel' making the ladies of the court 'ashamed to be dressed and painted like peacocks'. The contrast was particularly sharp on the occasion of the reception at court of Mary of

Lorraine, the Queen Dowager of Scotland, in October 1551. The arrival of the Franco-Scottish party, Aylmer tells us, triggered a revolution in *coiffure*, which turned all heads but Elizabeth's. 'All the ladies went with their hair froused, curled, and double curled, except the Princess Elizabeth.'[8]

Even her taste in architecture was plain. When she finally arrived at a settlement of her estates with Edward's government, she acquired as her townhouse the Protector's former palace of Somerset House. This was not only the largest and most magnificent private residence in London, it was also the newest and most fashionable. It was built throughout in the new, Italian style, and the frontage to the Strand was the first attempt in England at a harmonious, symmetrical classical façade. It was also singularly free from ornament.[9]

Elizabeth, who always knew a good deal when she saw one, may have been more interested in the building's substance than its appearance. But her account book for 1551–2 shows an interest in the pure classical style. Account books were often decorated with calligraphic flourishes and 'historiated' – that is, pictorial – initials. The penmanship of these decorations was frequently splendid, but the illustrations tended to be disappointing. This reflected the provincial backwardness of contemporary English artists. But Elizabeth's account book is a shining exception. It contains five capital letters, each illustrated with a pen-and-ink drawing. The drawings themselves are accomplished, and their artist was either an Italian or a confident imitator of the latest Italian mannerist art. The subjects are Italianate, too. Each shows the personification of one of those entities which so occupied the Renaissance mind: *Tempus* (Time); *Dolor* (Grief); *Natura* (Nature); Temperance and Justice. It is easy to guess how some of these might have applied to Elizabeth at that time. Grief confronts a shipwreck, but the inscription confidently asserts that afflictions are in vain. Nature restores dismembered limbs, and Justice has her sword and scales broken. In the case of the remaining two personifications there is no need to guess. Elizabeth herself invoked Time in one of her off-the-cuff comments at her eve-of-coronation procession; while Edward VI, according to William Camden, Elizabeth's contemporary biographer, was in the habit of referring to Elizabeth as his 'Sweet Sister Temperance'.[10]

In so describing his sister, Edward was not only praising her taste and life-style but her religion. Or rather, he was praising her life-style as the *consequence* of her religion. Similarly, when, in a notorious letter, Edward asked his stepmother Catherine Parr to dissuade his other sister, Mary, from the 'foreign dances and merriments which do not become a most Christian princess', he was implicitly denouncing Mary's conservatism in religion which led, in his view, to an equally old-fashioned and deplorable mode of life. The thought of the lugubrious Bloody Mary as a bit of a goer and the exuberant Gloriana as a Puritan Maid rather taxes the imagination. But certainly the contrast was there for those with eyes to see in the early 1550s.[11]

And Elizabeth did her best to reinforce her side of the image. Protector Somerset had informed her, probably late in 1548, that the custom of giving New Year's gifts was to be abandoned. They were expensive; they were also part and parcel of the superstitions of Merry, Catholic England which the Reformers were determined to root out. A handful of the ladies and gentlemen of the court, headed by Somerset's own wife, the redoubtable Duchess Anne, ignored the prohibition (one hopes out of kindness for the lonely little boy who was their king) and gave Edward gifts. Elizabeth was not one of their number. Instead, on 2 January she wrote a letter of explanation to her brother. She began by endorsing Somerset's proposed ban on expensive gifts as 'a very wise arrangement'. And she further excused herself from sending even her by-then accustomed translation of an improving work on the grounds of her lengthy indisposition, which had 'so wasted' her studies that the task, 'formerly not difficult', had become impossible. Terrible, when she herself occupied Edward's throne, would have been the fate of one of her own courtiers who offered so miserly a token to her.[12]

The proposed ban on New Year's gifts was quickly forgotten. Gifts Rolls, recording the presents given and received by the King, are known to have existed for 1552 and 1553 and in December 1551 the Privy Council ordered the payment of £300 towards the cost of the King's presents. Nor did Elizabeth hold out. In January 1552 she spent £32 3s. 10d. on gilt plate 'for New Year's Gifts' and she also received a substantial present from the King. What she gave in return is not clear. But probably she continued with her practice of giving her own translations of religious texts.[13]

One of Elizabeth's New Year's gifts survives. It is her translation from the Italian into Latin of Bernardino Ochino's 'Sermon on the Nature of Christ', prefaced with a Latin dedicatory letter to Edward dated at Enfield on 30 December. The year is not stated. Many of Ochino's sermons were on controversial points of the new theology, including predestination. Probably in 1551 fourteen of these theological sermons were translated by Anne Cooke, who was to become sister-in-law of Elizabeth's future minister William Cecil and wife of her Lord Keeper, Nicholas Bacon. But Elizabeth avoided such vexatious issues and instead selected for her own translation something at once more straightforward and passionate: Ochino's sermon on Christ and His sufficiency for our salvation. It was a perfect choice for a king who by Christmas 1552 was already mortally sick.

Elizabeth explained her choice more circumspectly. 'Since', she wrote to Edward, 'the subject is the nature of Christ, [it is] a matter well becoming to you who speak daily of our Lord, and who is closest of all to him in the position and dignity you have upon earth.' Here, at least, Elizabeth is absolutely consistent. She had referred to her father (in a similar dedicatory letter) as 'a god upon earth'; now her little brother was next to Christ 'in position and dignity'. In the fullness of time she would inherit their throne and, in her own eyes at least, their quasi-divinity.[14]

Elizabeth also received such presents from others. About this time, for instance, Jean Belmain, who was French tutor to both Elizabeth and her brother, gave her a French translation of one of St Basil's Greek works. It was an epistle on the 'Virtues of the Single Life'. The saint explained that only the single life enabled the soul to devote itself to God; marriage, on the other hand, enmeshed it in the world and its cares. It was, apparently, an odd choice of gift for a young Protestant princess, since Protestants were enthusiasts for marriage, including even the marriage of priests. But no doubt Belmain knew his pupil's mind. And, according to Robert Dudley, who should have known, that mind was already made up. Much later Dudley was reported as saying: 'I have known her, from her eighth year, better than any man upon earth. From that date she has invariably declared that she would remain unmarried.'[15]

But, as Dudley was the first to hope, on that issue, above all, a woman was entitled to change her mind.

We normally think of teenage years as a period of chaotic experiment. But Elizabeth's, thanks to the forcing-house of her royal status, were all of a piece. And the piece was uniformly plain. Should we follow Creighton and see this as play-acting? The temptation to do so is considerable. Elizabeth was a born actress. And art had perfected nature. Her education focused on rhetoric: that is, on the use of language as a vehicle to communicate thoughts and feelings, and, at least as importantly, to conceal them. Elizabeth had become a mistress of all its mysteries: she could not only perform a script, she could write and direct it too. And it is easy to see the purpose of the part she was playing in these years. The Seymour affair had presented her to the world as no better than she should be and much worse than her friends hoped. How better to turn away the slings and arrows of outraged opinion than by playing the Puritan Maid, whose plain dress and plainer manner proclaimed her invincible virtue?

This indeed is part of the story. But it cannot be the whole. For Elizabeth's style was not unique to her. Instead, it was typical of almost all the leaders of intellectual and political opinion in the brief reign of Edward VI. Power politicians such as Seymour, Somerset and John Dudley, Earl of Warwick, and intellectuals such as Thomas Smith, William Cecil and John Cheke, who made the leap from Cambridge to the court, were progressives and thought of themselves as such. Their progressivism embraced every field of activity: like Elizabeth, they were Protestants in religion, classicists in art and amateurs of what C. S. Lewis pejoratively called the 'drab' style in literature. This range led to some strange juxtapositions (the austere Cranmer was painted with an improbably erotic classical relief in the background). Elizabeth, less constrained than the Archbishop, was a leader of the movement and her household was one of its principal academies. The movement proved short-lived. But its collapse was due, not to its own internal contradictions or to the hypocrisies of its practitioners (though they had their share). Instead, it was overthrown by the pressure of external circumstances. And these were to bear particularly heavily on Elizabeth herself. The result changed both her and her opinions, and meant that the adult Elizabeth in some ways presented the antithesis of her youthful, Edwardian self.[16]

CHAPTER 14

The Dudleys

Elizabeth's retreat behind the university-like walls of her household was a measure of the damage done to her public status by the Seymour affair. But both the damage and the consequent retreat have been greatly exaggerated. Historians, hampered by the uncertain chronology of Edward's reign, have talked of a two-year rustication from the court. Actually, Elizabeth's enforced exile lasted only for a few months. For she had been fortunate in her choice of enemies.

It was Protector Somerset who had acted as a hard judge of her behaviour. But the Protector's destruction of his brother Seymour proved fatal to his own reputation. He had come to power on a wave of golden opinion. But his popularity had been quickly dissipated by his arrogance, his courting of the lower orders and his monopolization of power. Fratricide was the last straw. In October 1549 Somerset panicked in the face of hardening opinion on the council and tried to play the royal card: he retired to Windsor with Edward and proclaimed that the King was in danger. But Edward refused to co-operate. Instead he sulked, complaining loudly of a cold and his gloomy, old-fashioned quarters in the castle. The council, meanwhile, held its nerve.

The result was the first bloodless transfer of power in England for decades.

Both sides had been friends and colleagues. Even more importantly perhaps, they shared a common educational background. In their student days, they had studied the history and theory of Roman politics. Subsequently, their professional training had been in English law. Both predisposed them to a 'constitutional' view of power. This emphasized that political office was a public trust, and that duty to the State or 'commonwealth' transcended loyalty to any particular individual. The new tone is best caught in a letter from Sir Thomas Smith, the royal Secretary with the beleaguered Protector at Windsor, to his colleague Sir William Petre, the other royal Secretary, who had remained with the Council in London. 'Join you with us . . . that things may be brought to moderation,' Smith pleads. There had been a terrible revolt in the summer: 'Let Christian charity work with you . . . that this realm be not made in one year a double tragedy . . . and scorning stock of all the world'.[1]

This new, consensual politics was to have profound long-term implications for Elizabeth's own reign. Almost all her leading councillors had either been members of the Council under Edward or had shared the same educational background. For them, the fall of Somerset was a traumatic event, a sort of fall of 'political man'. In retrospect, it could be seen as the beginning of the end of the Edwardian régime. This had begun with high hopes for a reformed Church and State. But it was to end in squalid, self-destructive acrimony. It was a road down which Elizabeth's councillors were determined never to travel again. This determination, and the sense of a collective will which it embodied, were to put a very real brake on Elizabeth's own power when she became queen.

In the short term, however, Somerset's fall meant only liberation for Elizabeth. Somerset, the subject who had presumed to sit in judgement on her, a king's daughter, was deprived of the protectorate; he forfeited a tranche of his lands and was excluded, though only temporarily, from the Council. The protectorate was not revived.

Instead, John Dudley, Earl of Warwick, emerged as the first amongst equals on the council.

The history of the Dudleys and the Tudors was intertwined – like a tree and, Dudley's many enemies would have said, its parasitical ivy. John Dudley's father, Edmund, had been minister to Elizabeth's grandfather, Henry VII, and had been the most notorious agent of the King's rapacity in his later years. But when Elizabeth's father, Henry VIII, came to the throne he was determined to make a fresh start. And he signalled this by executing Edmund as a traitor. Thereafter John had worked slowly and cautiously for his rehabilitation. But now that he and his family stood once more on the threshold of supreme power, the memories of his father's treason were revived.[2]

Elizabeth was probably indifferent to such historical niceties (as she certainly was in the case of John Dudley's son, Robert). More immediately worrying was the fact that John Dudley seemed to be keeping risky company. In his attack on Somerset he had been supported by the most important surviving Catholic politicians. There had even been rumours that Mary, Elizabeth's Catholic sister, would be made Regent for her brother and that Catholicism would be restored or at least tolerated. From Elizabeth's point of view this would have been stepping out of the frying pan into the fire.

But John Dudley duped and threw over his Catholic allies during an extraordinary Christmas. Normally business was suspended for the twelve days of festivities; in 1549 political activity was unremitting.[3]

And it was on Christmas Day itself that the King and council issued a circular letter to the bishops which reaffirmed that it was full steam ahead with reform in the Church. The letter began by recalling that parliament had already agreed to a new, reformed Prayer Book in English. But, following Somerset's fall, 'divers unquiet and evil disposed persons . . . have rumoured that they should have again their old Latin service, conjured bread and water and such like superstitious ceremonial'. Using the new political language, the letter sharply reminded its addressees that the Prayer Book was not Somerset's private action; instead it was 'our act and that of the whole realm'. As such, it would stand.[4]

The Christmas Day letter was one sign of future policy. The other was Elizabeth's own arrival at court for the Christmas festivities. She had come on 19 December and had been received with much pomp. Her arrival was noted by the imperial ambassador, Van der Delft, who immediately spotted its significance. Hitherto Mary, as the King's elder sister and heiress to the throne, had been treated with especial distinction, while Elizabeth had been under a cloud. Now Elizabeth was preferred to Mary. This was because she shared the same religion as John Dudley and the new ruling group. Or, as the ambassador put it more quaintly, Elizabeth 'was more of their kidney'.[5]

CHAPTER 15

Property

Elizabeth reaped material as well as political benefits from the fall of Somerset. And the process shows her in a harshly acquisitive light. Perhaps the explanation was psychological. As a little girl, she had lost everything with her mother's death; thereafter, she was never to voluntarily let go of anything again. But her genes spoke as loudly as childhood memories. Rapacity was the besetting sin of the Tudors. It marred Elizabeth's great-grandmother, Lady Margaret Beaufort, for all her saintliness; her grandfather, Henry VII, for all his wisdom; and her father, Henry VIII, for all his *bonhomie* and magnificence. Elizabeth was no exception. But maybe the times spoke loudest of all. The reign of Edward VI was a bonanza. The reformed preachers might denounce ambition and avarice; but Edward's councillors, who were their patrons, helped themselves to lands and titles at the King's expense. Elizabeth showed, not for the last time, that a woman could play a dirty game just as well as any man.

For the first six months of Edward's reign the annual income of £3,000, due under her father's will, had been paid to her in cash, in irregular instalments of a few hundred pounds at a time. Then the process of assigning her lands of equivalent value began. In the case of her sister Mary, matters had been quickly and generously settled and her letters patent were issued on 17 May 1548. With Elizabeth, on the other hand, the haggling over which lands and on what terms went on for over two years.[1]

Late in 1548, for instance, Thomas Seymour asked 'if the Lady Elizabeth had had her letters patent out?' Elizabeth's cofferer, Thomas Parry, replied, 'No, for there were some things in them that could not be assured to her grace yet.' In fact, the chief obstacle had been not technical but personal, for the all-powerful Somerset was Mary's friend but lukewarm at best to Elizabeth. With his removal from the protectorate, matters were settled with sudden and suspicious swiftness. On 17 February 1550 the council issued instructions for the settlement of lands to Elizabeth to be completed and a month later (which is a mere twinkling of a bureaucratic eye) letters patent were issued in her favour on 17 March.

They granted her dozens of manors, houses and other parcels of land, grouped by county. Most were situated in an arc to the immediate north-west of London. One group centred round the old De La Pole manor-house at Ewelme in Oxfordshire, where the red-brick church and college still stand much as they were in Elizabeth's day. Another larger holding clustered round Ashridge, one of Elizabeth's childhood homes on the Buckinghamshire–Hertfordshire border, some two and half miles north of Berkhamsted. Berkhamsted itself, Hemel Hempstead, Great Missenden and Princes Risborough were included in this part of the grant.

Then there was a string of manors in Huntingdonshire, stretching up towards Northamptonshire, where Collyweston, the great country palace of Elizabeth's great-grandmother, Lady Margaret Beaufort, formed another nucleus. The group of manors included Uppingham and Preston, across the Rutland border, and Maxey, then in the Soke of Peterborough in Northamptonshire. These estates lay in a line below the town of Stamford in the extreme south of Lincolnshire. Stamford was the seat of the Cecil family, which had begun its ascent thanks to its connection with its powerful royal neighbours at Collyweston.

In Berkshire, Elizabeth's lands focused on Newbury, including the town and a ring of other manors around it which had belonged to Elizabeth's former stepmother, Jane Seymour, and to her stepfather (as husband of Catherine Parr) and suitor, Thomas Seymour. There were other, remoter estates in Dorset, Hampshire, and, remotest of all, another manor of Lady Margaret Beaufort's at Caistor in Lincolnshire.

Finally, Elizabeth was given Durham Place in London, the former London residence of the prince-bishops of Durham, as her town-palace, and Enfield manor and forest as her suburban retreat.

It was a princely endowment. The income totalled just over £3,000, as her father's will had stipulated, and payment of it was backdated to the Michaelmas (29 September) quarter-day before his death. But the grant would endure only 'until the councillors of Henry VIII, nominated by his will, shall provide [Elizabeth] with a suitable marriage'. In short, marriage for Elizabeth was like the stroke of midnight for Cinderella: with the sound of wedding bells, her principality would vanish. Is it any wonder that she viewed marriage with such mixed feelings?[2]

Now it is clear that Elizabeth had in fact been in informal possession of most of these lands and their income since much earlier in Edward's reign. For instance, when Seymour had been in discussion with Thomas Parry about the financial side of his proposed marriage with Elizabeth, the broad disposition of her estates seems to have been the same as in the 1550 letters patent. 'I told him,' Parry testified, 'where the lands lay, as near as I could, in Northamptonshire, Berkshire, Lincoln and elsewhere.' Even then Elizabeth was scheming to improve her property by exchange, and Parry explained that she was confident that Sir Richard Moryson, another scholar turned councillor, would 'help her to have Ewelme for Apethorpe'. In the event, as the patent makes clear, Elizabeth went one better and got her coveted Ewelme without having to surrender Apethorpe, which formed a useful part of her Northamptonshire territories.[3]

But clearly the jewel was Ashridge. Seymour noted approvingly that it 'was not far out of his way and he might come to see her in his way up and down' from his estates in the west country. Ashridge was situated on high ground, surrounded by the woods and rich hunting forests of Hertfordshire, much of which Elizabeth now owned. The house had been a monastic foundation before it had been confiscated in the Dissolution and it still retained most of its ecclesiastical buildings, including the church and cloister. Elizabeth's apartments were fitted into this structure. The great hall, measuring 66 feet by 28 feet, and the great chamber, measuring 46 feet by 26 feet, were on the ground floor. The presence chamber was above the great chamber and had the same dimensions; then, also on the first floor, came the privy chamber above 'a certain chapel'. Most unusually, however, the vast bedchamber (48 feet by 24 feet) seems to have been on the ground floor, under the old monastic infirmary. Or perhaps the infirmary had been on the second floor, which would allow for a more usual first-floor placing of the bedchamber.[4]

Despite the curious disposition of the rooms, however, Elizabeth inhabited a royal apartment, with the name and function of each room and the rules of access to it following the pattern which had been established in her father's palaces. The great hall, as in an Oxbridge college, was the main dining room. Here her servants ate every day, though Elizabeth would join them only on high days and holidays, such as Twelfth Night. The great chamber was the principal reception room, where her gentlemen and yeomen waited as a sort of guard of honour. The presence chamber was where she received important visitors. Beyond lay the privy chamber and the bedchamber. This was a female realm, presided over by the formidable Catherine Ashley. Most days Elizabeth would eat privately in the privy chamber, attended only by her ladies. Men were admitted here only on sufferance and – save for her tutors – they were excluded utterly from her bedchamber. That at least was the theory, though the Seymour affair shows how easily the etiquette could be breached.

Lodgings on this royal scale required to be royally furnished. Fortunately, the King's wardrobes were full to bursting point after decades of relentless, acquisitive accumulation under Henry VIII, and in December 1548 horses were hired to transport furnishings from three royal palaces – Oatlands, the More and Greenwich – and hand them over to 'the Lady Elizabeth her grace's officers' as a loan. This pinpoints the date at which Elizabeth set up house on her own. Previously, as we have seen, she had lodged with Catherine Parr and, after her falling out with her stepmother, with Anthony Denny. Then, after Catherine's death in September 1548, Catherine's widower and Elizabeth's suitor, Thomas Seymour, had offered Elizabeth furnished accommodation in his own townhouse. Wisely, Elizabeth had declined to accept without Denny's advice. This had not been forthcoming and instead the council, probably at Denny's prompting, must have agreed that Elizabeth should set up housekeeping on her own.[5]

The council's provision of furnishings was fairly generous. Best equipped, curiously, were Elizabeth's chapel and closet, where she heard divine service. She was given three lavishly embroidered and bejewelled pairs of altar frontals and five equally elaborate sets of priestly vestments for mass. Her theology (let alone her supposed taste for the plain style) should have rendered these repugnant. In fact, to the horror of her Protestant councillors and bishops, she continued to use rich chapel furnishings for the whole of her reign.

Most of the remaining household stuff also consisted of rich textiles. These included several sets of tapestry: five pieces showing hunting and hawking, four with the history of Hercules, six of the Triumphs of Petrarch, and another six with the City of Ladies. There were dozens of pieces of cheaper 'verdure' tapestries as well, worked with all-over patterns of foliage. Some of these had been cut to fit rooms. Several others were 'window-pieces', to serve as curtains. There were foot carpets, carpets to cover cupboards or sideboards, and a dozen Turkey carpets. For her bedchamber she was given a rich bed canopy, or sparver, of cloth of gold and silver and crimson velvet, with matching white, red and yellow curtains and two traverses. The latter were curtains for screening off parts of rooms and would have been used to humanize her vast bedchamber.

Another mark of her status was the two chairs, upholstered in cloth of gold and red velvet, with gilt pommels and roundels of the royal arms. One of these chairs would have stood in her privy chamber and the other in her presence chamber. She was also given thirteen rich cushions, which were used to make window-seats and chests more comfortable.

All of this was 'portable magnificence', that could be carried from house to house, while the heavy basic furniture of cupboards, trestle-tables, benches and stools was left *in situ*. So included in the loan were six bear hides to provide waterproof covers for as many carts, and four cloth sacks for trussing vulnerable objects.

There was, however, one significant omission. Elizabeth was given no cloth of estate. Mary, on the other hand, had been provided with two in *her* loan of furnishings from the royal wardrobe. The cloth of estate or canopy was the mark of royal status. One hung in the presence chamber over a chair or throne and was always treated with the same reverence as though the king or queen were actually sitting there. Mary, as heir to the throne, was accorded this symbol of royalty; Elizabeth, as only 'the second inheritor in remainder to the crown', was not.[6]

Materially, however, Elizabeth had benefited from her youth and junior ranking in the succession. Her income had been paid more or less in full from the beginning of her brother's reign. But her household, the principal object of her expenditure (costing, for example, £3,700 in 1551–2), had only been set up on a fully independent basis at the beginning of 1549. Before that, she had squatted on her relations and friends. There were costs in this arrangement, but they were nothing like the burden of running an independent establishment. The result was that Elizabeth had built up a substantial cash reserve. For example, in 1551–2 Elizabeth gave Parry, 'by her grace's own hands', £4,600 in cash to cover the deficit on the previous year and to meet the current year's expenditure. Then, at the end of the year, the cash surplus of over £1,500, which was specified to the last third part of a farthing, was once again 'delivered to her grace's own hands'. Elizabeth had learned early on to run a tight ship in her household. In the fullness of time, she was, Thatcher-like, to apply these same techniques of good housewifery to the finances of her kingdom. As with Thatcher, the consequences were mixed.[7]

Elizabeth also showed herself adept at exploiting her new political favour for further financial gain. When she had first set up housekeeping on her own, she had done so not at one of the many houses assigned to her, but at Hatfield in Hertfordshire. Hatfield had originally belonged to the Bishop of Ely and it had been magnificently rebuilt in brick by Henry VII's minister, John Morton, during his tenure of the diocese. Henry VIII had taken a fancy to it and had treated it as his own for twenty years before formally acquiring it from the then bishop in 1538. Subsequently, Elizabeth's friend and mentor, Sir Anthony Denny, was given responsibility for the administration of the property. And it was probably due to Denny's influence that Elizabeth was allowed to take up residence there at Christmas 1548. But on 6 January 1550, when Elizabeth was still revelling in her first visit to court since the Seymour affair, Hatfield was granted to the now dominant John Dudley, Earl of Warwick. Six months later, however, Dudley returned Hatfield to the Crown as part of a deal in which he got more lands in Warwick, the county from which he took the name of his earldom. And three months later still, in September, Hatfield was granted to Elizabeth, in exchange for her surrender of the most distant of her estates, at Caistor in Lincolnshire.[8]

Elizabeth's motives in acquiring Hatfield have been subsequently seen in largely sentimental terms. She had loved it, according to one writer, 'since childhood, with its gentle gardens and vast park stocked with fine deer'. This is possible. But it would require Elizabeth to have had a prodigious memory since she had last made an extended stay at Hatfield between the ages of three and six months! Building works had been undertaken to prepare the palace for her in 1533 and she had been moved there in December, with the reluctant and humiliated Mary in tow. In 1537 the two half-sisters, then on much better terms, had paid fleeting visits; while in 1546 Elizabeth seems to have spent some more time at Hatfield with her brother Edward. And that is all.[9]

So sentiment, as usual with Elizabeth, explains little or nothing. Elizabeth liked Hatfield because it was modern, convenient and commodious – unlike Ashridge, say, with its lodgings cobbled together out of a former monastery. She was also probably attracted to its site on strategic grounds. It lay in the heart of Hertfordshire, astride the Great North Road and only twenty miles north of London. And Hertfordshire was the frontier between Elizabeth's lands and those of her sister Mary. Elizabeth's estates, as we have seen, lay to the north and west of London; Mary's main landholdings, on the other hand, were to the east, in East Anglia and Essex. But both half-sisters had their main residence in Hertfordshire: Mary's was at Hunsdon, to the extreme east of the county, on its border with Essex, while Elizabeth's was supposed to be at

Ashridge, to the extreme west, on the Hertfordshire–Buckinghamshire border. The grant of Hatfield, plumb in the centre of the county, gave Elizabeth the upper hand. And at this point, though not before, Hatfield became her principal seat.

But as Elizabeth and her new friends grew rich, the young King, shorn of his lands for their benefit, sank into chronic poverty.

Even Seymour, who is usually seen as one of the most unthinkingly selfish politicians of the age, had had some compunction. While Catherine Parr had been walking in the grounds of Sudeley Castle, her husband's principal seat, she had remarked to Sir Robert Tyrwhitt (later Elizabeth's interrogator): 'You shall see the king, when he cometh to his full age, will call his lands again, as fast as they be now given from him.' 'Then,' Tyrwhitt replied, 'is Sudeley Castle gone from my [Lord] Admiral.' That, according to Catherine, had already been anticipated by Seymour. 'I do assure you,' she had told Tyrwhitt, 'he intends to offer them to the king, and give them freely to him at that time.'[10]

No such hesitations were shown by Elizabeth, though she was Edward's half-sister. Instead, she out-vultured the worst. When, for example, her father's old servant Stephen Gardiner was finally deposed from the wealthy bishopric of Winchester in 1551, the imperial ambassador reported that 'several persons appear to be watching for the prey'. Conspicuous among them was 'the Lady Elizabeth, the king's sister', who 'is in love with a certain manor, which she is said to be sure of acquiring'. In this instance, at least, she failed. She was also, as part of a comprehensive resettlement of her lands, forced at last to disgorge Apethorpe, which was immediately regranted to Sir Walter Mildmay, her future Chancellor of the Exchequer. It was a rare defeat.[11]

CHAPTER 16

Rival Sisters

It was Nell Gwyn, Charles II's rumbustious mistress, who cried, as the mob were about to stone her carriage: 'Good people, forbear! I am the *Protestant* whore.' In similar vein, Elizabeth might have proclaimed herself 'the King's *Protestant* half-sister' – as opposed to Mary, the King's Catholic half-sister and, as the law had been left by Henry VIII, heiress presumptive to the throne. The council's principal motive in facilitating Elizabeth's enrichment was to build her up as a rival to Mary; Elizabeth – as greedy for applause as for cash – was probably playing at the same game of sisterly rivalry.

But it was Mary who had first consciously staked out her position in religious terms. In autumn 1547 she issued her first protest at the religious developments of Edward's reign. It took the form, ironically, of committing herself to Henry VIII's anti-papal but doctrinally fairly conservative settlement of the last years of his reign. The council, she sharply told Somerset, had sworn to this settlement too; now they were going about to overthrow it. Somerset replied in reasoned terms, explaining that the religious settlement of the 1540s was incomplete and claiming that Henry VIII had died regretting that it was only a halfway house to Reformation. Neither side pressed matters to a conclusion.

The stakes were sharply raised two years later. In May 1549 the Act of Uniformity made the new Protestant Prayer Book the only form of worship legally permitted in England. Mary immediately played the foreign card – just as she had done under her father – and sought an assurance from the imperial ambassador that her cousin, the Emperor Charles V, would give her the protection of his mighty hand. The Emperor agreed, and instructed his ambassador to request that Mary be given written authority, 'in definite, suitable and permanent form', that she would be exempt from the law and be allowed to 'live in the observance of our ancient religion'. Somerset explained in an interview with the ambassador that for Mary to have a formal exemption from the laws by letters patent was impossible; on the other hand, he would be prepared to turn a blind eye to Mary and her household continuing to celebrate Mass, providing it was done 'quietly and without scandal'. But the Council, where Somerset was already losing influence, took a harder line, requiring Mary's submission to the law.[1]

There matters stood at the fall of Somerset in late 1549. Elizabeth's first reaction to this event was to invite herself to court for Christmas. Edward wrote to Mary to explain that Elizabeth had 'desired to remain with us all this Christmas holidays' and to invite Mary to join the family party. But, he continued, he knew that the notice was short and he was also unsure about the state of her health. Mary took what was probably a palpable hint and made her excuses. She was wise: it was on this Christmas Day that the King and Council publicly reaffirmed their commitment to continue the Protestant Reformation.[2]

For Elizabeth, the new dispensation was an opportunity, to be exploited to the full; for Mary it was a threat. In January Mary told the imperial ambassador that John Dudley, Earl of Warwick, was 'the most unstable man in England' and that the conspiracy against Somerset 'had envy and ambition as its only motives'. By May she was involved in a hare-brained scheme, once again, to flee abroad. Meanwhile, Elizabeth and Warwick were getting on famously – or, if some rumours were to be believed, infamously. In November 1550 the imperial ambassador reported the worst of the rumours: Warwick was to divorce his wife and marry Elizabeth, 'with whom he is said to have had several secret and intimate personal communications'. There seems to have been nothing in the story. But it was not the last time that Elizabeth's name was to be linked to the house of Dudley in compromising circumstances.[3]

The long-postponed showdown between Mary and the régime finally began at Christmas 1550. This time the family reunion, put off the previous year, actually took place, and Edward, Mary and Elizabeth all gathered together for the festivities. But, as so often, Christmas turned into a time for family quarrels. Edward openly upbraided Mary for hearing mass in her chapel. Humiliated, Mary burst into tears. Later she went on the attack and informed Edward that he was not yet old enough to make up his own mind about something as important as religion. It was now Edward's turn to dig in his heels and he demanded Mary's obedience to his laws as both his sister and his subject. It was stalemate.[4]

But as Mary's star set, Elizabeth's rose. The imperial ambassador, who was no friendly witness, described the scale of Elizabeth's arrival in London for the Christmas season. She came 'with a great suite of gentlemen and ladies, escorted by a hundred of the king's horse, [and] she was most honourably received by the Council'. The reason for rolling out the red carpet 'was to show the people how much glory belongs to her who has embraced the new religion and become a very great lady'.[5]

The point was driven home by the ceremonies on Epiphany. Epiphany, celebrated on 6 January, commemorates the presentation of gifts to the infant Christ by the Magi or Three Kings. It was the last of the festivities of the Twelve Days of Christmas; it was also, because of the royal associations of the Three Kings, the day of highest court ceremony in the whole year, when the King wore his crown and dined in public. Mary had by then gone home in dudgeon. But Elizabeth was the star of the festivities. She dined with her brother and the French ambassador and after dinner went with the King to watch a bear-baiting. Elizabeth was the King's half-sister; now she appeared as a sort of consort.[6]

Looked at in one perspective, Elizabeth's arrival in London with a great procession anticipates her many *entrées* into her capital when she was queen. But, from another point of view, it was a throw-back to an earlier age. Elizabeth rode into her brother's city at the head of scores of men. They were armed; they wore her livery and they were drawn from her own household and her vast, strategically sited estates. In short, they were retainers, and she was an over-mighty subject, who was made all the mightier and more dangerous by her claim to the throne. After all the efforts and blood-lettings of her father and grandfather, it looked as though the ingredients for civil war were in place once more. There was division in the royal family and faction in the kingdom. Arguably, it was even more dangerous than in the fifteenth century. Then, men had worn the badge of a lord as a sign of their allegiance; now, they wore the tokens of religious faith. Were the Wars of the Roses about to be followed by the English Wars of Religion?

The risk was made palpable by Mary's next visit to court. She came on a peremptory summons and it was Edward's intention this time to make her submit. But she came in force. On 15 March, Mary rode through London to her town-palace at St John's, the old headquarters of the Knights of St John of Jerusalem. Fifty knights and gentlemen rode before her in velvet coats and gold chains, and after her eighty ladies and gentlemen, 'everyone having a pair of beads' – that is, a rosary. The rosary was the badge of their faith. Two days later, on the 17th, Mary went from St John's, through Fleet Street and the Strand, to the court at Whitehall. Her retinue was now even more impressive, 'with many noblemen of lords and knights and gentlemen and ladies and gentlewomen'. At the court gate, the usual courtesy was done and the Controller of the King's Household came to meet her. She was then taken through the hall to the presence chamber and given a rich banquet. But in the middle of the festivities she was summoned to meet King and Council. They taxed her with disobedience. She replied that 'her soul was God's and her faith she would not change'. She was answered in the King's name that he 'constrained not her faith but willed her (not as a king to rule) but as a subject to obey'.[7]

Next day Mary gave a double answer. Alerted by her, the imperial ambassador came to court to threaten war if she were not given freedom of worship. Mary herself preferred deeds to the ambassador's words. After spending the night at St John's, she rode out of the city in state and withdrew to the heart of her estates at New Hall in Essex. It was like Achilles skulking in his tent. And Mary, too, had her myrmidons. Mary's gesture and the ambassador's ultimatum threatened a combination of foreign war and domestic insurrection. And the threat was very real: East Anglia had been in turmoil since the suppression of Kett's Revolt in 1549, while England's defences against invasion were demoralized and underpaid. Not surprisingly, despite the King's tears of frustration, the council backed off. Mary got her Mass.[8]

Almost a year later, in March 1552, Elizabeth made an even grander showing. She had come up from Hatfield to her own town-palace at Durham Palace; then on the 17th she rode through London on her way to take up residence in St James's Palace. She was proceeded by 'a great company of lords and knights and gentlemen' and followed by 'a great number of ladies and gentlewomen'. The total size of her escort was two hundred on horseback as well as yeomen, who presumably went on foot.[9]

St James's Palace – with its red-brick walls, twin-towered gatehouse and chapel with its ceiling decorated with the badges and ciphers of Henry VIII and Anne of Cleves – is little changed from its appearance on that day. It had been built by Henry VIII as a sort of guesthouse to Whitehall Palace, from which it was separated only by the park. Thomas Cromwell, Henry's minister, had used it and it was probably intended as Edward's London residence when Prince of Wales. Its use by Elizabeth was thus a mark of great favour.[10]

Elizabeth reposed herself for a day at St James's, where she may have spent her time looking at the pictures of her ancestors which hung on the walls. The portraits included her grandmother and namesake, Elizabeth of York, her grandfather, Henry VII and her father, Henry VIII. But there was none of Elizabeth's own mother, Anne Boleyn.[11]

Then, on the 19th, Elizabeth rode through St James's Park to the court at Whitehall. We can follow her route across the park towards Horse Guards' Parade, which was then the tiltyard. Between the park gate and the court, she had to cross the busy thoroughfare of King Street. This was the main highway between Westminster and the City and bisected the palace. The two parts of the building were joined at first-floor level by private galleries which were carried across the street by two arched gateways. The way at ground level, however, was frequently noisesome. But for Elizabeth's visit it was freshly 'strewn with sand fine'. In this procession, the great of the court joined her own following and even the two premier peers – the Duke of Suffolk and the Duke of

Northumberland (as John Dudley, Earl of Warwick, had become) – formed part of her train.[12]

Elizabeth's account book for this year shows something of the planning behind the pageantry. Five of her servants were paid considerable sums for 'riding for gentlemen when her grace went to the court'. These were the messengers who had been sent to summon outlying members of her following to join her train. Another officer was paid 'for going afore to Durham Place by the space of three days' to help get it ready. Elizabeth also paid out £9 15s. in 'rewards' or tips to various of the King's officers and other during her stay at St James's. But, apart from these incidentals, all the expenses of her household for the visit were met by the King, even down to £10-worth of wheat 'spent at Barnet as parcel of her grace's expenses going to the court'.[13]

In early 1553 the game of sisterly tit-for-tat continued. Mary began a sort of state visit to her brother on 6 February; while Elizabeth planned another sojourn at St James's at Candlemas, that is, 2 February. It is unclear, however, whether Elizabeth's planned visit took place.[14]

It is, however, a mistake to focus solely on the tensions and rivalries between the three royal half-siblings. Mary and Edward might clash publicly over religion but their personal relations never broke down and Mary continued to profess an affection for Edward, which, as his godmother, she probably felt. Similarly, despite their semi-religious rivalry, Elizabeth continued to condole with Mary on her frequent ill health and Mary responded politely.[15] But the warmest ties were between Elizabeth and Edward.

And it was to Edward that Elizabeth wrote perhaps her most natural and human letter. It begins with one of the irritating conceits which were to become habitual with her. But it rescues itself by the ingenuity with which the simile is pursued and the real feeling Elizabeth displays.[16]

She plunges straight into the figure of speech: 'Like as a shipman in stormy weather plucks down the sails, tarrying for better wind', so did she 'pluck down the high sails of my joy and comfort', in the hope that 'as the troublesome waves have repulsed me backward so a gentle wind will bring me forward to my haven'. Then follows the explanation. She had been on her way to see him when she had been ordered to turn back. The blow had been twofold: 'the one for that I doubted your majesty's health; the other because for all my long tarrying I went without that I came for'. She had been partly reassured on the state of Edward's health. But she would not be happy until she saw him. She is open about her frustration at having to make 'the half of my way the end of my journey'. Otherwise, her tone is resolutely cheerful. It is, in short, the sort of letter that you might write to an invalid who is not expected to live long.

The letter is dated 'from Hatfield, this present Saturday', while the incident it describes had occurred the previous Thursday. There is no indication of month or year. On the other hand, Candlemas Day, when we know Elizabeth intended to visit Edward in 1553, fell on a Thursday. It is possible that Elizabeth's letter refers to this aborted visit.

But why, in that case, was Mary allowed to see Edward on 6 February, only four days after Elizabeth had been turned back? Maybe Elizabeth's letter again contains a clue. For, she protests, 'whatsoever other folks will suspect, I intend not to fear your grace's goodwill, which as I know I never deserved to faint, so I trust will stick by me'. Were Edward and Elizabeth too close for some people's comfort?

CHAPTER 17

Exclusion: Edward VI's Will and Death

The order to abort her intended visit to Edward was not the only rebuff Elizabeth received in the spring of 1553. Hitherto Elizabeth and John Dudley had enjoyed a close working relationship. But now their interests clashed.

Dudley had been promoted to the dukedom of Northumberland in October 1551. In order to give substance to his title, he set about accumulating a princely land-holding in the north-east. It drew on two main components: the Percy earldom of Northumberland, which had reverted to the Crown in dubious circumstances under Henry VIII, and the prince-bishopric of Durham, whose holder, Cuthbert Tunstal, was deprived in October 1552. The jewel of the bishop's vast estates had been his London palace known as Durham Place. It stood on the site of The Adelphi, and occupied the whole area from the Strand to the Thames. The hall was particularly sumptuous, with a lofty roof, Purbeck marble columns, and three bays of vast, traceried windows. It was, in short, too desirable for the bishop's good and Tunstal had been forced to exchange it with Henry VIII. Elizabeth's mother and Boleyn grandfather had lodged there before Anne's marriage to Henry VIII and, in the fullness of time, it had been granted to Elizabeth as her townhouse. Now Dudley decided that, as Duke of Northumberland and heir of the prince-bishops, he must have their former seat as his own London palace.[1]

The task of broaching the matter with Elizabeth was delegated to the Chancellor of the Court of Augmentations. As its name indicates, the court had been established to administer the enormous increase in the Crown's property which followed Henry VIII's dissolution of the monasteries. Under Edward VI, however, it was mockingly suggested that its name should be changed to the Court of Diminutions, since it spent most of its time giving the King's property away! Elizabeth had been a prime beneficiary of this process and she did not at all relish having to surrender any of her gains. She was also irritated that the decision had been taken 'without first knowing her mind'. But she had the sense to acquiesce with reasonably good grace. The pill was further sweetened by Northumberland's offer of Somerset House, the townhouse of his fallen rival, as compensation. Somerset had spent almost £10,000 on the building; now a further £900 was expended in fitting it out for its new mistress. Curiously, the keeper of Somerset House was Northumberland's son, Robert Dudley.

In all of this, Northumberland was careful to protest that he would 'not offend willingly' Elizabeth. But still he had taken her for granted.[2]

The point is heightened by Northumberland's treatment of Mary. Once he had been her enemy. Now nothing seemed too good for her. The attempt at forcing her to submit to the new religion had long been abandoned; instead, she was treated with all possible consideration. While Elizabeth was turned away from court, Mary was welcomed with honour. On 10 February 1553 she rode from St John's to Whitehall, and Northumberland headed the group of councillors who received her at the gate and conducted her to the presence chamber where the King 'met and saluted her'. Susan Clarencius, one of Mary's oldest and most faithful ladies-in-waiting, and a fervent Catholic, was given a land grant. Even Mary's hawks received special protection and the council ordered two men suspected of stealing them to be put to the torture if they would not confess.[3]

It was, as both Elizabeth and her half-sister must have reflected for different reasons, curiouser and curiouser.

The explanation, as it was for almost everything in this year of upheavals, was the state of the King's health.

The brother Elizabeth had grown up with was a healthy, vigorous child. True, the rigours of his education had perhaps made him too bookish and sedentary. But this had changed when John Dudley inaugurated a fresh régime for the young King after Somerset's fall in 1549. Study and training in statecraft continued. But alongside them Edward was introduced to knightly sports and pastimes. He took to them like a duck to water. Indeed, much the liveliest passage in his *Chronicle* is a description of a mock tournament held at Greenwich in May 1551. Edward led his team in black and white, while the opposing side, captained by the Earl of Hertford, Somerset's eldest son, wore yellow. Edward's side lost and the King proved a bad loser. 'My band tainted [hit] often – which was counted as nothing – and took never – which was very strange – so the prize of my side was lost.'[4]

But almost a year later there is a very different entry in the *Chronicle*. 'I fell sick of the measles and the smallpox,' Edward wrote on 2 April 1552. He seemed to make a quick recovery and Elizabeth sent her congratulations.[5] But the recovery was deceptive. By the winter he was seriously ill and in the spring of 1553 it was clear to those nearest to him that he was suffering from consumption. Consumption, or tuberculosis, was incurable in the sixteenth century. It was only a matter of time.

The succession suddenly became the question of the moment. The rules were still those which had been laid down by Henry VIII's will, which in turn rested on the authority of parliamentary statute. If Edward died without heirs, the crown should pass to Mary. If Mary died without heirs, it should go to Elizabeth. And if Elizabeth also died childless, it should descend to the issue of Henry VIII's younger sister Mary.

The problem was religion. The established religion of England was now Protestantism – and, with the second Prayer Book of 1552, unadulterated Protestantism that explicitly rejected the central Catholic doctrine of transubstantiation. The new establishment of Edward's reign was Protestant too. But Mary, the heir presumptive, was unequivocally and vocally Catholic. Her succession thus threatened to be a burning issue indeed. It would be a bonfire of the vanities of office and wealth for the men of the new establishment, whom Mary held personally responsible for the apostasy of England and the wreckage of her own life. And their bodies might be at risk of burning, too, if Mary as queen decided to revive the old heresy statutes as well as the old religion. There was also, for those in the know, a real doubt over Mary's patriotism. She had never hesitated to appeal to the Habsburg emperor against her own father and brother. And she had even procured a threat of war against England. And yet, if Edward died, she was Henry VIII's chosen, legal heir.

So the situation of 1553 placed four of the strongest and most fundamental values of the sixteenth century – religion, patriotism, due legal process and dynastic loyalty – at war with each other.

Only Edward himself seems to have found it easy to choose between them. He plumped unhesitatingly for religion. Mary was Catholic, so Mary must go. It was Edward who had been determined that Mary would submit to his laws and his religion. He had been frustrated then. Now he was resolved she should not, as his successor, undo everything that had been done in his reign to establish the true worship of God.

But what of Elizabeth, his 'Sweet Sister Temperance'? She was Protestant and she was 'the second inheritor of the crown'. But she was also passionately loyal to her father's memory and wishes. And she had an unshakeable sense of dynastic legitimacy. Probably without asking – which would anyway have been dangerous – the King knew that she would refuse to have any share in his scheme. Elizabeth would not agree to be the usurper of Mary's claim. So Elizabeth had to go as well. Edward's 'good will', which she had been so confident would stick by her, evaporated in the face of his burning conviction.

For us it is hard to believe that a fifteen-year-old boy could arrive at such momentous decisions. And it is even harder to accept that he could force them on a hard-bitten, experienced and generally reluctant Council. But the evidence is clear. Edward had a handful of co-operators, but the driving will was his.

Edward's first scheme was set out in a paper, with the Adrian Mole-like title of 'My device for the succession'. The paper was 'wholly written in his most gracious hand'.[6] Edward's writing, of which the 'Device' is a characteristic example, is a clear but rather wooden italic, with none of Elizabeth's style or flair. On the other hand, the 'Device' shows Edward's intelligence and his clear understanding of the history of the English succession up to his own times. Until then no woman had succeeded to the throne. But women *had* transmitted their claims to their male descendants. Indeed, all Edward's ancestral families – York, Tudor and Lancaster – traced their title to the throne through the female line. The 'Device' turned these facts into principles.

Edward's other starting point in the 'Device' was Henry VIII's will. Under the terms of the will, the next in line to the succession after Edward and his half-sisters were the descendants of Edward's aunt Mary. Henry VIII had conceived of this as a remote possibility. Edward's decision to eliminate both his half-sisters made it a matter of immediate importance.

But here Edward's historical principles came into play. Edward recognised only his aunt's *male* descendants as his heirs. Her female issue instead were accorded only a transmissory role: they could beget a male heir and, if they did, the mother was to act as regent or governess of the realm until her son came of age at eighteen.

These provisions now seem odd. But they are easily explained, not only by the previous history of the succession, but also in terms of Edward's own experiences. Like the good, advanced Protestant he was, he disapproved on theological grounds of the exercise of public, political authority by women. The most infamous statement of this doctrine is John Knox's pamphlet, 'The First Blast of the Trumpet against the Monstrous Regiment [that is, rule] of Women', written in 1558 against the rule of Henry VIII's other great-niece, Queen Mary Stuart, in Scotland. Coincidentally, John Knox was in England while Edward was composing the 'Device', and Edward would have sympathized with his later views. Edward also had personal experience of 'the monstrous regiment of women' in the more modern sense, with his two formidable sisters, the one twenty years older than he, and the other four. Like many other boys in such circumstances, he probably came to feel hen-pecked and resentful. All this meant that excluding women from the succession was a pleasure as well as a duty. On the other hand, Edward clearly felt even more resentful at the sort of exploitative bossing around he had received from his male Protector, Somerset, and pseudo-Protector,

Northumberland. Hence his decision to copy the French device of a female regency.

Unfortunately for Edward's planned gender revenge, *all* the available heirs of Henry VII, the patriarch of the Tudor dynasty, were female. This was the case not only for the excluded, such as Edward's half-sisters, Mary and Elizabeth, and his cousin, Mary, Queen of Scots; it was also true for the privileged line of Mary, Edward's younger aunt. Mary had first been married to King Louis XII of France. Louis was a prematurely aged roué, who is supposed to have died of his exertions with his pretty young English bride. Despite his efforts, there were no children and Mary declared that she would rather be torn with wild horses than have another such husband. To exclude the possibility completely, she ordered Charles Brandon, Duke of Suffolk, her brother's dashing favourite who had been sent to escort her back from France, to marry her on the spot. Ever obliging to the ladies (which is why he got through four or five wives, one of whom he married twice and another not at all), Suffolk obeyed. But only two daughters of the marriage survived: Frances, who married Henry Grey, Marquess of Dorset, and Eleanor, who married Henry Clifford, Earl of Cumberland. In turn, Frances and Henry (who was created Duke of Suffolk in 1551 in right of his wife) only had daughters: Jane, Catherine and Mary; as, for that matter, did the Earl and Countess of Cumberland, whose single surviving child was Margaret.[7]

With time, no doubt, one of the Grey or Clifford girls would have had a son. But time, with the state of the King's health, was in short supply. More drastic measures were needed.

The first step was to get some of the royal brood-mares married and breeding. Margaret Clifford was in her late teens; Jane, the eldest Grey daughter, was seventeen, and Catherine, her next sister, was a year or two younger. All three were married or betrothed within a few days of each other in May 1553. On the 21st, Jane was married to Lord Guildford Dudley, Northumberland's fourth surviving son. The ceremony took place at Durham Place, so recently filched from Elizabeth, and was celebrated 'with a display truly regal'. Indeed, Edward himself supplied robes for the occasion. The celebrations at Durham Place continued without interruption to the 25th, when two more weddings took place there: Catherine Grey was married to Lord Herbert, then aged about fifteen, the heir of the Earl of Pembroke; while Lady Catherine Dudley, Northumberland's daughter, was married to Henry, Lord Hastings, the heir of the Earl of Huntingdon. Catherine, of course, had no claim to the crown, but her husband, as the direct descendant of George, Duke of Clarence, Edward IV's brother, who met his end in a butt of Malmsey wine, did. Finally, Margaret Clifford was betrothed to a man at least thirty years her senior: Northumberland's brother, Sir Andrew Dudley, KG, one of the Chief Gentlemen of the King's Privy Chamber. Edward issued him too with robes and jewels for his forthcoming wedding in June. But events overtook it and it never happened.[8]

A few days after the celebrations at Durham Place, the King's health took a sudden turn for the worse as his tuberculosis entered an acute stage. 'The sputum which he brings up is livid, black, fetid and full of carbon; it smells beyond measure.' Clearly, there was no time to wait for one of the May brides to have a son; instead, an immediate successor had to be found.[9]

Edward, despite his illness and the fact that he could sleep only with opiates, acted with characteristic decisiveness. The 'Device', with its intended exclusion of female succession, had originally left the crown to 'the Lady Jane's heirs males'. With a few strokes of the pen, to delete two letters and to insert two words, the King altered this to leave it to 'the Lady Jane and her heirs male'. It remained only to give effect to the decision.[10]

Legally, the position was clear. As the order of succession in Henry VIII's will was based on parliamentary statute, only another parliament could change it. Writs were duly sent at the beginning of June to summon a new parliament in September. But, almost immediately, it became clear that the King's life was to be measured in weeks, not months.

Once again, Edward took the lead, and on 12 June summoned the Chief Justice and the law officers to his bedside at Greenwich and ordered them to draw up a will in the form of letters patent to give effect to the terms of the revised 'Device'. They withdrew to consult among themselves before delivering a reasoned refusal. Letters patent, they reminded the council, could not overturn an act of parliament, and to seek to change the succession without parliamentary authority was treason. Three days later they were summoned to the presence once more and Edward, whose will seemed to strengthen as his poor, broken body weakened, commanded them, with 'sharp words and angry countenance', absolutely to obey on their allegiance.

Reluctantly (the Chief Justice said he was 'in great fear as ever I was in my life before'), they acquiesced and drew up the necessary instruments. These fleshed out the legal details of Edward's scheme; they also added the propaganda justification. Elizabeth and Mary were bastards, of the half-blood and might marry abroad. The Grey girls on the other hand were legitimate, of the whole blood and 'very honourably brought up and exercised in good and godly learning and other noble virtues'. In other words, like Edward, they were learned Protestants.

On the 21st the councillors and other notables countersigned the letters patent. The group which had been bound together by the marriage alliances of May was enthusiastic. The rest shared, more or less vehemently, the lawyers' doubts. Or at least later they said they did, as they were trying to save their skins. But they signed none the less.[11]

The reason for the sudden change in the relative standing of Elizabeth and Mary is now clear. Elizabeth could be disregarded as she had been eliminated from everybody's calculations. Mary, on the other hand, was sweet-talked and cherished to lull her into a false sense of security so that she could be picked off at the critical moment. It was not a foolish calculation. Why add to the already fevered and dangerous atmosphere of May and June by arresting the heir to the throne, when she was in residence at Hunsdon under the nose of the government and a few hours' riding distance from London?

But to work, the scheme needed secrecy. In similar circumstances, the deaths of Edward's father and grandfather had been successfully concealed until the necessary arrangements for a smooth transition of power were in place. But the coalition round Edward's bedside was too divided and shaky. Inevitably, news leaked.[12]

On the evening of 6 July Edward died in the arms of Sir Henry Sidney, Northumberland's son-in-law. His last words were: 'I am faint; Lord have mercy upon me and take my spirit.' He was at peace. But he had left turmoil behind him.[13]

Probably before Edward was quite dead, Lord Robert Dudley had ridden off with a detachment of the guard to seize Mary. But when he reached Hunsdon the bird had flown. Mary had been warned, and had fled north-east, to Kenninghall, in the heartland of her East Anglian principality.[14]

Without control of Mary, her obvious rival, it was that much harder to impose the succession of Jane. But a serious attempt was made. On the 7th the Tower was reinforced; on the 8th the City government was informed of the King's death and sworn to the new succession and on the 10th Jane was brought to the Tower and proclaimed Queen.[15]

But the next day news arrived that Mary had also proclaimed herself Queen and that men were flocking to her standard in East Anglia. At the core of Mary's following were the Catholic officers of her household. They both organized her *coup* and supplied the initial military leadership. They were joined by other landowners from her estates, and neighbouring lords and gentlemen who were sympathetic to her cause. All this was typical of a bastard-feudal revolt. But there was also a strong popular element, as Mary's troops were reinforced 'with innumerable companies of the common people'. They were turning out for her because she was a Tudor, and Jane was not. Mary, for once reading the runes of public opinion correctly, played on this sentiment by insisting that her title rested on statute and her father's will. The numbers prepared to fight for King Henry's daughter redoubled and a separate gentry rising broke out in the Thames valley.[16]

When the first word of all this reached the Tower, the Council decided that the Duke of Suffolk, Jane's father, should lead the expeditionary force against Mary. Jane, in tears and on her knees, forced a change of plan. Her father was to stay and Northumberland was to lead the army. With this, she sealed the fate of her régime and, it proved, signed her own death warrant and that of her leading supporters. Northumberland was the best soldier in England. But he was hamstrung by his own ill health and the melting away of his troops in the field. Meanwhile, without the presence of Northumberland's dominating personality, the Council in London disintegrated, and a *sauve qui peut* ensued.[17]

On the 19th, Mary was proclaimed in London and on the 21st Northumberland, having retreated from Bury St Edmunds, himself proclaimed Mary at Cambridge. There were no heralds or trumpeters, so he threw up his cap in the air. He might as well have thrown up his head. He was arrested that night and brought to the Tower on the 25th, 'all the streets full of people, which cursed him and calling him traitor without measure'.

There were bells, bonfires, banquets and *Te Deums* for the new Queen.[18]

We know nothing of Elizabeth in all this, save that she was at Hatfield. As soon as the immediate dust settled, she wrote to Mary to congratulate her and to ask whether, when they met, she should be wearing mourning for Edward. What either of the sisters really felt about their brother, who had so callously brushed them aside in his last months, we can only speculate.

But Elizabeth had survived. And with her power intact. This was formidably displayed on 29 July when she entered the city of London and rode through Fleet Street to her new town-palace at Somerset House. Her escort numbered two thousand horses, 'with spears and bows and guns'. Sir John Williams and Sir John Bridges, her chamberlain, were the captains of her host and they and all her men wore green, garded, or trimmed, with white, in fabrics graduated according to their rank – from velvet, down through satin to taffeta. Green and white were the Tudor livery colours. Elizabeth, like her sister, was advertising her dynastic legitimacy; she was also showing that she too was able to recruit a bastard-feudal following and that 'right strongly'. But, for the moment, like Elizabeth herself, her following was placed in Mary's service.[19]

The next day, the point made, Elizabeth reduced the number of her followers by half as she rode back through the city to meet her sister. Mary had remained in her own princely house at New Hall (Beaulieu) in Essex while her opponents were mopped up and put under arrest. Now she was ready to enter her capital. She came to Wanstead, where Elizabeth met her. Then, on 3 August, they rode to Aldgate, where they were greeted by the city. Mary wore the royal purple with a great gold and jewelled chain, while 'next to her followed the Lady Elizabeth's grace, her sister'.[20]

The events of these weeks made a profound impression on Elizabeth. The faction in London had proclaimed her a bastard as well as Mary. In particular, Dr Nicholas Ridley, the Bishop of London, had preached two sermons at Paul's Cross, on the successive Sundays, 9 and 16 July, in which he declared 'the Lady Mary and the Lady Elizabeth, sisters to the king's majesty departed, to be illegitimate and not lawfully begotten in the estate of true matrimony according to God's law'. 'Which,' according to an observer, 'the people murmured sore at'.[21]

Elizabeth took equal resentment at the words and, years later, threw them in the teeth of her bishops, most of whom were proud to call themselves Ridley's disciples. 'I do not marvel, though *Domini Doctores*, with you my Lords, did so use themselves therein, since after my brother's death they openly preached and set forth that my sister and I were bastards.' There were other scars as well. She never forgave the Grey sisters, who had been preferred to her in the succession. Even Margaret Clifford, who seems to have been a rather silly dabbler in prophecies, was to feel her wrath with a spell of imprisonment in the Tower. Nor, finally, could Elizabeth ever abide again the sort of advanced Protestantism which had led Edward to decide that only men were fit to rule. It was, in short, a defining moment.[22]

CHAPTER 18

Queen Mary

The woman whom Elizabeth rode to meet at Wanstead had also been through a defining experience. For twenty years – the whole of her adult life – Mary had known neither security nor much happiness. Everything she loved – her mother, her religion and her friends – had been condemned and rejected. Often she thought of self-imposed exile; always she was marginal to English life and politics. Now suddenly her enemies had melted away and she had been carried to the throne of England on the shoulders of her cheering people. The transformation was prophetic, miraculous, messianic. No wonder that Queen Mary, according to the Venetian ambassador, always had a medley of quotations from the Missal and scripture on her lips: 'In thee O Lord I trust, that I be not confounded forever: if God be for us; who can be against us?'[1]

Who indeed? The answer in those heady summer days was very few. Mary had been supported by the great bulk of the people, whatever their religious persuasion: Protestants, Catholics and the merely conformist majority had all flung themselves behind her cause. But it is important to understand their reasons. People supported Mary because she was Henry VIII's daughter and the lawful heir. They had also come to hate Northumberland and his friends as a selfish, greedy, oppressive clique. And they were increasingly mistrustful of the speed and radicalism of religious change, as a handful of clergy, many of them foreign, rushed to make England into a new Strasbourg or even a new Geneva.

All this was a clear mandate both for Mary's accession and for stopping the Edwardian reformation in its tracks. But it was much less clear what positive policies might have kept the initial, broad-based coalition together. And whatever they were, Mary was not interested in them. Instead, she embarked on an increasingly lonely voyage that was to leave her at the end as isolated and unpopular as Northumberland and as marginal to English life as she had ever been in her unhappy youth.

But in the summer of 1553 it did not look like that. 'Money was thrown out at windows for joy. The bonfires were without number, and what with shouting and crying of people and ringing of the bells, there could no one hear almost what another said, besides banquetings and singing in the street for joy.'[2]

And Mary's first actions seemed of a piece. Inevitably, the leading figures in the attempt to change the succession were rounded up, tried as traitors and condemned. But only the most obviously guilty – Northumberland himself and his two henchmen, Sir John Gates and Sir Thomas Palmer – were executed. Shamefully, all three died as apostates to the Protestant faith which they had professed so glibly and exploited so ruthlessly in life. Otherwise, Mary showed herself the 'merciful princess' she proclaimed herself to be. Her rival, Jane Grey, and Jane's intended consort, Guildford Dudley, were kept in honourable imprisonment in the Tower; while the Duke of Suffolk, Lady Jane's father, was pardoned and released.

This should not surprise too much, as Mary was always benign when merely temporal matters were concerned. More striking is that fact that on 18 August she issued a conciliatory proclamation on religion. She made no bones about her own faith or her wish that her subjects would embrace it 'quietly and charitably'. But there was to be no compulsion – at least not 'until further order by common assent may be taken therein'. Meanwhile, in words that recalled her father's great speech to parliament in 1544, she exhorted everyone to live in peace with their neighbours and to avoid opprobrious terms such as 'Papist' and 'heretic'.[3]

Her words were good and well chosen – if not as idiosyncratic or effective as those of her father or sister in similar crucial circumstances. But actions spoke louder. And what spoke loudest of all about the Broad-Church nature of Mary's early support was the presence at her side of her sister Elizabeth.

Publicly, relations between Mary and Elizabeth continued to be excellent for the remainder of the summer. But privately the pressure on Elizabeth to convert to Catholicism became intense. It was a rerun of the earlier sibling clashes between Edward and Mary, with Mary this time playing Edward's part as well as occupying his throne. Like Edward, Mary found the disobedience of her nearest relative a personal affront. And she was just as concerned that her designated successor, as Elizabeth still was, differed from her in religion.

Here, as so often in this story, it is Mary's extraordinary lack of self-awareness that strikes. She, clearly, could see no parallel between her former situation and Elizabeth's present dilemma. And she certainly showed no sympathy for it. The explanation, probably, is simple. Edward, Mary would have said, had been wrong then and Elizabeth was wrong now, while she, Mary, had been right all along.

When it came to the crunch, however, Elizabeth's reaction was very different from Mary's previous pious obstinacy. Unlike Mary, Elizabeth never toyed with the idea of martyrdom. Nor was there an imperial ambassador to threaten war on her behalf. So she temporized. At the beginning of September she sought an audience with her sister at Richmond. Kneeling and in tears she implored Mary's understanding. She acted as she did, she claimed, not out of obstinacy but ignorance, 'having been brought up in the creed which she professed, without having ever heard any doctor who could have instructed her in any other'. To facilitate her conversion, she begged for books and a priest. And meanwhile she promised she would attend the Chapel Royal.

The moment came on 8 September. This was the Feast of the Nativity of the Blessed Virgin, after whom Mary had been named. Elizabeth duly appeared but she was conspicuously indisposed, complaining 'all the way to church that her stomach ached, and wearing a suffering air'.[4]

For the time being, this was enough to satisfy Mary, though it deceived few others. At the coronation celebrations at the end of the month, relations between them appeared as good as they had ever been. Elizabeth accompanied her sister to the Tower and rode immediately after her in the eve-of-coronation procession. On this occasion she shared a carriage with that other great survivor, Anne of Cleves. And the two ladies also sat with the newly crowned Queen Mary at her coronation banquet.[5]

Four days after the coronation, the Queen opened parliament after hearing the traditional mass of the Holy Ghost. At the same time convocation, the assembly of the clergy, also met. In the opening session of convocation, Elizabeth was prayed for immediately after the Queen, as the heir presumptive. But the actions of the two assemblies inevitably made her already difficult position worse. Parliament took the first step to the legal restoration of Catholicism by repealing all the religious innovations of Edward's reign. It also rehabilitated the memory of Mary's mother, Catherine of Aragon: her marriage to Henry VIII was declared valid and Mary herself legitimate.[6]

It was after the passage of this act, the Venetian ambassador noted, that Mary's attitude to Elizabeth became overtly hostile. The deliberations surrounding the act seem to have revived all Mary's memories of the humiliations she and her mother had suffered at the hands of Anne Boleyn. At the same time, Elizabeth's continuing prevarications over religion confirmed that she was indeed her mother's daughter. So the old bitterness returned and the Queen's mind started to harp, dangerously, on the parallels: Elizabeth was heretical and illegitimate, 'characteristics in which she resembled her mother'. 'Her mother had caused great trouble in the kingdom,' and Mary feared that Elizabeth would do the same. Was she even, Mary wondered, her half-sister? After all, she looked rather like Mark Smeaton, the musician who had been executed for adultery with Anne Boleyn.[7]

The upshot was that Mary had serious discussions about removing Elizabeth from the succession. But her confidential advisers were divided on the issue. And in any case it was clear that parliament would not as yet wear the idea. At court, however, Mary was less trammelled and she made her feelings plain by demoting Elizabeth's precedence.

Her choice of instrument displayed an exquisite feminine malice. Lady Margaret Douglas had long been resident at the English court. She was daughter of Henry VIII's elder sister, Margaret, Queen of Scots, by her second marriage to the Earl of Angus. In 1544 she had been married to the Earl of Lennox, a renegade Scottish nobleman in the English service, as part of the price for his treason. The wedding had taken place at St James's and Elizabeth had been present at court for the occasion. Now Elizabeth was sometimes made to give place to her. She was outraged. It was not only a blow to her pride; it was also a threat to her place in the succession, as Tudor blood flowed through Margaret's veins.[8]

Finally, in early December Elizabeth asked her sister's permission to leave court. The relief was probably mutual. Outwardly, they parted on affectionate terms and Mary gave her a coif of rich sables. *En route*, Elizabeth paused to write to Mary for 'copes, chasubles, chalices and other [Catholic] ornaments for her chapel'.[9]

Six weeks later, the sisters' elaborate fencing was to turn into a duel almost to the death.

CHAPTER 19

The Spanish Marriage

When Elizabeth left the court to spend Christmas at her house at Ashridge it was less then six months since Mary's accession. But the atmosphere in the capital had changed beyond recognition. The euphoria of summer had gone and in its place there was a poisonous winter fog of paranoia and plotting. Part of the reason (though a surprisingly small part) was hostility to the return of Catholicism. But the principal problem was the response to Mary's intended marriage.

Mary was eager to marry. She was thirty-seven. By contemporary standards, it was unlikely that she would bear children. If she were to stand any chance at all, she had to marry quickly. The need for haste was one element in the rapidly mounting tensions. Another was Mary's deeper attitude to marriage. She wanted a husband, not merely to engender an heir or to complete her womanhood. She also seems to have felt that her queenship itself was defective without a king.

She made this clear even at the moment of her coronation. The tradition was that, on the eve of the coronation, the monarch created a group of knights of the bath. Mary followed the tradition and created a rather distinguished group of fifteen knights, headed by the restored earls of Devonshire and Surrey. But instead of dubbing them herself, she delegated the task to the Earl of Arundel. The Earl's powers were given to him formally, by letters patent. In the case of the knights of the bath, the delegation was understandable on grounds of female modesty. According to the ritual, the knights were stripped and bathed and given the admonition of knighthood by the sovereign while still naked in the bath. But Arundel's powers were granted for two days. And he exercised them again on the day after the coronation when he created another, larger group of knights on the Queen's behalf. Here – as these were knights of the carpet rather than the bath and were fully clothed throughout – there was no question of female modesty. Instead it can only have been a matter of female incapacity: Mary felt that, as a woman, she could not exercise the chivalric or military aspects of kingship.[1]

These were less crucial than they had been. But they were still important, both practically and symbolically. Large expeditionary forces, or armies royal, were usually led by the king. The king appeared as a knight, mounted and in armour, on the obverse of the great seal. And the male rituals of the joust were a major point of contact (in every sense of the word) between the king and governing class. Some of this a woman could not physically do. She could not – at least under sixteenth-century conditions – command in the field or joust. But the rest were symbolic functions, symbolically discharged. By delegating even the dubbing of knights, Mary was saying she could not do these either. The effect was to declare a central part of monarchy out of bounds to a woman. Instead, the functions of her future husband would have to expand to fill the vacuum. This made his role, already great, greater still. And the greater it was seen to be, the more contentious became the choice of man who would discharge it.

There were two candidates. Both were part of Mary's extended family and both were royal. Edward Courtenay was, like Mary and Elizabeth, a great-grandchild of King Edward IV. His father, the cousin and boyhood playmate of Henry VIII, had been executed for his support of Catherine of Aragon and Mary herself. Courtenay had been imprisoned in the Tower since his father's execution. When Mary took possession of the Tower after her accession, Courtenay had knelt at her feet along with Stephen Gardiner, the former Bishop of Winchester, who had also been incarcerated since early in Edward VI's reign. 'These be my prisoners,' the Queen exclaimed expansively. Courtenay was restored to the earldom of Devon and Gardiner to his bishopric. Gardiner was also made Lord Chancellor. His principal aim then became to marry Courtenay to Mary as the English candidate for her hand.[2]

The trouble was that Mary did not think of herself as particularly

English. Ever since her troubles began under Henry VIII she had looked to her Habsburg cousin, the Emperor Charles V, for advice and protection. Now she looked to him for a husband. Many years ago, Charles himself had been betrothed to Mary. Now, he was a widower once more, but he was too old and weary to take up matrimony again. Instead he offered his son, Prince Philip, who ruled Spain as regent for his father. Philip was also a widower and the father of an heir. But he was only twenty-six as opposed to Mary's thirty-seven. And he was blond, bearded and rather lascivious, with a notable collection of tastefully pornographic paintings by Titian. Mary, squat, mannish and deeply religious, was not an appealing prospect.

But there were powerful dynastic issues at stake. Charles had spent his life in two titanic struggles, with heresy and with France. A marriage alliance with England would be a trump card in both. Henri II, the King of France, had gained control of Scotland by bethrothing his son, the Dauphin François, to Mary, Queen of Scots. With the marriage of Philip to Mary Tudor Charles would riposte by adding England, the much richer, larger and more strategically important part of Britain, to the Habsburg agglomeration. But the religious aspects of the union were just as important. A heretical England had been the ally and experimental laboratory for Charles's rebellious Protestant subjects in Germany. The recatholicization of England, which the marriage of Philip and Mary would guarantee, would strike a powerful blow against the rebels. So Charles was insistent and Philip, eventually and unenthusiastically, was obedient.

Mary approached the issue in a very different spirit. As she had done for the past twenty years, she made the imperial ambassador her confident as a proxy for her cousin the Emperor. The ambassador, Simon Renard, used all his diplomatic skills to persuade her that Philip was her only possible choice as consort. He preached to the converted. And this is more than a metaphor as Mary considered the question, as she did everything else, religiously.[3]

Her final decision was taken in a scene of high devotional drama. On 29 October Renard was summoned to the palace. He found the Queen alone with her favourite attendant, Susan Clarencius. The reserved sacrament, as usual, was on an altar in the room and all three – the Queen, the ambassador and the lady-in-waiting – flung themselves on their knees before it and sang 'Veni Creator Spiritus': 'Come Holy Ghost, eternal God'. This was the great hymn which had been sung over Mary before her anointing at her coronation. Now, as it ended, Mary was ready for another sacrament. She rose from the ground and spoke as one inspired. As the ambassador wrote: 'Philip was the chosen of Heaven for her, the Virgin Queen. If miracles were required to give him to her, there was a stronger than man who could work them. She would cherish him and love him and him alone.'[4]

Whom God hath joined, let no man put asunder – and, Mary would have added, let no man come between their union either.

The decision, once taken, was pursued with implacable urgency. On 8 November, Mary, by her personal intervention, jumped her deeply divided council into agreement, and on 16 November it was the turn of the commons. The Speaker was admitted to present their petition, which begged her to marry within the realm. Mary brushed aside Chancellor Gardiner and replied herself. She gave formal thanks to the commons for their wish that she should marry. But then, passionately and without reserve, she rejected their specific advice:

> The English parliament has not been wont to use such languages to their sovereigns. And where private persons in such cases follow their own private tastes, sovereigns may reasonably challenge an equal liberty. We have heard much from you of the incommodities which may attend our marriage; we have not heard from you of the commodities thereof – one of which is of some weight with us, the commodity, namely, of our private inclination. We have not forgotten our coronation oath. We shall marry as God shall direct our choice, to his honour and to our country's good.[5]

It was magnificently imperious. But at a price. For the speech placed a dangerous weight on the monarch's private wishes. Mary claimed to know, by particular and divinely inspired insight, that her personal inclination and the public welfare were compatible. But what if it turned out to be otherwise?

In December, the terms of a draft marriage treaty were agreed by the English council and Charles's advisers. They were very one-sided: the English asked and Charles, by and large, agreed, so keen was he for the marriage. On 2 January, a grand embassy, headed by the Count of Egmont, arrived in London to conclude the treaty and on the 14th Gardiner announced the terms agreed to a specially convened meeting of 'the lords [of the council], nobility and gentlemen' in the presence chamber at Whitehall.[6]

Mary, with the mere force of her personality and the power of her office, had been able to coerce the court élite into accepting the Spanish marriage. Public opinion proved more difficult to control.

Word of her intentions spread very quickly. And the reaction was uniformly hostile. The lower orders in England hated all foreigners and foreign ways (what's changed?). But the élite had tended to be more cosmopolitan. Here, however, Henry VIII's reign had brought about a sea-change. Henry's emphasis on England as an island empire, standing independent and alone against a papal Antichrist on the one hand and Protestant rebels and heretics on the other, helped make xenophobia intellectually and politically respectable among the élite as well. So any foreign king-consort would have had problems. But Philip's perceived Spanishness aroused a particular, virulent hatred. Why is unclear, but it did. And that is what matters.

As the outriders of the special imperial embassy rode into London, the boys pelted them with snowballs. When the noble ambassadors made their entry the following day, the English lords appointed to greet them excelled themselves in courtesy and magnificence. But the people – the mothers and fathers of the snowball-throwing boys – 'nothing rejoicing, held down their heads sorrowfully'. Their silence was more ominous than their sons' snowballs.[7]

In his address in the presence chamber on 14 January, Gardiner tried, with characteristic boldness, to confront the concerns directly. 'No Spaniard,' he insisted, 'should be of the Council . . . neither should have the custody of any forts or castles; neither bear rule or office in the queen's house, or elsewhere in all England.' Here was a master-rhetorician conjuring with language, trying to draw the sibilant poison of the word 'Spanish'. He succeeded only in making matters worse.

'These news', the best-informed contemporary diarist notes in his eyewitness account, 'although before they were not unknown to many and very much misliked, it being now in this wise pronounced, was not only credited, but also heavily taken of sundry men, yea and thereat almost each man was abashed, looking daily for worse matters to grow shortly thereafter.'[8]

They did not have to wait long. Within days rebellion had engulfed the south and south-east.

CHAPTER 20

Rebellion

The rebellion of 1554 – known from the leader of its most important sector as Wyatt's Revolt – brought Elizabeth to her nadir. It led to the most dangerous and difficult time of her life when she often feared imminent execution or murder. She even expressed a preference as to how she should die: like her mother, by the sword, rather than by the axe. The rebellion also threatened both Mary's throne and her life. Which is why, in turn, her rage against Elizabeth was so deep and long-lasting.

The original plan was that the revolt would raise the whole of southern and midland England in a series of separate but co-ordinated regional risings, each under the leadership of local grandees: Sir Thomas Wyatt in Kent; Lady Jane Grey's father, the Duke of Suffolk, who had been released from the Tower on the previous 31 July, in the midlands; the Carew family in the south-west; and Sir James Croft in the Welsh Marches. The rebels would then march on London from the four points of the compass and overthrow the government. The Spanish marriage would be blocked; Mary dethroned; and Elizabeth, married to Courtenay, who had been alienated by Mary's rejection of him as her husband, made Queen and King Consort. As well, in the grander version of the design, there would be French help to prevent England being turned into a Habsburg colony.[1]

But the plots leaked. In early January, Sir Peter Carew's refusal to obey a summons to court alerted the government to the preparations for the south-western rising. On the 18th, ambassador Renard, having heard that a French fleet was assembling off Normandy, rushed to court to tell Mary what he knew about the plot to make Courtenay and Elizabeth King and Queen. And on the 21st Gardiner himself interviewed

Courtenay. The Chancellor soon extracted the truth from his blubbering protégé, and delivered an edited version to the Queen. At this point the rebels were panicked into premature action, which seems to have been Renard's intention all along. Suffolk fled to his country estates after a summons to court on the 25th while Wyatt had already left for Kent after hearing of Renard's interview with the Queen on the 18th.[2]

Three of the four provincial revolts – in the south-west, the midlands and the Welsh Marches – now went off at half-cock and fizzled out harmlessly.[3] But Wyatt, who had shown an interest in advanced military organization under Edward VI, managed to raise a substantial force in Kent. On Sunday, 28 January, the Duke of Norfolk, who was then eighty years old, was sent against the rebels with a detachment of the guard and five hundred City whitecoats. At the first opportunity, the whitecoats and part of the guard went over to the rebels, crying joyfully, 'A Wyatt! A Wyatt!', while their leaders and a few others fled ignominiously back to the capital. The sight of the remaining yeomen of the guard returning, bedraggled, weaponless and with their coats turned inside out to conceal the embroidered royal crowns and ciphers, symbolized the seriousness of the Queen's plight.[4]

It was indeed desperate. The issue was not the size of Wyatt's army but the loyalty of London. For the desertion of the whitecoats threw in doubt the allegiance of the whole city. If Wyatt had pressed forward immediately resistance would probably have collapsed. But he delayed, giving Mary a chance to regain the initiative.

On Thursday, 1 February, she rode from Whitehall to the City, with all the guard in armour. In the procession, her demeanour was studiously sombre. But in her speech in the Guildhall to the City government she roused herself to oratorical heights.

She was their Queen and at her coronation she had been 'wedded to the realm and the laws'. At this point she showed them her coronation ring. 'I have [it] on my finger,' she said, 'which never hitherto was, nor hereafter shall be, left off.' She was the true inheritor of the realm, her father's daughter. Above all, she loved her people. 'On the word of a prince,' she swore, 'I cannot tell how naturally the mother loveth the child, for I was never mother of any.' Here we can imagine the pathos, part real, part acted, in her voice. Then her tone steadied and swelled. 'But certainly, if a prince and governor may as naturally and earnestly love her subjects, as the mother doth the child, then assure yourselves, that I being your lady and mistress, do as earnestly and tenderly love and favour you.'[5]

This is extraordinary. Mary had fumbled over female monarchy at her coronation with her failure to dub the knights herself. Here, however, she gave it magnificent and original voice. The ideas came from her father's parliamentary speech of 1545. But the invocation of a wedding ring and motherhood supplied a characteristically feminine twist. It was both tough *and* tender, as only a woman could make it. No man could have given the speech; and no woman, not even Elizabeth herself, could have given it better.[6]

Hearts and minds won over, Mary ended by giving solemn under-takings on her marriage. They were carefully noted down by one of her listeners. 'She never intended to marry out of the realm,' she protested, 'but by her council's consent and advice. And that she would never marry but all her true subjects shall be content. Or else she would live as her grace has done hitherto. But that her grace will call a parliament as shortly as maybe.'[7]

This too was superb. But it was a lie. Or rather, it was a whole series of deliberate, unequivocal lies. Mary had already sworn to marry Philip. She had stampeded the council into acquiescence and she had repudiated parliament's interference. And she had no intention of changing tack now.

But the lies worked – as barefaced lies often do when they are uttered by someone convinced, as Mary was, of the higher justice of their cause. The crowds cheered and nerves were steadied.

On the 3rd, Wyatt appeared in Southwark at the southern end of London Bridge, only to find the drawbridge up and guarded with cannon. After a three-day stand-off, he marched his army upstream, to the next Thames bridge, at Kingston. The bridge had been broken down. But Wyatt made it passable and panic broke out again in the capital when news came early on Ash Wednesday, 7 February, that the rebels were approaching from the west.[8]

Once more Mary was a tower of strength. In the crisis, her niceness about the military side of monarchy was thrown overboard and 'many thought she would have been in the field in person'.

And, indeed, she almost found herself, willy-nilly, in the midst of battle. Beyond Knightsbridge lay St James's Park, which was surrounded with a strong, high wall. Confronted with this obstacle, the rebel forces split, most making for Charing Cross by the north of the park, along the line of the modern Pall Mall. But a few went to the south, via Westminster. Here the road turned into King Street, the narrow thoroughfare which lay between Whitehall Palace, to the east or river side, and the park wall, to the west. Mary herself was in the privy gallery of the palace, by the Holbein Gate. There she could both see and be seen. Below, were the yeomen of the guard, some in the street in front of the open gates, and others in the great court; while the gentlemen pensioners, the élite bodyguard, were in the hall, where they covered the entrance to the royal quarters.

As the rebels, who numbered only a handful, approached, the guard behaved as disgracefully as they had in Kent. Those in the street turned tail and fled back through the gates, crying, 'Treason, treason!' The Lord Chamberlain and Captain of the Guard, Sir John Gage, led the flight, falling over himself in his haste and fouling his garments in the mire. As the fleeing guards pushed into the court, those who were already there tried to take refuge in the hall, but were denied entrance by the gentlemen pensioners. 'There should ye have seen running and crying of ladies and gentlewomen, shutting of doors, and such a screaming and noise as it was wonderful to hear.'

In the midst of the confusion, somebody had the presence of mind to close the gates, and the rebels, after shooting arrows harmlessly at the gates and windows, marched on towards Charing Cross.

The immediate danger past, it was time for heroics and the gentlemen pensioners petitioned the Queen that the gates be opened for her honour. She agreed, but on condition that they did not leave her sight. For, she said, with a pretty flutter, 'her only trust is in you for the defence of her person this day'.

Meanwhile, Wyatt and his band had reached Ludgate. But the gates, like London Bridge, were firmly closed against them. The spell was broken. Now Mary's troops had confidence as well as numbers on their side, and they fell on the rebels. Wyatt surrendered, and his followers fled or were captured. At 5 p.m. he was taken to the Tower by water. He had failed, and there could be only one outcome. 'It is no mystery now,' he said as he entered the fortress.[9]

When the news of his capture came to the palace, the pensioners were ushered into the presence and every man kissed the Queen's hand as she lavished thanks on them and promises of reward (which were to be unfulfilled). But it had been a close-run thing. If Wyatt himself, with his larger force and greater presence of mind, had taken the southern route by the Palace, Mary's personal danger would have been very real. Maybe the divinity that doth hedge a queen would have protected her, for certainly her inadequate, incompetent and cowardly guards could not. But who knows?

In any case the real measure of the 'trouble and fright' given to the Queen and her advisers was the vengeance they exacted. The three days after the failure of the revolt saw a stream of prisoners admitted to the Tower as the remaining leaders were rounded up. On the Saturday, the Duke of Suffolk was brought in under heavy guard. And on the Sunday the fate of the rebels was indicated by Gardiner's Lenten sermon, preached at court before the Queen.

'We beseech you,' Gardiner began, applying to England the words that the Apostle Paul had addressed to the erring Corinthians, 'we beseech you . . . that ye receive not the grace of God in vain.' Then he turned to the Queen and admonished her directly. Previously she had 'extended her mercy, particularly and privately'. But 'familiarity had bred contempt' and she had been repaid with rebellion. Now it was time to be 'merciful to the body of the commonwealth . . . which could not be unless the rotten and hurtful members thereof were cut off and consumed'.

'Whereby,' the diarist noted, 'all the audience did gather that there should shortly follow sharp and cruel execution.'[10]

They were correct. But the government's targets extended beyond the leaders of the rebellion. One intended victim was Lady Jane Grey, the other was Elizabeth.

CHAPTER 21

Retribution

At 10 o'clock on Monday, 12 February, the first of Gardiner's 'sharp and cruel' executions took place, as Guildford Dudley, husband to Lady Jane Grey, was beheaded on Tower Hill. Afterwards, the body – 'his carcass thrown into a cart and his head in a cloth' – was brought back for burial in the chapel of St Peter ad Vincula in the Tower. His wife watched both his departure for the scaffold and the bringing back of his mangled remains – 'a sight to her no less than death'. She knew it was her turn next.[1]

The scaffold on which she would die had been built on Tower Green, within the walls and directly in front of the chapel. The lieutenant led her out. She was wearing the same black dress trimmed with black velvet in which she had been tried and she prayed all the way from a book. The two ladies attending her were in tears but she remained calm. She made a brief speech, in which she acknowledged her fault in loving 'myself and the world'; then knelt and recited the penitential psalm 'Misere mei Deus', in English. 'Have mercy upon me, O God. Wash me thoroughly from mine iniquity and cleanse me from my sin.' She continued 'in most devout manner' to the end. 'The sacrifices of God are a broken spirit, a broken and contrite heart, O God, thou wilt not despise.'

The psalm ended, she stood up; gave her gloves and handkerchief to one of her ladies and her book to the lieutenant; and began to untie her dress. The executioner went to assist her. Revolted, 'she desired him to let her alone' and turned to her women. They helped her off with her dress and head-dress and handed her a fine handkerchief.

The executioner knelt down to ask her forgiveness. Then he told her to step on the straw. For the first time, she saw the block. For a moment, the steadfast martyr became a frightened, bewildered girl. 'I pray you despatch me quickly,' she begged. After she knelt she asked: 'Will you take it off before I lay me down?' 'No, madam,' he replied. She tied the handkerchief round her eyes and groped for the block, in a deadly game of blind-man's buff. 'What shall I do? Where is it?' A bystander guided her. She laid her head on the block and stretched out her body. 'Lord, into thy hands I commend my spirit,' she said. And the axe fell.[2]

She was seventeen.

Of all the terrible things done in that terrible place, this was one of the worst. It is also an indelible stain on Mary's reputation. Only a little while before, Jane had described her as 'a merciful princess'. She at least died knowing the truth.

For Elizabeth, the event was searing. She was only three years older than Jane. And the rebels had acted in her name, not Jane's. She feared, with good reason, that Mary would show her the same quality of mercy as she had shown Jane.

On 26 January, the day after the outbreak of Wyatt's rebellion, Mary had written to Elizabeth in firm but friendly terms. She had informed her of the rebellion and summoned her to court for her own safety; she could stay as long as she wished and would be assured of a hearty welcome. Elizabeth had declined to bite. She protested her ill health and the dangers of the way because of the disturbed state of the countryside. Thereafter, she bided her time at Ashridge as the drama of the rebellion turned into the tragedy of failure.[3]

Until that point, the government had had too much on its hands to worry about Elizabeth's disobedience. But the moment the revolt was over, her fate moved to the top of the agenda. On the 8th (the day after the collapse of the rebellion) Renard wrote to the Emperor that Mary had decided that the heads of both Courtenay and Elizabeth would fall. The next day a posse was sent to bring her willy-nilly to court. It consisted of three privy councillors and a troop of horse. When it arrived at Ashridge Elizabeth continued to plead that she was so weak that she could not make the journey 'without peril of life'. But the two royal doctors, who seem to have been in attendance on her since her first summons to court in January, certified, on the contrary, that she was fit enough to travel. Cornered, Elizabeth had to give way, and the party began its slow journey to London the next day. It was 12 February, the day of Jane Grey's execution.[4]

The journey to London, in which Elizabeth was carried in a litter, was planned to take five days, travelling at a leisurely pace of six or seven miles a day. In fact, it took much longer as Elizabeth fell ill in good and earnest. The nature of her disease is not known. But the strange lumps and swellings which disfigured her might be consistent with acute psychosomatic illness brought on by stress. Not until the 23rd did her cavalcade enter London.

It was a striking sight. A hundred horsemen in velvet coats rode in front of her and a hundred behind, in scarlet cloth trimmed with velvet. Surrounded by this sea of red, Elizabeth herself, dressed all in white, made a dramatic contrast. Malicious rumour, eagerly reported by the imperial ambassador, had interpreted her swollen body as a pregnancy resulting from 'some vile intrigue'. Elizabeth was determined to face down the rumours. So she had the curtains of her litter flung open and defied the world with an expression that was 'lofty, scornful and magnificent'. Renard was impressed, against his better judgement.[5]

She was carried through Smithfield and Fleet Street to Whitehall. There the Queen refused to see her and she was lodged in a remote, secure corner of the palace near the privy garden. On the day of her arrival, the latest victim, the Duke of Suffolk, Lady Jane's father, was beheaded on Tower Hill.

Meanwhile, the evidence against Elizabeth was mounting: Sir James Croft had called in at Ashridge on his way to raise the Marches and had advised her to move to Donnington, her castle two miles north of Newbury. Wyatt himself had sent her similar advice, and she had replied, orally and in non-committal terms, by her servant Sir William St Loe. Unfortunately, St Loe had undercut the careful ambiguity of his mistress's answer by going on to appear with two of the rebel leaders, Isley and Knyvet, in their demonstration at Tonbridge. Lord Russell, the son and heir of the Earl of Bedford, had carried Wyatt's letters to her. Finally, there was the fact that a copy of her letter, excusing her failure to come up to court from Ashridge in response to Mary's first summons, had been found enclosed in the seized despatches of the French ambassador.[6]

It looked very black. But it was not quite black enough. Clearly, Elizabeth had been kept informed by the conspirators. But there was no evidence – so far – that she had approved of their objectives or even that she knew of them. The government's interrogators now got to work. Wyatt was the key. On 25 February, two days after Elizabeth had been brought a virtual prisoner to Whitehall, Secretary Bourne reported to the council from the Tower: 'We have this morning travailed with Sir Thomas Wyatt touching the Lady Elizabeth and her servant Sir William St Loe.' Under examination, Wyatt implicated Croft, and Croft in turn implicated St Loe, who was immediately taken prisoner to the Tower. St Loe 'came in', it was noticed, 'with a wonderful stout courage, nothing at all abashed'. And his courage stood up under interrogation. That line of enquiry had drawn a blank.[7]

Nothing else seems to have been wrung out of Wyatt before his trial. This took place on 15 March in Westminster Hall and began with the reading of his indictment. In his reply, Wyatt continued to deny conspiring to cause the Queen's death and he put the mildest construction on his contacts with Elizabeth. He had indeed sent her a letter and she had replied, 'but not in writing'. Instead she had sent St Loe with the message 'that she did thank him much for his good will, and she would do as she should see cause'.[8]

It was no more than the council had known for weeks. But Wyatt's public testimony against Elizabeth – innocent though it was – provided the pretext for a sudden escalation in the pressure on her. On Friday, 17 March, two days after Wyatt's trial, the council came in a body to Elizabeth and formally charged her with involvement in both Wyatt's and the Carews' conspiracies. Elizabeth vehemently denied the charges. She was then informed that the Queen had determined she should be sent to the Tower, pending further investigations. As they withdrew, the council failed to uncover, and left instead 'with their caps hanging over their eyes'.[9]

In a moment, Elizabeth had descended from the second person in the kingdom to a suspected traitor. An hour or so later, she was sequestered from most of her servants, and armed guards were set to watch overnight in the courts, gardens and hall. On the Saturday, two councillors, the Marquess of Winchester and the Earl of Sussex, arrived to escort her to the Tower by water. She begged for a delay. She was told – in the proverbial expression that long antedated Shakespeare's use of it – that 'time and tide' waited for no one. She then asked for permission to write to her sister. That, too, was denied. But eventually Sussex relented and Elizabeth sat down to write the letter of her life.[10]

She began with the words of King John of France when he had returned to England to spend the rest of his life as a hostage: 'that a king's word was more than another man's oath'. Mary, she reminded her sister, had promised her that she would not be condemned without answer and due proof. Let Mary fulfil that promise by seeing her before she was sent to the Tower.

So far her tone had been measured, almost stately. But the writing of the word 'Tower' reminded her of what confronted her. It also reminded her of the Seymour affair, when those she had loved had last had to do with that terrible place. 'In late days I heard my lord of Somerset say that if his brother had been suffered to speak with him he had never suffered.'

'But,' Elizabeth continued, 'persuasions were made to him so great that he was brought in belief that he [Somerset] could not live safely if the Admiral lived.' This was, of course, what Renard and a faction of her council were insinuating to Mary. 'If [the council] do not punish [Elizabeth] now that the occasion offers, the queen will never be secure,' Renard had just written to Charles V. In her letter, Elizabeth immediately underscored the accuracy of the parallel with the Seymour brothers by denying it. 'Though these persons are not to be compared to your majesty,' she wrote, 'yet I pray God as evil persuasions persuade not one sister against the other.'

She ended by a direct rebuttal of the only two substantial charges against her. 'And as for the traitor Wyatt, he might peradventure write me a letter, but on my faith I never received any from him. And as for the copy of the letter sent to the French king, I pray God confound me eternally if I ever sent him word, message, token or letter.' By this time, Elizabeth was on the second page, of which she had used up only a quarter. She then filled the remainder of the page with strong, oblique strokes. These, like the lines drawn on a cheque, were a security device to prevent the addition of any forged matter. Then, at the foot of the page at the left, she added a sort of postscript: 'I humbly crave but only one word of answer from yourself.' And, at the right, her signature: 'Your highness's most faithful subject, that hath been from the beginning and will be to my end. Elizabeth.' The signature was as firm and flourished as ever, with the square, looped ornament after it that so resembled the 'R' of her father's sign manual.[11]

Elizabeth was writing not only for her life but also for time. The latter, at least, she got. By the time the letter was finished the tide had turned. The next tide was not until midnight. It was too dangerous to take her then, as an attempt could be made to rescue her under cover of darkness. So Elizabeth had bought a few, precious hours.

But they availed her nothing. Mary refused to see her and upbraided Sussex for allowing her to write.

The following morning, Winchester and Sussex came for her again. As she was brought through the garden, she looked up at the windows of the Queen's lodgings, in the hope of glimpsing her sister. But there was no one there. Similarly the river was empty. It was Palm Sunday and the citizens had been enjoined to take part in the revived ceremonies of carrying palms.[12]

CHAPTER 22

The Tower

Palm Sunday is the beginning of Passion week. Did Elizabeth perhaps spend the journey from Whitehall to the Tower reflecting on the parallel between Christ's agony and her own plight? Death also seemed near when the boat struck the bottom while passing under London Bridge. Between the state of her mind and the state of the tide, she was near breaking point.[1]

But we must not exaggerate, or let others exaggerate for us. Much of our knowledge of these grim moments comes from a tract entitled *The Miraculous Preservation of the Lady Elizabeth, now Queen of England*, which was appended to John Foxe's *Acts and Monuments*, otherwise known as the *Book of Martyrs*. The tract is broadly reliable. But, like most martyrologies, it exaggerates its subject's sufferings. The present moment is one example. According to the *Miraculous Preservation* Elizabeth at first refused to get out of the boat at Traitor's Gate and, when she did, she made an affecting speech. 'Here landeth as true a subject, being prisoner, as ever landed at these stairs,' she is supposed to have said.

The sentiments were certainly Elizabeth's. But the actual words cannot have been delivered, as the contemporary Tower diarist makes it clear that Elizabeth did not go into the Tower through Traitor's Gate. Bearing in mind the lowness of the tide, it would anyway have been impossible, since Traitor's Gate was a water-gate. Instead, Elizabeth landed at Tower Wharf and entered the Tower across the drawbridge, to the west of the fortress. This was scarcely less dreadful, however, as the sequence of narrow causeways and drawbridges took her past the Tower menagerie, with its roaring lions. Just as alarming, probably, were the guards lining the route. 'What! Are all these harnessed men here for me?' she asked Sir John Gage. 'No, madam,' he replied. 'Yes,' she said, 'I know it is so. It needed not for me, being, alas, but a weak woman.' Some of the guards are supposed to have behaved with unsoldierly deference, doffing their caps, kneeling and crying out: 'God save your grace!' Their enthusiasm meant little, however, when Elizabeth passed under the Bloody Tower and glimpsed, on the other side of the court, the scaffold on which Lady Jane Grey had been executed. Then she was led through the Coldharbour Gate.[2]

The *Miraculous Preservation* also misleads as to Elizabeth's accommodation when it refers to her 'dungeon'. This has led to pretty stories about her imprisonment in the Bell Tower – and clandestine meetings between Elizabeth and Robert Dudley, also a condemned prisoner in the Tower, on the leads between the Bell and the Beauchamp towers.

Alas for such romance, Elizabeth was imprisoned in neither a tower nor a dungeon but in the royal palace which lay in the inner ward of the Tower, to the south-east. There she had generous accommodation consisting of four chambers, and the attentions of a dozen servants. Other servants, of her hall and kitchen, were accommodated outside the Coldharbour Gate, which formed the main entrance to the inner ward.

However, if Elizabeth's physical privations were less than has been thought, her mental ones were probably even greater. For the area where she was accommodated had been rebuilt by her father for her mother's coronation. Anne Boleyn had stayed there then, on the eve of her greatest triumph; she was also lodged there three years later when she was brought to the Tower as a traitor and adulteress. Anne had left the apartment only for her trial and execution. Elizabeth had boasted that she was her father's daughter. Was she now, instead, to follow in her mother's footsteps?[3]

As the councillors left her, two of them turned the heavy keys in the doors. Whereupon Sussex, who had been uneasy throughout, wept and warned his fellows: 'What will ye do, my lords? . . . She was a king's daughter and is the queen's sister. And ye have no sufficient commission so to do. Therefore go no further than your commission.' Another version adds the prudent reminder: 'Let us use such dealing, that we may answer it hereafter.' Such concerns among Mary's councillors lent Elizabeth a glimmer of hope.[4]

She had need of it when, probably on Good Friday, she was subject to formal examination by the council. The object was to prove an overt act by Elizabeth that would implicate her in the rebellion. There were still hopes that Wyatt would provide evidence of this, and offers of pardon were being made to him and his wife, in return for implicating Elizabeth. But still he hesitated. Otherwise, the most promising leads were the known contacts between Elizabeth and the rebel leaders about her removal from Ashridge to Donnington.

What were her intentions, the council sternly demanded, in proposing to go to Donnington? Elizabeth's first defence was evasion. Had she got such a house? After all, she had so many. Then, under prompting, she recovered her memory. 'Indeed, I do now remember that I have such a place, but I never lay in it in all my life. And as for any that hath moved me thereunto, I do not remember.' Her memory was jogged on this point, too, by confronting her with Croft himself. The reminder worked. 'And as concerning my going unto Donnington castle, I do remember that Mr Hoby and mine officers and you Sir James Croft, had such talk,' she said. Then she went on the attack. 'But what is that to the purpose, my lords, but that I may go to mine own houses at all times?'

At this point, according to the *Miraculous Preservation*, one of Elizabeth's interrogators, the Earl of Arundel, threw himself on his knees to beg her forgiveness for having 'troubled you about so vain matters'. Croft joined in. Kneeling also, he declared that he was sorry to see the day that he was brought in as a witness against Elizabeth. But, he assured her: 'I have been marvellously tossed and examined touching your highness.'[5]

But Elizabeth had not totally bamboozled the council. An important despatch from Renard on 3 April shows that they were well aware of the truth. 'From day to day,' Mary had assured him, 'they are finding new proofs against [Elizabeth]. That especially, they had several witnesses who deposed as to the preparation of arms and provisions which she made for the purpose of rebelling with the others, and of maintaining herself in strength in a house [Donnington] to which she sent the supplies.'[6]

This indeed seems to be what had happened. Modern historians have failed to give it due weight because they misunderstand Elizabeth's position. It is so romantic, is it not, to see her as a 'weak woman', vulnerable and defenceless? The fact that Elizabeth used these words herself when she entered the Tower should put us on our guard. For Elizabeth was *not* defenceless. She was the second largest landowner in the kingdom. She had coffers bursting with cash. And she had a devoted household and following that was capable of raising thousands of knights and gentlemen. So far, they had only paraded with her in procession. But they were certainly able and willing to fight for her in the field. In those circumstances, their numbers would have multiplied, with each gentleman bringing his little army of scores or hundreds of his own servants and followers.

And preparations *had* been made, as Mary said, to muster these forces, with Donnington as their operational centre.[7] It was well chosen. It was a castle and fully defensible. It had strategic significance, too. It commanded the valley of the River Kennet, which linked the Thames lowlands with the main road to Marlborough and the west. What Elizabeth intended to do there is a mystery. Probably her reply to Wyatt ('That she would do as she should see cause') is as good an indication as any. She would wait on events. If Wyatt had won, or half won, in London, she would have had to intervene to protect her own position. A victory by any of the other provincial risings would have presented her with a similar dilemma – or opportunity. Donnington would have been an ideal vantage point for this game of wait-and-see.

But the game never took place. Mary's first letter summoning Elizabeth to court had made a pointed reference to Donnington, 'whither, as we understand, you are minded shortly to remove'. This was a shot across the bows. Much more important was the collapse of all the planned risings. This made armed resistance, even if Elizabeth had been inclined to it, suicidal. There was no choice but to submit. She had to hope that she had been discreet enough to cover her tracks. She had also to hope that her servants would remain steadfast under questioning. On the whole, both hopes were fulfilled.

The final serious threat to Elizabeth came with Wyatt's execution on 11 April. The government had strung him along to the last, and even on his way to the scaffold he seems to have hoped for a reprieve. None came. Instead, he made a speech exonerating Elizabeth and Courtenay. 'And whereas it is said and whistled abroad,' he began, 'that I should accuse my lady Elizabeth's grace and my lord Courtenay; it is not so, good people. For I assure you neither they nor any other now in yonder hold or durance was privy of my rising or commotion before I began. As I have declared no less to the queen's council. And this is most true.'

Wyatt's disclaimer was deeply unwelcome to the official representatives on the scaffold. Dr Hugh Weston, Dean of Westminster, was there as a priest, to minister to Wyatt. But he was also in the confidence of the régime. Trying to undo the effect of Wyatt's words, he butted in, saying: 'Mark this, my masters, he sayeth that that which he hath showed to the council in writing of my lady Elizabeth and Courtenay is true.' Someone else seems to have plucked Wyatt by the gown. At all events, he said no more but immediately prepared himself for death.[8]

He was beheaded between 9 and 10 a.m. Within the hour, his body was quartered on the scaffold; his bowels and private parts were burned nearby, while the head and quarters were put in a basket and taken to Newgate by cart. There they were parboiled, as a primitive preservative, before being nailed up. The head was placed on top of the gibbet at St James's. But it was stolen away within the week.[9]

The day after Wyatt's execution, a group of councillors again confronted Elizabeth in the Tower. What took place at the meeting is not known. But probably they tried to get Elizabeth to agree to some generalized admission of guilt to save everyone's face. They failed.[10]

In fact, it had been clear to Renard for a fortnight that Elizabeth was likely to escape the worst. For one thing, there was not enough evidence against her. Standards at Tudor treason trials were loose. But by this time the government was finding that juries, sickened by the blood-letting, were becoming more demanding. On 17 April, Sir Nicholas Throckmorton was acquitted, 'whereat many people rejoiced'. And on the 27th only eight jurors were willing to condemn Sir James Croft. So four others, more compliant and of inferior quality, were empanelled to secure his conviction. These were straws in the wind. To put Elizabeth on trial in these circumstances was to risk a hurricane. And, in any case, there was no real political will to do so. The council had divided and redivided on the subject. Before the rebellion, Gardiner, the Lord Chancellor, had been Elizabeth's best friend, and Paget, the Lord Privy Seal, her most dangerous enemy. After the revolt, the two exchanged positions. And as her council bickered and debated, Mary dithered.[11]

The result was that the terms of Elizabeth's incarceration were quickly softened. She already had permission to walk in the privy garden of the palace, mornings or afternoons, or in the great chamber adjacent to her lodgings. On 25 April, she asked for the further privilege of being allowed to walk in the great gallery. The matter was referred to the Queen. But it was overtaken by larger events.[12]

On 4 May, Sir Henry Bedingfield was appointed Constable of the Tower and directed to raise a hundred troops. The appearance of these fresh guards gave Elizabeth another bad fright. She asked 'whether the Lady Jane's scaffold were taken away or no'. For, as the *Miraculous Preservation* explained, she 'fear[ed] lest she should have played her part'. The real reason for the reinforcements became apparent when, on 19 May, Elizabeth was informed that she was being removed to house arrest.

She left the Tower as she had arrived, by water.[13]

CHAPTER 23

Prisoner's Progress

Elizabeth was popular in London. On 12 April two men were set in the pillory at Cheapside for 'saying that Wyatt had cleared my lady Elizabeth'. And, a month earlier, the very stones had seemed to speak on her behalf. When the people said, 'God save Queen Mary!' in front of a wall in Aldgate, there was no reply. But when they cried, 'God save the Lady Elizabeth!', an angelic voice answered 'Amen!' On 15 July a young woman stood at St Paul's Cross to confess her part in the pious fraud.[1]

Others met far harsher punishment. Gallows were set up at every gate of the city and at crossroads, open spaces and landmarks both within the city and in the suburbs. Batches of rebels suffered exemplary punishment at each one. Some were hanged, others hanged in chains, others quartered and beheaded. The gallows, festooned with bodies and body parts, were left standing for four months, in awful warning.

Now Elizabeth was deliberately being taken away from all this. Her barge moved swiftly. But word of Elizabeth's removal from the Tower spread faster still. Public rejoicing broke out, as the people mistakenly thought she was at liberty. As her barge passed the Steelyard, the Thames-side depot of the privileged Hansa merchants, guns fired a salute. But still the barge with its attendant craft rowed on, not landing till it came to Richmond, a dozen miles upstream. Richmond, with its fantastic, clustering turrets and onion domes, had been rebuilt and renamed by Henry VII, Mary's and Elizabeth's grandfather. Mary revered Henry VII, because he was ultra-orthodox, and Richmond, his particular creation, was her favourite residence. For Elizabeth it was foreign territory.[2]

It felt doubly strange with Bedingfield and his men surrounding

Elizabeth and watching her every move and she spent one of the most disturbed nights of her life. 'For this night,' she told her gentleman usher, 'I think to die.' It is hard to know how seriously to take her repeatedly expressed fears of assassination during these months. It is unclear even how seriously she took them herself.[3]

But they did have some basis in fact. Once it became clear that there was insufficient hard evidence to proceed by law against her, Gardiner and Renard discussed sending her to Pontefract Castle in Yorkshire. Richard II had met his mysterious end there, and the (probably) unspoken thought was that a similar fate could be arranged for Elizabeth. Renard was still nursing the scheme when he reported Elizabeth's removal from the Tower and stated in his despatch that Woodstock was only a staging-post for Pontefract. On the other hand, Bedingfield's correspondence with the Council, which survives in full, contains no trace of the project. Maybe, like much else in this period, the Pontefract scheme existed largely in Renard's fertile brain; in any case it sunk amid the divisions of Mary's council.[4]

As both the council and the imperial embassy leaked like sieves, Elizabeth was probably aware of these discussions and their outcome. But, sensibly, she did not relax her guard. And her vigorously expressed fears had their purpose, too. By constantly calling public attention to the possibility of assassination, she was making it doubly difficult for the government to do the deed itself and get away with it.

After staying only a single night Elizabeth was removed from Richmond, which lay on the south bank, and ferried across the Thames to begin her journey north to Woodstock, which had been chosen as her place of imprisonment. A group of her own servants was waiting to catch a glimpse of her. As she landed, she sent a message to them consisting of the single biblical phrase: 'Tamquam ovis.' The *Miraculous Preservation* glosses Elizabeth's words as a reference to the verse in the Acts of the Apostles: 'He was led like a sheep to the slaughter.' Humble passivity in the face of danger and even death might be Elizabeth's public stance. But, as her guardians were soon to discover, she took her real text from St Matthew's Gospel: 'Behold, I send you forth as sheep in the midst of wolves; be ye therefore wise as serpents, and harmless as doves.'[5]

Elizabeth, moreover, had a temperamental preference for the wisdom of the serpent – and, occasionally, its bite and venom as well.

Her journey from Richmond to Woodstock took four days. On the first night, she stayed in the dean's lodgings at Windsor; on the second, at Sir William Dormer's house at West Wycombe in Buckinghamshire; and on the third at Lord William of Thames's princely residence at Rycote in Oxfordshire. Everywhere people gathered to watch her: men, the scholars of Eton, and, above all, women. At High Wycombe women had prepared cakes and wafers and loaded them in her litter. She was overwhelmed with the quantity of gifts and the smell of the herbs and spices in the wafers, and had to ask them to stop.

For, despite the presence of the intensely Catholic Dormer family, the rich, wooded landscape of the Chiltern Hundreds was Protestant country. Or, as Bedingfield put it in the first of his reports: 'Men betwixt London and these parts [West Wycombe] be not good and whole in matters of religion . . . be[ing] fully fixed to stand to the late abolishing of the bishop of Rome's authority.' Elizabeth herself was a great landlord in Buckinghamshire, and Lord Russell, the Earl of Bedford's son, 'and certain other gentlemen of his sect hath procured and practised' the advance of Protestantism in both High Wycombe and neighbouring Wooburn. The result was real conviction, that knew both its religion and its princess. A quarter of a mile outside Wooburn, Christopher Cook, 'a plain husbandman [farmer] . . . awaited on a hill to see her grace . . . and the said Sir Henry talking with him found him a very Protestant'. And at Aston Rowant, four men rang the church bells as Elizabeth passed and were put under arrest for their pains.

The last leg of the journey lay through Oxfordshire, skirting the city to the east. Here Bedingfield felt more confident: 'The sure hope, as I can learn, consisteth in Oxfordshire men,' he reported. The country was more remote from London; while the university, the home of old causes if not lost ones, had remained a bedrock of Catholicism. But even in the little towns and obscure villages of Oxfordshire, to Bedingfield's evident disappointment, Elizabeth's magic worked. At Wheatley, the whole population turned out to watch the cavalcade, crying out: 'God save your grace!' The same happened at Stanton St John. At Islip, Elizabeth was an appreciative audience for some rural merrymaking by the men and children and was greeted by the women on the bridge. Gosford also turned out to welcome her.[6]

Elizabeth's prisoner-progress from the Tower to Woodstock was the strangest of her life. But it was by no means the least successful.

As for Bedingfield, his relief at getting her to Woodstock was palpable. But his troubles, as he soon discovered, had just begun.

CHAPTER 24

Imprisonment: The Politics of Protest

Sir Henry Bedingfield, Elizabeth's gaoler, was the head of an ancient Norfolk family, based at Oxburgh Hall near Swaffam. The Bedingfields' history went back to the Normans. But, for the last three generations, their story too had been intertwined with the Tudors'.

Henry Bedingfield's grandfather, Sir Edmund, had rebuilt Oxburgh Hall, much as it stands today, with its moat, towered gatehouse and red-brick battlements. There he had entertained Elizabeth's grandfather, Henry VII, in 1487, after fighting alongside him in his victory at the Battle of Stoke. Forty years later, Henry Bedingfield's father, also called Sir Edmund, had been steward and gaoler of Mary's mother, Catherine of Aragon, in her last unhappy years at Kimbolton. Now Bedingfield himself was appointed to fulfil the same role for Elizabeth. We can imagine that Mary signed the warrant with a grim smile, if her piety allowed her such an ironic pleasure.

The earlier years of Bedingfield's career had been of purely local significance. But the course of his life had changed on a single day when, in July 1553, he had been one of the first to rally to Mary's standard in Norfolk when she levied war against Queen Jane. Mary never forgot the loyalty of her Norfolk partisans, and their leaders, including Bedingfield, were immediately sworn members of her council. The decision does more credit to Mary's heart than her judgement. Bedingfield had only a basic education. His handwriting was slow and bad. And anything more than schoolboy Latin defeated what he called – with justifiable unpretentiousness – 'his Norfolk understanding'. But he was loyal, to her and to the faith. And that was enough for Mary.[1]

But it was not enough for the task of watching over the wittiest, best-educated princess in Europe.

Bedingfield kept a careful record of his custodianship, in the form of a letter-book with copies of his correspondence. Its purpose was to justify his actions, should he ever be called to account. From our point of view, however, it acts as a blow-by-blow account of the grossly unequal duel between him and his prisoner.[2]

Most of Elizabeth's biographers have played Bedingfield's testimony for laughs. And there are indeed some very funny episodes. But it has a serious message, too. It shows how far Elizabeth's household had become a government in (internal) exile. It also shows how far Elizabeth, supposedly so risk-averse, was prepared to pursue a bold, high-risk strategy when she thought the circumstances were right.

Bedingfield's instructions were strict – both for himself and for Elizabeth's other attendants-cum-goalers. Elizabeth was to be securely kept, but treated 'as may be agreeable to [Mary's] honour and her estate and degree'. She was to have no conversation with any suspicious person out of Bedingfield's hearing. Neither was she to send or receive any 'message, letter or token to or from any manner of person' whatsoever.[3]

Those were his orders; enforcing them was another matter.

His difficulties began with Woodstock itself. It was one of the most ancient royal estates and kings had hunted its park since Anglo-Saxon times. The structure of the palace was largely the work of Henry II, who had enjoyed its rural bliss with his mistress, the fair Rosamund. But it had been extensively rebuilt and modernized by Henry VII. Bedingfield encountered the missing slates, defective lead-work and broken windows which are inevitable in large buildings which are only irregularly occupied. But the palace was far from the medieval ruin described by most historians. The problem, rather, was its size and the lack of security. The outer court was 200 feet square and the inner court only a third smaller. But in all of this, as Bedingfield explained, there were 'three doors only that were able to be locked and barred'.[4]

A gaol almost without locks is one problem. A gaol where half the warders are the devoted servants of the prisoner is another. For Elizabeth had been allowed to keep six of her own personal attendants: three women and three men. Her women were alone with Elizabeth in her lodgings for long periods of time, while the men came and went freely in the rest of the house. Any one of them could whisper a message or secrete a letter. And it is clear that they did.[5]

Mary and her council had sought to minimize the difficulties by removing the most offensively Protestant of Elizabeth's women. The worst offender in their eyes was Elizabeth Sandes. Mary herself described her as a woman 'of evil opinion and not fit to remain about our sister's person'. Bedingfield was instructed to replace her at once with Elizabeth Marbery. He effected his instructions at 2 p.m. on 5 June, to loud demonstrations of 'great mourning' by both Elizabeth and Mrs Sandes. Mrs Sandes was packed off, with Parry's assistance, either to her uncle in Clerkenwell or to her father in Kent. Soon she made her way abroad in the great Protestant diaspora.[6]

Mrs Sandes travelled in a group that was the epitome of the Elizabethan connection. Its leader was her cousin, Sir William Stafford. His first wife had been Mary Boleyn, Elizabeth's maternal aunt, while his second was Dorothy (also née Stafford) who was both Elizabeth's lady-in-waiting and her cousin as the great-granddaughter of George, Duke of Clarence. Dorothy, as a 'noble lady', was accorded special privileges of citizenship in her final foreign refuge of Basel, before returning to a lifetime of service in Elizabeth's bedchamber. Whilst in Basel, she hired the 'little house' next to the Clarakloster, where most of the poorer English exiles led a sort of communal life. One of their leaders was the martyrologist John Foxe, who, as we have seen, appended the *Miraculous Preservation* to his *Book of Martyrs*. The effect of this highly flattering memoir of Elizabeth's 'sufferings' under Mary was to make Elizabeth seem to have shared in the experiences of exile, while, in reality, she had ostentatiously conformed to Mary's Catholicism. My guess would be that Foxe's neighbour in the Clarakloster, Dorothy Stafford, was his prime source for the highly circumstantial details of the *Miraculous Preservation*. Dorothy had not been there. But she knew, intimately, everybody who had, including Elizabeth herself.[7]

Mrs Sandes's exile and the company she kept amply confirm Mary's verdict on her 'evil opinion'. Mary was also right to reject both Elizabeth's two preferred alternatives for Sandes. These were either the long-serving Dorothy Bradbelt or Elizabeth Norwich. Dorothy's family connexion are hard to establish. But Elizabeth Norwich was to marry Sir Gawain Carew of the rebellious Carew dynasty. This was not the only connexion between the family and Elizabeth's women. Lady Tailboys, another of Elizabeth's ladies, had married Sir Peter Carew in 1545 after the death of her husband, George, Lord Tailboys. Carew had fled abroad after his part in Wyatt's Revolt and his wife, as Bedingfield noted disapprovingly to the council, was living as a 'sojournant', or guest, in the house of one of Elizabeth's gentleman ushers, Mr Winter.[8]

With a pool so tainted with this combination of family connection, Protestantism and the politics of protest, it is not surprising that the council failed to find three properly orthodox ladies to attend on Elizabeth. Probably their choices were simply the least obstreperous in dissent. Bedingfield continued to worry about all Elizabeth's servants. Some refused point-blank to attend Catholic services; the rest, he feared, did so 'for form only'. His solution was to ask for some learned men to be sent 'to preach and talk with them in the matter of their religion' – either to convert them or to expose them as arrant Protestants. There is no evidence that the council even bothered to try.[9]

But at least the problem of Elizabeth's women was well known and, to that extent, containable. Moreover, their gender greatly limited their freedom of movement. Much more intractable were the difficulties which flowed from the apparent technicality of financial responsibility for the household at Woodstock.

For another oddity of this Through-the-Looking-Glass gaol was that it was the prisoner who paid her gaoler and his staff. The council supplied funds only for Bedingfield's soldiers. Otherwise, all the costs of the establishment were borne by Elizabeth. So she provided food and drink, which was much the most expensive single item, not only for herself and her servants, but also for Bedingfield and *his* servants. This put Bedingfield in a position of humiliating financial dependence. And Elizabeth's accounting officers drove the fact home mercilessly, first by quibbling about the cost and then by demanding a formal royal warrant to authorize the expenditure. More importantly, these financial arrangements led Bedingfield to a major miscalculation.[10]

Bedingfield had two principal characteristics: punctiliousness, on the one hand, and a fear of taking either initiative or responsibility on the other. Generally, the two, to Elizabeth's mounting irritation, reinforced each other. But in this case they were at war. His first instructions from the council had been to isolate Elizabeth by getting her factotum and *éminence grise*, Thomas Parry, as far away from Woodstock as possible. Normally Bedingfield would have fallen over himself to comply. But, he quickly realized, Parry's removal meant that he, Bedingfield, would have to take financial responsibility for the household. And that was more than he was prepared to do: 'I neither will nor dare intermeddle myself with [it],' he wrote bluntly. Instead he finessed a compromise: Parry should leave the palace but stay in the town. Lord Treasurer Winchester, extending a protective hand over Elizabeth as usual, helped fix the deal.[11]

From Mary's point of view, the decision was a disaster. Bedingfield was no match individually for either Elizabeth or Parry. But now he had to deal with the two of them: Elizabeth in the palace and Parry in the town. Combined, they ran rings round him and often barely bothered to disguise the fact.

Parry made his headquarters in the Bull inn in Woodstock. This offered the perfect cover. Or, in Bedingfield's own phrase, it was 'a marvellous colourable place to practice in'. Forty people a day, Bedingfield complained, repaired to Parry in his own livery, as well as Elizabeth's own servants, who seemed to find that the business of provisioning the household required a large number of visits to the Cofferer. The Bull must have done a roaring trade. But were these people there as customers or as conspirators to undo the State? Bedingfield had his suspicions. But he found it frustratingly difficult to make them stick.[12]

For Parry covered his tracks well. So well, indeed, that we too find it difficult to know just what he was up to. But he seems to have had three main aims, only one of which was legitimate from the point of view of the government. His first, legitimate, objective was to keep Elizabeth's land-holding intact as a functioning administrative and financial unit. For its very survival was imperilled by Elizabeth's imprisonment. She had been a hands-on manager of her own moneys. That, in the circumstances of her imprisonment, in which she was not even allowed to issue a warrant over her own signature, was impossible. But worse than the loss of management control was the risk of a tenants' revolt. Charges of treason normally signalled confiscation of property. When the landlady was about to go to the block, why should the tenants pay up? Parry moved swiftly to enforce Elizabeth's rights. Rents were demanded and stocks of deer in her parks were protected.[13]

Mary was consulted on these moves and, up to a point, supported them. But Parry's two other aims were clandestine and subversive. The first was to defy Mary's specific instructions by keeping Elizabeth in touch with the outside world. And the second was to mount a general challenge to Mary's régime by keeping Elizabeth's political connection, which in the nature of things was oppositionist, alive and kicking. The two aims overlapped and frequently involved the same incidents and individuals. And the stories of three of them will illustrate Parry's success and poor Bedingfield's difficulties.

Soon after Elizabeth's arrival, a student at Oxford, which lay only five miles down the road, sent presents of three books for Elizabeth, each with a covering letter. Bedingfield's suspicions were immediately aroused and the books and the covering letters sent off to the council for further investigation. The books were discovered to be harmless but one of the letters seemed to contain code. What did the writer mean by the phrase that 'he lacked a compass to set a pen in'? And another seemed to mingle matters of love and religion. Bedingfield was instructed to examine the student. The boy was summoned. He turned out to be a plausible young man who blagged himself out of an awkward corner, partly by blinding Bedingfield with his 'diffuse words . . . in the Latin phrase', and partly by luck as, in its haste, the council had forgotten to return the incriminating letters, so he could not be held to the actual words he was accused of writing. To cut a long story short, Elizabeth got her books, while the plausible young man was John Fortescue. He was Parry's stepson and became his political heir. But as an important but rather colourless figure in both Elizabeth's court and council, he never quite lived up to the romance of his early escapade in her service.[14]

Another Greek bearing gifts was Christopher Edmonds, who turned up with freshwater fish for Elizabeth and then with 'two dead pheasant cocks'. On both occasions, he spent a long time talking to his fellow-servants. Like Fortescue, Edmonds had the gift of the blarney (perhaps learned from Elizabeth, who was no mean mistress of the art) and swore that 'he neither did nor would do anything contrary to his allegiance'. Bedingfield was not deceived. He had already encountered him at Lord Williams of Thame's house at Rycote, where Elizabeth had stayed on the last night of her journey from Richmond. Edmonds also lived at Rycote, and had the distinction of carrying a dish to the door of Elizabeth's dining chamber. Edmonds, it turned out, was Williams's stepson.[15]

Much about Williams's behaviour as Elizabeth's host must have now become clear to Bedingfield. Williams was a good Catholic and a Marian loyalist. Yet he had gone out of his way to entertain Elizabeth lavishly and treat her royally. For through his wife and her son, Christopher Edmonds, he had a foot in both camps. Many other families, to Mary's distress, had taken out similar reinsurance. And Williams was to cash his in, to triumphant success, at Elizabeth's accession when he became, as Lord President of the Council of the Marches, the chief figure in the government of his native Wales. His stepson did well too, as did his wife Dorothy Edmonds, who was to be one of the most influential of Elizabeth's ladies.[16]

But most extraordinary is the case of Francis Verney. The Verneys were a rich and influential Buckinghamshire gentry family. Buckinghamshire, as we have seen, was one of the areas where Elizabeth's landholdings clustered most thickly. It was thus the most natural thing in the world for Francis, a younger son, to become the servant to his powerful royal neighbour. But, as well as the traditional reasons of spheres of magnate influence, Verney had been attracted into Elizabeth's orbit on the new grounds of religious sympathy. He was a Protestant; he was also a plotter. Actually, it seems rather difficult to find gentlemen Protestants in this period who were *not* plotters. And Verney was no exception. One of Bedingfield's informants put it succinctly: 'If there be any practice of ill, within all England, this Verney is privy to it.' Verney had a meeting with a servant of the executed rebel leader, the Duke of Suffolk, at the Bull, and he opened council letters which had got into his hands. Within three years, he and his eldest brother Edmund were convicted traitors. Bearing in mind the almost pathological flamboyance of his behaviour, this is hardly surprising. What is surprising, however, is the extent to which Parry was prepared to countenance him. Parry had him to stay with him at the Bull, and, with staggering impudence, used him as Elizabeth's special messenger to solicit for her at Mary's court.[17]

Yet it was this same Elizabeth who, on 26 August, after hearing confession, summoned Bedingfield to witness and hear her solemn declaration. Bedingfield and one of Mary's ladies knelt in front of her while Elizabeth spoke. 'That [she], in all her life, had done nothing, nor intended to do, that was perilous to the person of the queen's highness or the commonwealth of this realm, as God, to whose mercy she then minded to commit herself, was judge.' She then sealed her words by taking the sacrament 'in Catholic form'.[18]

It was a breathtaking performance. Elizabeth sailed on, swan-like and serene, while her servants, male and female, paddled away furiously in submarine plots, heresies and treasons.

Perhaps only Mary, who had known her half-sister from her precocious infancy, really had the measure of her; certainly Bedingfield did not.

CHAPTER 25

Imprisonment: Personal Resistance

There were two ways in which Elizabeth could escape from her custody at Woodstock. The first was by a *coup d'état*. This would have required one of the multitude of plots, in which Parry and his agents and acolytes were involved, to be successful. Elizabeth was too hard-headed to have had much faith in either the plots or the plotters – though she did want to know what was going on. She was also, if her subsequent behaviour is any guide, reluctant to bring about directly her sister's dethronement and death. She had scruples about the act. She also had serious concerns about the impact of repeated usurpations on the standing and security of the monarchy.

Such scrupulosity left a negotiated settlement with Mary as the only way out. The question was the terms on which she would settle. Mary thought she held all the cards and could force Elizabeth into an humiliating climb-down. At the least, Elizabeth would have to acknowledge her guilt and sue for mercy. At best, she would be removed from the English political scene entirely, by marriage to a foreign princeling. These were plausible outcomes, that were debated by serious politicians. But instead, Elizabeth, by magnificently overplaying a weak hand, bluffed her way to an almost total victory.

Elizabeth's first problem lay in opening negotiations at all. Bedingfield, with characteristic obtuseness, interpreted the ban in Elizabeth's sending or receiving letters as applying to communications with the Queen herself. So Elizabeth's initial request to write to Mary was refused. But at least he mentioned her wish to the council. The council's reply came in the form of a postscript that was clearly the result of direct royal intervention: Mary was 'pleased' that Elizabeth should write. There is almost a sense that she was looking forward to the encounter. And certainly, to begin with, she more than held her own.[1]

Elizabeth's letter does not survive. But we know its broad outline from Mary's summary in her reply. Elizabeth had protested she was innocent. As for the matters she was charged with, either they had been misunderstood, or the persons making the charges deserved no credit. This was the position she had taken in the Tower and was to stick with throughout. It cut no ice with her sister.

Mary wrote that she was 'most sorry' to have had any occasion of suspicion. But copies of Elizabeth's letters *had* been found in the French ambassador's bag. And the leaders of Wyatt's revolt *had* used her as their figurehead. This 'we can hardly be brought to think that they would have presumed so to do, except they had had more certain knowledge of her favour towards their unnatural conspiracy than is yet by her confessed'. Subsequently, Mary had treated her with 'more clemency and favour' than was usual in cases of treason. But Mary was not to be abused with fair words. 'Conspiracies be secretly practised, and things of that nature be many times judged by probable conjectures and other suspicions and arguments, where the plain direct proof may chance to fail.'

Judged in this light, Elizabeth's actions spoke of guilt, however loudly her letters might protest her innocence. Mary's conclusion was brutal. She wished to be troubled with no more of 'her disguise and colourable letters'. Instead, Elizabeth should compose herself to behave properly towards God, which in turn would improve her behaviour towards her sister and Queen.[2]

The letter was harsh enough. What made it far worse was that Mary had not deigned to write directly. Instead, her reply had been addressed to Bedingfield, who was instructed to communicate its contents orally to Elizabeth.

Understandably, he did so with some trepidation. He made a summary of the Queen's letter so that he could read it word for word. And he waited for Elizabeth to broach the subject. She did so after hearing Mass in her chamber. Had he received a reply, she asked.

Bedingfield said he had and asked leave to get his prompt sheet, which he read over to her twice after dinner. He knelt as she listened. But instead of the expected explosion, Elizabeth uttered only carefully modulated expressions of regret that her letter had not taken better effect. And she immediately, terrier-like, returned to the attack. Granted *she* was forbidden to write, could not *Bedingfield* write a summary of her petition to the council?

Bedingfield refused; Elizabeth pressed. State prisoners in the Tower could communicate with the council via the governor; even, as she pointed out when she returned to the same theme next day, friends were allowed to sue on behalf of ordinary villains in Newgate Gaol. Was she to be denied even this, she asked, as she walked in the garden with Bedingfield, and to continue thus without any worldly hope? Then, she protested, she must look only to God and the truth to vindicate her. The effect of this impassioned declaration was rather spoiled by a sudden shower of rain. She broke off exclaiming, 'It waxeth wet', and scurried indoors.[3]

Once again, as a result of Mary's direct intervention, Elizabeth was eventually given permission to approach the council, using Bedingfield as an intermediary. And, once again, Elizabeth increased the stakes. Since her avowals of innocence were not believed, she begged that she should be put on trial on the charges as they stood. Alternatively, she wanted a personal interview with the Queen to put her case. Or, if even that were denied, that a delegation of the council be sent to Woodstock to hear her replies.[4]

Not that the council were left to make up their own minds. Instead, Elizabeth showed herself an expert lobbyist. The essential feature of Mary's council was its divisions. It has been rather fashionable of late among historians to play these down. But informed contemporaries, like ambassadors, had no doubt, either that the divisions existed or that exploiting them was the quickest and surest means to obtain their ends. Elizabeth, at least as well informed as any ambassador, concurred.

So she highlighted personal differences. She sneered at her enemy (who was also an enemy of her mother, Anne Boleyn) Sir John Gage, for his age and forgetfulness; she protested her dependence on her friend Lord Treasurer Winchester. More importantly, she put her finger on the main structural faultline in the council. This lay between the long-serving councillors who had held office under Henry VIII and Edward VI, and the johnny-come-latelies, who had been appointed by Mary. The former tended to be moderate and 'politique' over religion; the latter were Catholic enthusiasts. Elizabeth touched this nerve by appealing directly to the councillors who were 'parties and privy to and for the execution of the will of the king's majesty her father'. This was unexceptional, as Henry had appointed his executors as her especial guardians. But it was also tactically shrewd, as it pitted the Henrician councillors against the Marians.[5]

Elizabeth even applied the same technique to the dangerous territory of religion. She had made the minimum necessary gesture towards her own Evangelical faith by asking for the Bible in English. Otherwise she conformed to the forms of Catholic worship without fuss and, when it suited her, with apparent devotion. But she continued to use the Litany and Suffrages in English. Mary then made a specific request that Elizabeth conform to usage of her own Chapel Royal and 'the ancient and laudable custom of the church' and employ Latin for these prayers instead. Elizabeth hastened to comply. But she also offered an exquisitely calculated defence of her previous practice.[6]

She had begun hearing the Litany two or three times a week, she explained, while she was in the Tower. This was because its sequence of petitions to God for 'mercy upon us miserable sinners' was so appropriate for her situation.

> 'Remember not, Lord, our offences, nor the offences of our forefathers; neither take Thou vengeance of our sins; spare us, good Lord, spare thy people whom Thou has redeemed with thy most precious blood, and be not angry with us forever', her chaplain intoned.
>
> 'Spare us, good Lord,' Elizabeth replied.
>
> 'From all sedition, privy conspiracy, and rebellion; from all false doctrine', he continued.
>
> 'Good Lord, deliver us,' she answered.

But, Elizabeth insisted, she had used the Litany only after consulting Gage and he 'would not say nay to it, because it was used in the king [her] most noble father's days.' One can imagine Gage, whom she detested, squirming at the uncomfortable memory.

But it was not only tactics. Elizabeth had memories – indeed, a whole complex of memories – too. The Litany had first been authorized in May 1544. This was the triumphant year when Elizabeth had been restored to the succession, to her father's bosom and to an honoured place at court. In October, Catherine Parr's clerk of the closet had bought a copy of the Litany for the Queen Regent's use and Elizabeth had knelt beside her stepmother as the priest had recited the Litany's prayers for the King: 'that it may please Thee to be his defender and keeper, giving him victory over his enemies'. As Henry was away fighting the French, Elizabeth had replied with especial fervour: 'We beseech Thee to hear us, good Lord.'[7]

The words stuck in Elizabeth's mind, not only because of their message but also because of the sonorous musicality with which Cranmer, the compiler and translator of the Litany, had endowed them. Cranmer was Elizabeth's godfather. As Elizabeth heard his English text recited for the last time, Cranmer was only five miles away in his prison in Oxford. Bedingfield mentioned in one of his reports to the council that the presence of Cranmer and his fellow prisoners, Ridley and Latimer, 'hath done no small hurt in these parts, even among those that were known to be good afore'. The council replied promptly, acknowledging Bedingfield's concern, and adding ominously that 'we shall shortly cause such order to be taken with them as shall be convenient'. The 'convenience' was their trial and execution for heresy. Ridley and Latimer were burned first, on 16 October 1555. Cranmer followed five months later, on 21 March 1556, after the drama of his recantation and retraction.[8]

Cranmer's goddaughter was more fortunate. By September 1554, Elizabeth's lobbying won permission for her to write to the council herself and to send the letter by one of her own servants. She made a famous to-do about the whole affair. Bedingfield told her of the council's permission on Monday, 17 September. To his great surprise, Elizabeth then let the matter sleep until the following Sunday, when she ordered him to prepare writing materials for her the next day. Bedingfield obeyed and on the Monday morning delivered to one of her ladies a standish, or writing desk, with five pens, two sheets of fine paper and one of coarse. In the afternoon, Elizabeth announced that she could write no more because she had a headache. On the Tuesday morning, she washed her hair. Then, in the afternoon, she summoned Bedingfield and delivered her bombshell: her royal autograph was too grand for the council, to whom she only ever wrote by a secretary. As 'I am not suffered at this time to have none,' Elizabeth explained, 'therefore you must needs do it.' Bedingfield desperately tried to excuse himself, but Elizabeth would have none of it. 'At her importune commandment and desire', he wrote to Elizabeth's dictation, 'as she read unto me of her handwriting, which she retaineth as a minute [copy].'

Bedingfield retaliated by delaying handing over the letter to Verney, Elizabeth's chosen messenger. There followed another contest of wills between Bedingfield and his prisoner, which Elizabeth, as usual, won. Eventually, the letter was handed over and Verney despatched. Bedingfield had to console himself by reclaiming the writing desk, what was left of the paper and the pens 'wanting one'.[9]

The mountains had laboured over Elizabeth's letter to the council. But there issued, not a ridiculous mouse, but a poisoned pen in the form of Mary's reply. It was brief and devastating. Her sister would have enjoyed her favour long ago, Mary wrote, if her defence had 'so well . . . satisfied indifferent ears as it seemeth to satisfy her own opinion'. It was a put-down worthy of Elizabeth herself at her best: the biter was bit.[10]

Elizabeth now gave up any hope of a quick resolution of her situation. Instead, she petitioned only that she might be moved to a house nearer London. Even this was denied, and she spent the winter at Woodstock, protected from the chill of the Oxfordshire countryside only by the scratch repairs to broken windows and leaky roofs which had at last been carried out.

Wood for fuel, Parry protested, was in short supply. But Elizabeth warmed herself more effectively by venting her rage on Bedingfield.

Following the lead of Francis Blomefield, the great eighteenth-century county historian of Norfolk, later writers have sought to soften, even to sentimentalize, the relationship between Bedingfield and Elizabeth. Bedingfield might be crusty and Elizabeth tempestuous, but, they seem to suggest, the two developed a *modus vivendi*, even, perhaps, a grim mutual affection in adversity.

Bedingfield's own record offers not a scrap of evidence in support of this view. Rather to the contrary. Bedingfield satirized Elizabeth's airs and graces by referring to her as 'this great lady'. Elizabeth responded by treating Bedingfield with open contempt. She acknowledged that he meant well. But as for everything else, 'knowledge, experience, and all other accidents' required for his present position, she told him to his face, he lacked them all. It was crushing and it crushed. Bedingfield petitioned desperately to be released of his position: news of his discharge would be, he assured the council, 'the joyfullest tidings that ever came to me'. But the news did not come. Instead Elizabeth continued the abuse. It was as if she vented her temper by kicking the dog.[11]

Elizabeth's frustration during these long, dreary months of apparently interminable detention is understandable. So, too, is her irritation with Bedingfield's wooden obtuseness: he was a fool and she never suffered fools gladly. But, equally, her treatment of Bedingfield shows her in her worst light. Like her father and her mother, she could charm when it suited her. When it did not, she was, again like both her parents, an unmerciful bully, who lashed with words and later with fists. Henry VIII had the physique of the bully: he was bigger and stronger than any of his victims. But the real issue was the superiority of status rather than strength. Kings and queens could hit out with impunity because to lay a finger on them was treason. Of course, the imprisoned Elizabeth was royalty under a cloud and one of Bedingfield's first acts was to forbid her Gentleman Usher to put up the cloth of estate to which she was entitled as heiress presumptive. But, despite the shadow of treason, Elizabeth remained royal and whenever Bedingfield spoke with her indoors, even to reprove her or deny her wishes, he did so kneeling. Kneeling is not a strong negotiating position. Elizabeth exploited her advantage shamelessly and vindictively.[12]

CHAPTER 26

A New Dynasty?

Elizabeth's detention at Woodstock lasted just under a year. During these months, time, as it does for prisoners, stood still. Outside the park gates, on the other hand, events moved at breakneck speed. These months were the apogee of Mary's reign when, for a glorious moment, it seemed as though history were to be rewritten and the words of the *Magnificat*, spoken by Mary, Queen of Heaven, were about to be fulfilled for Mary, Queen of England.

> He hath shown strength with his arm; He hath scattered the proud in the imagination of their hearts. He hath put down the mighty from their seats, and exalted them of low degree. He hath filled the hungry with good things and the rich He hath sent empty away.

Mary and her mother had been despised; now they were exalted. Her relations with her protector, Charles V, had been treasonable; now Charles's son, Philip, was to become her husband and king consort. Her religion had been trampled under foot; now it was to be restored.

But in this counter-revolution there was no place for Elizabeth. The achievement of Henry VIII, the mighty father of whom Elizabeth was so proud, would be swept aside; her mother's marriage would be an aberration, while Elizabeth herself would be written out of this revised history of new beginnings and unlooked-for happy endings. Or at least she would be written out if Philip and Mary had a child.

Elizabeth understood this perfectly and for once she let her feelings show. The English Litany, which Elizabeth used as long as she was permitted, contained a noble prayer for the monarch.

'That it may please Thee to keep and strengthen in the true worshipping of Thee, in righteousness and holiness of life, Thy Servant MARY, our most gracious queen', her chaplain had prayed.

'We beseech Thee to hear us, good lord', Elizabeth had fervently replied. After Mary's marriage, Elizabeth's chaplain, 'according to his most bounden duty', altered the text to pray for King Philip and Queen Mary together. Bedingfield placed himself so that he could see Elizabeth's response. He could hardly contain his glee when he reported her obstinate silence. Since the inclusion of Philip's name, he wrote, 'her grace hath never answered word to that article, that could be heard or perceived by any man, being marked of very purpose by . . . me'. Elizabeth took kingship very seriously. But she was not going to pray for her own irrelevance.[1]

News of the fulfilment of Mary's dreams of marriage reached Elizabeth, partly through official letters to Bedingfield, and partly through Parry's clandestine communications network.

The last legal obstacles to the marriage had been cleared even earlier, while Elizabeth was still in the Tower. Reflecting Mary's own failure to assume the quintessentially male, chivalric attributes of monarchy, two lawyers had argued that Mary's sovereignty was both conditional and temporary. As a female, she had only 'a woman's estate' in the Crown. This meant that, when she married, she lost her title to the crown and her husband became King. The argument was based on an analogy with the succession to noble estates and titles. If the male line of a noble house died out, the heiress transmitted the landed estate to her husband, while the title was normally recreated for him by the so-called 'courtesy of England'. The heiress, that is to say, had not become duke or earl herself, but she had been the agency through which the dukedom or earldom had passed to her husband. Mary, the lawyers were arguing, was in a similar position with regard to the crown. This sort of reasoning was dangerous and it was quashed by a brief act of parliament, passed in April 1554, which declared that 'the regal power of this realm is in the Queen's majesty as fully and absolutely as ever it was in any of her most noble progenitors, kings of this realm.'[2]

What was Mary's absolutely was hers also to bestow as she pleased and, after the crushing of Wyatt's revolt, the momentum for the Spanish marriage became unstoppable. As Elizabeth completed her journey from the Tower, through the Thames valley to Woodstock, Mary began a slow royal progress west to meet Philip at Winchester. First she took up residence at Richmond; then, on 12 June, the council informed Bedingfield that Mary intended to move to Oatlands near Weybridge as news had come that Philip was definitely on his way from Spain. He landed at Southampton on 20 July and Mary arrived in Winchester two days later. On the 25th they were married in Gardiner's cathedral by the bishop himself.[3]

The arrangements for the wedding were modelled on the precedent of Catherine of Aragon's marriage with Prince Arthur. It was not a happy augury. But Mary brushed such thoughts aside. First the groom and then the bride walked the length of the elevated walkway which stretched from the west doors to the steps of the sanctuary. Mary was given away by four nobles, 'in the name of the whole realm'. Her wedding ring, at her insistence, was a plain gold band, unornamented by any stone. 'For she said she would be married as maidens were in the olden time'. She also used the old Catholic rite, in which the woman promises 'to be bonny and buxom [or compliant], in bed and at board'. The marriage was consummated that night, to Mary's ecstatic satisfaction at least.[4]

The royal couple then danced and hunted their way back towards London, with a lavish reception at Lord Treasurer Winchester's great house at Basing and a more formal visit to Windsor for Philip's installation as knight and co-sovereign of the Garter. This time Mary shared in the knightly rituals: Arundel, once again acting as the Queen's lieutenant, invested Philip with the robe, but Mary herself placed the collar round her husband's neck. Then came the new King's entry into London.[5]

Most of the pageants made the usual plays on Philip's name and the parallels with his great namesakes, from Philip of Macedon to Philip the

Fair, his own grandfather. But 'the most excellent pageant of all' was an animated genealogy, showing the descent of both Philip and Mary from Edward III of England. Curiously, Philip had the more impressive English lineage, descending twice over from John of Gaunt, Duke of Lancaster, Edward's third and favourite son, by two of Gaunt's legitimate daughters, Philippa and Catherine. Mary also had a double descent from Gaunt, but hers lay through two of the children, later legitimated and surnamed Beaufort, of the Duke's long-term liaison with Catherine Swinford. On the Tudor side, she descended from John Beaufort, Earl of Somerset, and on the Yorkist, from Ann Beaufort, Countess of Westmorland. Another version of this genealogy appears on the back of the title page of the book of commemorative verses that the scholars of Winchester College presented to the King and Queen on the occasion of their marriage.

The message of the London pageant was summarized in the following verses, which were written in letters of gold on a ground of silver:

> England, if thou delight in ancient men
> Whose glorious acts thy fame abroad did blaze,
> Both Mary and Philip their offspring ought thou then
> With all thy heart to love and to embrace,
> Which both descended of one ancient line
> It hath pleased God by marriage to combine.[6]

This, of course, is not history as we know it. We know, and Wyatt's rebels feared, Philip the Spaniard. The pageant genealogy, on the other hand, presented him as an Englishman come home. Nor was Mary the familiar half-Spanish daughter of Catherine of Aragon. Catherine did not appear at all; instead, Mary too was presented as mere English. Likewise, both spouses were shown to descend from the ancient Lancastrian house, and with their marriage the fractured, divided descent was reunited once more.[7]

The parallel with that other great Tudor genealogical reconstruction of history, the union of York and Lancaster, is obvious. That had been designed to knit up the rift of the Wars of the Roses; this was intended to heal the wounds of the Reformation. And if the marriage of Philip and Mary had been as fruitful and successful as that of Henry VII and Elizabeth of York, the new myth, of the royal and orthodox reunion of the house of Lancaster, would certainly have displaced the old legend of the union of York and Lancaster.

Both myths depended on exclusion as well as inclusion. The union of York and Lancaster entailed the demonization of Richard III. The awkward figure in the new myth was, of course, Henry VIII. As Mary's father, he could hardly become a non-person, like Richard III. Instead, his *actions* had to be airbrushed from the picture. And that, with starling literalness, is what happened.

Another of the pageants showed a representation of Henry VIII with, in his right hand, a book inscribed '*Verbum Dei*', that is, 'The Word of God'. The image was taken from the title page of the English Great Bible of 1539, which had shown Henry distributing the Bible, thus inscribed, to his grateful subjects. But the English Bible was a *bête noire* of the Catholics and the target of Gardiner's especial detestation. Gardiner was contemptuous of English as a recent and unstable language, fit only for the rootless novelty of heretical opinion, in contrast with the magnificent, millennial fixidity of Latin, which made it the natural vehicle for the eternal verities of the Faith. He also felt that to translate the Scriptures from the learned tongue into the vernacular was to lose control of their interpretation. For Gardiner, therefore, the appearance of the reference to '*Verbum Dei*' in the pageant was like a rude word in church, challenging all the hopes of the conservative counter-revolution which the marriage of Philip and Mary was intended to bring about. The bishop-chancellor poured out his wrath on the artist and ordered him to paint out the offending book in the King's hand and replace it with a pair of gloves![8]

CHAPTER 27

A New England?

In the autumn after Philip's entrée into London, the painting-over of the recent national past shifted to a larger canvas. Hitherto parliament had resisted any attempt to take the English Church back to Rome or to restore the Church's coercive powers over heresy and dissent. But, with the euphoria over Mary's marriage, the resistance collapsed.

The new parliament was opened on 12 November and its first act was to repeal the attainder or condemnation for treason of Cardinal Reginald Pole. Pole was papal legate or ambassador, charged with the mission of reconciling England to Rome. Since the beginning of Mary's reign he had been kicking his heels in the Netherlands. He had been kept there partly by English hostility to Rome and partly by the Habsburgs' determination to do first things first and get Mary safely married to Philip before allowing her government to embark on the messy and risky business of religious counter-revolution. Pole, with the mellifluous eloquence that came to him almost too easily, described his situation in more dignified terms: he had been knocking, he wrote to King Philip, on the doors of England like St Peter at the house of the Virgin Mary. Mary had opened to Peter; when would England open to him? Now, with the repeal of his attainder, it had; and Pole started his journey immediately in the company of the high-ranking English delegation that had been sent to escort him home.[1]

Home. For Pole the words 'going home' had the bitter-sweet meaning they do for all exiles. But both the bitterness and the sweetness were intensified by the enormity of the issues which had led to his rupture with his king and country. For Pole was no ordinary exile. He was grandson of George, Duke of Clarence; Henry VIII was grandson of

George's elder brother, Edward IV. So Reginald Pole and Henry VIII were cousins and boyhood companions, and Pole had worshipped Henry with the intensity of the younger for the elder and the weaker for the stronger personality. Pole turned out to be the intellectual of the family. Henry had sponsored Pole's higher education at the fashionable university of Padua, and Pole in turn had played an important part in Henry's negotiations to get the European universities to deliver a favourable verdict on the invalidity of his first marriage to Catherine of Aragon. But Pole had become increasingly unhappy at the King's proceedings and, with the decision to renounce obedience to Rome, he broke with them completely. Henry responded by denouncing Pole as an ingrate and arch-traitor and putting a price on his head. Pole's mother and his elder brother Henry, Lord Montagu, were executed and Pole himself was forced to remain in exile for twenty years.[2]

But now all this was over and Pole was coming home to England. Still better, England was coming home too – back to its obedience to Rome. On 28 November Pole explained his mission to both houses of parliament at Whitehall; the next day lords and commons each passed a resolution in principle to return to Rome, and on the 30th, the day of St Andrew who first brought his brother Peter to Christ, Pole pronounced the formal act of absolution.

> 'We', he intoned, 'by apostolic authority given unto us by the most holy Lord Pope Julius III, [God's] vicegerent on earth, do absolve and deliver you, and every of you, with the whole realm and the dominions thereof, from all heresy and schism, and from all and every judgement, censure and pain for that caused incurred; and also we do restore you again into the unity of our Mother the Holy Church, in the name of the Father, of the Son, and of the Holy Ghost.'[3]

It was a moment of high ceremony and high emotion. Philip and Mary knelt in their robes of estate while Pole stood in his cardinal's scarlet; the crowd cried 'amen, amen', and tears flowed. But there was sense amid the sentiment. For Pole's speeches outlined, fully on the 28th and in summary form on the 30th, a new providential history of England.

God, he explained, 'by providence hath given this realm prerogative of nobility above other', since 'this island first of all islands received the light of Christ's religion'. Other nations had been converted bit by bit and individual by individual. But England, Pole claimed, had embraced Christianity 'at once as it were in a moment'. The claim is based on a strange mixture of fact (the mission of St Augustine of Canterbury to the Anglo-Saxon kingdom of Kent) and fiction (the legendary apostolate of Joseph of Arimathea to the Britons). And it is utterly spurious.[4]

It is as spurious, indeed, as Henry VIII's own opposed providential English history. This argued that England's pre-eminence consisted, not in being the eldest daughter of the Roman Church, but in properly being independent from it. 'The realm of England', the act of appeals of 1532 asserted, 'is an empire'. This is not the modern meaning of 'empire'. By 'empire' we mean the *extent* of rule over several states or territories. For the Tudors, on the other hand, 'empire' was the *intensity* and *completeness* of royal rule over the particular territory of England. Internally, they claimed, the English King was supreme. So he was Head of the Church as well as the State. And externally, England was independent too. So the English Church was as self-governing and autonomous as the English State.

These rights and preprogatives of England and her king were, the act went on to state, 'manifestly declared and expressed' in 'divers sundry old authentic histories and chronicles'. Wisely, the act refrained from naming these supposed sources. But royal research and propaganda relied on a rag-bag of legends and fictions: the English descent of the Emperor Constantine from his mother, Helen; the continental conquests of King Arthur and, once again, Joseph of Arimathea.[5]

Pole, who had the reputation of sound scholarship, would have been fully justified in dismissing all this as a farrago of nonsense and insisting instead, as a matter of true and sober fact, that for seven hundred years England had been a faithful daughter of the Roman church and its kings suppliants at the throne of St Peter.

But Pole did nothing so radical. Instead, he chose to argue on Henry VIII's own ground, and with the sort of 'evidence' which Henry had used. To fight on your enemy's territory is always a dangerous decision. Pole only took it because he had to. He might have devoted his life to obedience to Rome. But when he returned to England he, like everyone else, felt obliged to kowtow to English nationalism.

For Henry VIII's reinvention of England had been extraordinarily successful. The sense of England alone, uniquely separate from Rome, had struck a chord. Instead of being a source of anxiety, England's isolation was a matter of pride. 'If God be for us, who shall be against us?' had become a new national motto and, to be welcome once more on English shores, the eternal verities of the Universal Church had to present themselves in terms of the particular national history of England.

This was what Pole was trying to do when he boasted of his patriotism; talked of the supposed unique mass conversion of the English to Christianity and claimed that it gave the realm 'a prerogative of nobility above all other'. The Reformers' vision of the English as a peculiar Protestant people had of course to go. But it would be replaced by the English as the chosen Catholic nation. The religion might change, but England's manifest destiny remained.

Gardiner returned to the same theme of Englishness a few days later in his sermon at St Paul's Cross on 2 December, the first Sunday in Advent. He took his text from the Epistle to the Romans: 'now it is high time to awake out of sleep'. St Paul had been calling on the Romans to wake out of the sleep of paganism and self-indulgence. Gardiner was trying to rouse the English from the twenty-year slumber of heresy and schism. The result was one of the great set-pieces of sixteenth-century pulpit oratory. It also contained a series of extraordinary insights into high policy, when Gardiner drew on his almost unique knowledge as a privy councillor to reveal how close England had come to returning to Rome on several previous occasions: after the Pilgrimage of Grace of 1536; again in 1541; in 1549 after the fall of Somerset, and, most recently, at Mary's coronation. 'But the time was not then,' he said repeatedly, after the narrative of each incident. 'But now,' he thundered in a carefully orchestrated climax, 'the time *is* come.' Now a reconciled English people join with the shepherds and angels in hailing a new advent, the return of England to the faith. Now all sing 'Glory to God in the highest and on earth peace, good will toward men.'[6]

But, under the rhetoric and the revelations, Gardiner was confronting a deeper task. Pole was trying to redefine England history; Gardiner was fighting no less than a struggle for the soul of England. What was England? What was essential Englishness? Was it the England of the Ages of Faith? Or was it the England of Henry VIII?

And who were the heroes of England? Were they King Arthur, King John and King Henry, who had fought for England against Rome? Or were they the saints and martyrs and kings, who had risked death, exile and imprisonment for the Catholic faith: from Edmund and Edward under the Anglo-Saxons to Mary, Pole and Gardiner in their own time?

And it is as three new English worthies that Mary, Pole and Gardiner appear in one of the most remarkable propaganda works of the period: John Elder's *Letter*. Dated New Year's Day 1555, it describes the events of the previous year in ecstatic terms, as England's 'calling home' to the safe haven of Catholicism. In this voyage, it claims, there had been three 'lode-stars and chief guides': the Queen, 'a virgin, and immaculate from all spot of heresies'; the cardinal, so long an unjustly reviled exile and traitor, but now 'restored to his blood and to the honour of his house'; and Gardiner, who had miraculously escaped snares, traps and long imprisonment in the Tower, all for his faith. But now, Elder concluded, their sufferings were over, their faith was vindicated and England was restored to 'the unity of Christ's religion' – 'which', he adds, 'shall be to the glory of God, the wealth of England, and to the perpetual peace, love and quietness of this most noble and whole isle of Britain.' The balance of the remark is characteristic: the glory of God wins one reference, national destiny two.[7]

The conclusion is surprising: in one aspect at least, Marian Catholicism turns out not to be very different from Elizabethan Protestantism. We think of the former as foreign and the latter as quintessentially English. In fact, both faiths set out to present themselves as national religions, and Catholics blew the patriotic trumpet as noisily as did any Puritan. The similarities went beyond tone to the detail of thought and language. Pole, Gardiner and Elder talked of the history of England and the destiny of the English, as confidently as did Shakespeare or John Foxe; they even hailed Mary as the Virgin Queen. All was set for a successful Marian England which would have gone down to posterity as an 'Elizabethan Age'. But it would have been Hamlet without the Prince and Elizabethan without Elizabeth.

Sir Thomas Wyatt the Younger led a Kentish revolt against the Spanish marriage of Mary to Philip II. But the revolt collapsed and he was executed.

After Wyatt's revolt Elizabeth was arrested. On 17 May 1554, she wrote this desperate letter to her sister Mary. The cross-hatching over her signature was to prevent any forgeries or additions. Despite the letter, Elizabeth was imprisoned in the Tower under threat of execution herself.

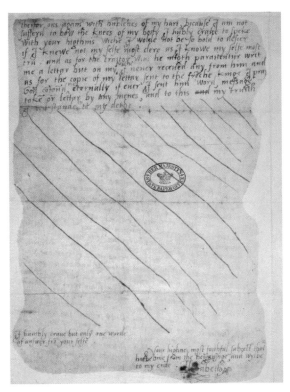

Sir Henry Bedingfield, Elizabeth's gaoler at Woodstock where she was detained for over a year. There Elizabeth tried to have her cake and eat it: her servants spent the time conspiring against Mary; while Elizabeth sought to persuade her sister of her innocence.

Philip and Mary. Philip II was son and heir of the Emperor Charles V. His marriage to Mary in July 1554, followed by her supposed pregnancy, immensely strengthened Mary's hand. In the euphoria, parliament restored Catholicism and the Papal obedience. But things fell apart when Mary's pregnancy proved to be false and even Philip started to turn to Elizabeth as the rising sun.

Thomas Cranmer, Archbishop of Canterbury. He became close to Elizabeth in 1544 when he was principal adviser to Catherine Parr as Regent, and Elizabeth learned to love his English Prayer Book. Mary treated him as her personal enemy and had him burned at Oxford in 1556.

The martyrdom of Latimer and Ridley in October 1555. Elizabeth regarded their burnings as a serious mistake and determined never to repeat Mary's persecution.

Stephen Gardiner, Bishop of Winchester. Irascible, intelligent and a superbly funny writer, he correctly identified Elizabeth as the most serious threat to the Catholic restoration and planned her execution. But Elizabeth – and events – outflanked him.

Cardinal Reginald Pole, himself of royal blood, was exiled under Henry VIII but under Mary returned in triumph to welcome England back to Rome. Elizabeth loathed him; but he smoothed her path to power by conveniently dying on the same day as Mary.

Robert Dudley, Earl of Leicester, and Elizabeth were contemporaries. They had known each other since childhood and were together imprisoned in the Tower under Mary. Elizabeth was deeply attracted to Dudley, who was of the same buccaneering type as Thomas Seymour, and, in other circumstances, would probably have married him. Instead, he became Master of the Horse and, after a brief but intense fling with the Queen, a sort of honorary consort.

Sir Nicholas Throckmorton, conspirator and diplomat, was involved in every plot against Mary. He advised Elizabeth on her strategy for her accession. She rejected his scheme for her Council as too Machiavellian. But she followed his suggestions for concealing her true intentions about religion.

Sir William Cecil. Elizabeth called Cecil her 'spirit' and their lives were indeed intertwined like body and soul.
Cecil came from a Tudor service family, studied at St John's and was surveyor of Princess Elizabeth's lands. At her accession she made him Secretary and pledged him her confidence. The partnership lasted till his death forty years later.

Elizabeth's coronation robes were a hand-me-down from Mary, but with the bodice remade to show off her elegant figure.

The coronation ritual was a clever compromise, too. She was crowned with the Latin text and by a Catholic bishop. But parts of the service were also said in English and Elizabeth absented herself from the Mass. As she left Westminster Abbey, she was almost laughing with triumph.

Sir Nicholas Bacon was Cecil's brother-in-law and a surprise appointment as Lord Keeper or acting Chancellor (he was given the junior rank because of his humble social origins). But he quickly established a rather thuggish authority and delivered, as the Queen's mouthpiece, the opening speech of Elizabeth's first parliament. This acted as a mission statement for the reign and Elizabeth kept its promises faithfully.

Mary, Queen of Scots, with Lord Darnley.
Mary was Queen of Scots in her own right and, through her great-grandfather, Henry VII, had a claim to England as well. Her arrival in England in 1568 as a refugee from her Scottish subjects was a disaster for Elizabeth, as Mary provided the focus that Catholic opposition had hitherto lacked. Mary had a finger in every plot against the Queen of England until Elizabeth was reluctantly driven to execute her in 1587.

Elizabeth: the Armada Portrait.
The defeat of the Spanish Armada was a personal victory for Elizabeth: in her speech to her land-forces at Tilbury she deployed a rhetoric that presented a woman as a convincing war-leader.

Robert Devereux, Earl of Essex, was Dudley's step-son and heir to his position in the Queen's favour. But he was young enough to be Elizabeth's son – or her gigolo. And he entered into a dangerous competition for popularity with her. Finally, when the Queen decided he had gone too far and stripped him of office, he tried to seize power by force. He failed pathetically. Elizabeth half blamed herself, but she still executed him. Thereafter, her own decline was rapid.

CHAPTER 28

A Royal Pregnancy?

Only one thing remained to complete this reconstruction of England's past, present and future and that was for Mary to follow the example of her namesake, the Virgin, and conceive a child. On 24 November 1554, when Pole met Mary at Whitehall for the first time since his exile, he addressed her with the opening phrases of the *Ave Maria*: 'Hail, thou that art highly favoured, the Lord is with thee: blessed are thou among women'. These are the words of the Annunciation, in which the Angel tells Mary that she has conceived a child, Jesus, by the Holy Ghost. Shortly after Pole had left Whitehall for his own palace at Lambeth, Mary sent a messenger in hot pursuit with the wonderful news. Pole's prophetic salutation had been answered: 'The babe had leapt in her womb.'[1]

Rumours of Mary's pregnancy had already been circulating for several weeks. On 9 November, Sir John Mason, the English ambassador to Charles V, reported his discussions with the Emperor on Mary's condition. Charles had asked him whether it were true that Mary was pregnant. Mason replied that he had heard nothing formally from the Queen, but that others had told him 'that her garments wax very straight'. Charles, delighted, said he was sure that it would be a boy. Mason answered bluntly that he didn't care; 'Be it man,' he said, 'or be it woman, welcome shall it be; for by that we shall be at the least come to some certainty to whom God shall appoint by succession the government of our estates.'[2]

The same sentiment was shared by all supporters of the new régime. Even in far-away Woodstock, Bedingfield caught the atmosphere of almost millennarian expectation when he ended his letters to the Queen on 4 October by praying for 'progeny of your excellent person' so that 'we may, as holy Simeon did for the birth of Christ, praise God for the same'. Elizabeth, we can safely guess, did not join in these prayers. But she was, willy-nilly, caught up in the pattern of events as the splendid rituals of a royal pregnancy unrolled once more.[3]

Mary, hesitant so long, took the prophetic spasm she felt after Pole's salutation as a divine sign that she really was pregnant. The news was proclaimed in the City and *Te Deums* were sung. Christmas at court was unusually magnificent. Finally, on 4 April the King and Queen moved to Hampton Court for Easter and the ceremony of the Queen's 'taking to her chamber'. She should now have remained secluded for the weeks preceding the birth. But her confidence and pride in motherhood led her to break with tradition by showing herself a few days later at a window of the palace on St George's Day, 23 April, to watch her husband lead the celebrations for the patron saint of the Garter, accompanied with all the pomp of the restored Church. The Queen, it would seem, carefully presented a sideways view to the crowd below, so that 'hundreds' witnessed the full glory of her swollen figure.[4]

At this supreme moment, Mary spared a thought for those members of her family who had been so brutally (if necessarily) excluded from her triumph. The Earl of Devon was recalled to court from his internal exile at Fotheringay Castle. And on 17 April Elizabeth herself was summoned from Woodstock. Mary's letter to Bedingfield breathes an easy confidence: bring up Elizabeth immediately, he was told, and don't bother with reinforcements. We can (and should) imagine the Queen's dismissive pity as she thought of her sister. Elizabeth was doubtfully legitimate, unreliable in her faith, unmarried, and likely to remain so. She was the sterile issue of an abortive union, while she, Mary, was heavy with destiny and the child who would guarantee the Catholic future of England.[5]

Elizabeth arrived at Hampton Court between the 24th and the 29th. She entered privately, by the back or garden side, and was accommodated in the Prince of Wales's Lodging, which had been built for her brother, Edward. Her rooms were secure and had secret access to the royal apartments. She was just in time to be a witness of the great event and to be informed, no doubt, of the part she was expected to play in the christening ceremonies. For on 30 April came the news that the Queen was safely delivered and of a son. Bells were rung and anthems sung.

But the next day the tidings turned out to be false. Mary herself remained invincible in her hopes and believed, like her mother forty years before, that notwithstanding her miscarriage (or whatever unnamed thing had happened) she remained pregnant still.[6]

Elizabeth, meanwhile, was left to her own devices for a fortnight. Clearly fearing that she had been forgotten, she finally took the initiative and asked to see the council. A delegation of councillors duly arrived, headed by Gardiner. He required her to submit to Mary's grace. Elizabeth refused. Gardiner returned next day with a message from the Queen: Elizabeth's refusal to acknowledge her guilt made it 'seem that the Queen's majesty had wrongfully imprisoned her grace'. Elizabeth was not to be caught. 'It may please her to punish me as she thinketh good,' she replied. Then Gardiner came nearer home: if Elizabeth did not submit, he told her, kneeling again, 'then your grace hath the vantage of me and other lords [of the council], for your wrong and long imprisonment'. Elizabeth did not dispute the fact, but loftily denied any desire for vengeance.[7]

A week later, at ten at night, Elizabeth was summoned to the Queen. Surprised by the lateness of the hour and the peremptory nature of the message, Elizabeth feared an assassination attempt. Instead, by torchlight she was taken across the garden to the foot of the privy stairs that led directly to the Queen's bedchamber. Then, accompanied only by Mrs Clarencius, Mary's favourite attendant, she was ushered into the presence. The two sisters had not met for a year. Elizabeth knelt, while Mary chided her with her refusal to submit.

'You [would say] that you have been wrongfully punished.'

'I must not say so, if it please your majesty, to you.'

'Then belike you will to others.'

Mary's sharp retort revealed her frustration. For a year she had been trying to force Elizabeth to admit her self-evident guilt and for a year Elizabeth had steadfastly refused. Now she had to reward her sister's obstinacy by taking her back into her good grace, on Elizabeth's terms, not hers.[8]

For there was another witness to the scene, apart from the confidential Clarencius. According to Foxe, Philip himself stood behind the arras to overhear the encounter between the two sisters. His reactions are unknown, but they cannot have been unfavourable to Elizabeth. Before his marriage, the Habsburg agent Renard had been her leading opponent, keen to have her out of the way to smooth his master's path to the English throne. But recently Philip's attitude had undergone a transformation. One explanation for this is romantic. Philip, it was rumoured, had transferred his interest from Mary to Elizabeth. Elizabeth's striking pallor was not to everyone's taste (the Venetian ambassador even described her as 'sallow'). Compared, however, with the ageing and dumpy Mary she was a thing of beauty and Philip probably did have unbrotherly thoughts about her.

But his head was at least as involved as his heart. Philip's interest in England had always been strategic rather than romantic. He had married Mary to bring England on side in the Habsburg struggle with France. So long as there was a prospect of offspring of the marriage, Elizabeth remained an obstacle. But now that it was clear to him (though not to his wife) that Mary's 'pregnancy' was a product of disease if not deception, Elizabeth's position changed and she became more useful to Habsburg aims alive than dead. For if Mary died childless, as seemed increasingly likely, the legitimate, Catholic heir to England was Mary, Queen of Scots, the granddaughter of Henry VIII's elder sister Margaret. Mary had been brought up in France and was betrothed to the heir of the French throne, the Dauphin François. Her succession to England would thus be the fulfilment of the worst Habsburg nightmare: she would unite Britain and fuse both island-kingdoms into a joint Franco-British realm that would control the Channel and rupture sea links between the eastern and western halves of the Habsburgs' European empire. Better Elizabeth on the English throne than that – bastard though she was and heretic though she might turn out to be. Better anyone.[9]

Philip's motives are the easiest to read of the three principals' in that strange, nocturnal meeting in the royal bedchamber. Mary's were much more confused. When she had summoned Elizabeth back to court it had been in a mood of easy, superior patronage. Now she was not so sure. *She* knew she was still pregnant. But others, even her husband, dared doubt. So suddenly, as so often happened in her relations with her younger sister, the ground had shifted beneath her feet and the advantage, which had all been on her side, started to shift as well. And, besides, her husband had insisted on a reconciliation and was hidden there to make sure she fulfilled her promise.

But what of Elizabeth herself? With her usual tactical skill she had sensed the shifting mood and exploited it to the full, against both the Queen and the council. But there was also, I suspect, womanly, sisterly pity. Certainly, Mary deserved compassion as, for weeks, then months, she persisted with the illusion of her pregnancy. At first her doctors, who were really little wiser than she, announced lamely that they had miscalculated her term. But early in May, the French ambassador tapped into the much better informed source of two of her intimate female attendants: 'the midwife and an old maid who had attended her from her youth'. Theirs was a much more pessimistic verdict: 'that the Queen's state was by no means of the hopeful kind generally supposed, but rather some woeful malady, for several times a day she spent long hours sitting on the floor with her knees drawn up to her chin'. Wise after the event, the Venetian ambassador reported the symptoms of this 'malady' in gruesome, circumstantial detail. 'From her youth' she had suffered from a retention of her menstrual fluids and a strangulation of her womb. Her body swelled and her breasts enlarged and sent out milk. It was these symptoms, he concludes, 'which led to the empty rumour of her pregnancy'.[10]

More and more people, within the court and increasingly without, were convinced of this explanation. But Mary, doubled-up in her agony, which was both physical and a sort of acute yearning, persisted in expecting a child. At the beginning of June it was reported that her delivery could not be delayed much longer. But by the end of the month Renard was contemplating the unthinkable. 'The entire future', he wrote to the Emperor, 'turns on the accouchement of the queen of which, however, there are no signs. If all goes well, the state of the country will improve. If she is in error, I foresee convulsions and disturbances such as no pen can describe.'[11]

'*If she is in error*'. After another month of delusory hopes, even Mary had to accept her mistake. Four months previously she had 'taken her chamber' with pomp and publicity. Now, at the beginning of August, she slipped out, unannounced, and removed to Oatlands some four miles away. As she was going through Hampton Court park to take her barge, the kind of miracle which had been refused her was vouchsafed to a common beggar. When he saw the Queen, 'for joy he threw his staves [crutches] away, and ran after her grace'. Moved, the Queen ordered that he should be given a reward and details of the event were carefully noted by the pious. Was it a happy augury or was Mary only mocked once more by a vulgar trick?[12]

Throughout these months Elizabeth was at her sister's side. She witnessed both her physical pain and her mental anguish. But what probably shocked her more, as a true believer in monarchy, was the damage done to the Queen's public reputation. This became worse with each week that the pretence of pregnancy was kept up. The gossips said that she had never been pregnant at all. That the foetus had been a pet monkey or a lap-dog. That there was a plot to pass off another's baby as the Queen's own (Foxe fingered Lord North as the agent who had tried to procure a suitable child). Thus the tide of rumour rose. Posters were nailed to the palace door and abusive papers thrown into the Queen's own chamber. But Mary's cup of sorrow was not yet full. God had denied her a child; now He was about to take away her husband.[13]

Philip had not made a bad start in his first year as King of England. He had shown characteristic Habsburg graciousness. He had revived the chivalric life of the court in a magnificent series of tournaments. Wisely (and in contrast to his wife's anguished outrage), he had appeared unruffled at the clashes between his Spanish followers and the xenophobic English. And he had established good working relations with most of the leading councillors. Nevertheless, neither the council nor parliament was prepared to give him any independent authority or even the appearance of it. The only thing that would change this would be the birth of a child. As father of the heir to the throne Philip's position would be immeasurably strengthened, both during Mary's life and especially after her death when he would be regent. This is why the hope of a child, which had largely been discounted before the marriage because of Mary's age, now loomed so large in Habsburg calculations. And it is why the eventual humiliating conclusion to Mary's phantom pregnancy was so bad a blow to Philip. There was now no serious chance of making his nominal English monarchy real. Moreover, other real kingdoms beckoned as his father, Charles V, had decided to implement his long-considered plans for the succession and abdicate. So Philip resolved to cut his losses and go.[14]

He broke the news to Mary gently but firmly. He also practised the innocent deception of assuring her that he would be gone for only a few weeks and that she might wait for his return at Dover. It would have been a long wait. Fortunately, the state of her health prevented the realization of the plan. Instead, Mary and Philip returned to Whitehall on 23 August. Three days later they went in solemn procession through the City. It was five months since the Queen had been seen in the capital and rumours were rife that she was dead. To quell them Mary was carried in an open litter. At Tower Wharf the King and Queen took the royal barges and went to Greenwich. First they prayed together in the Observant Friary which Mary had refounded. They were received solemnly with torches and a hundred torch-bearers accompanied them back to the palace, where they spent their final hours together. The parting came on the 29th. Philip set sail and Mary spent endless hours sitting in her chamber window, staring at the river which had borne her husband away.[15]

Once again, Elizabeth was present throughout these extraordinary events. On the 26th, she went directly by water from Hampton Court to Greenwich. And on 4 September, when Mary decided to seek the consolations of religion during her husband's absence, Elizabeth joined in the observances. The royal sisters led the court in fasting and receiving the Pope's jubilee and solemn pardon for sins. Mary was a passionate participant; Elizabeth was merely going through the motions, however convincingly.[16]

But she had plenty to think about during the *longueurs* of the service. Until the moment of her own release from Woodstock, her sister's reign had been a story of triumphant, almost miraculous success. Against all the odds, Mary had gained and held her throne; married the man she wanted, and restored the Catholic faith and the Roman supremacy. So much for those who said that a woman could not rule. But then, before Elizabeth's own eyes, her sister's triumph had turned to disaster. First there was the protracted humiliation of Mary's bogus pregnancy. Then came her husband's desertion, which was soon compounded by rumours of Philip's promiscuous womanizing.[17]

There was, in short, a paradox. A woman could take power. But the greatest risks to her holding of it would come from the fulfilment of her ordinary, womanly functions of wife and mother. Maybe Elizabeth, with her sharply logical mind, drew the obvious consequences there and then, as the candles guttered and the priests chanted promises of spiritual redemption she did not believe. Maybe she had drawn them already.

CHAPTER 29

Parliamentary Revolt

Elizabeth finally obtained permission to leave court for her own estates on 18 October 1555. Her way lay through London and the crowds cheered loudly. Their enthusiasm was gratifying but politically dangerous. So Elizabeth instructed some of her gentlemen to linger behind and quieten the people.[1]

Eventually she reached Hatfield, and there the former tenor of her life resumed. She studied with Roger Ascham, who, like her, had bent to the prevailing wind and conformed outwardly to Catholicism. And her old servants gathered round her, too. Bedingfield had been discharged of his responsibilities for her – to his relief and hers – and Parry had resumed his old duties. The following month Catherine Ashley, too, was released from protective custody in the hands of Sir Roger Cholmondeley and rejoined her mistress. It was just like old times.[2]

For Mary there could be no such retreat into a familiar world. So long as she was wrestling with the personal agonies of her false pregnancy and her husband's departure, Mary seems to have found solace in Elizabeth's company. But now that the Queen had once more to take up the burdens of government – and to take them up alone – her sister became a distraction, even a danger. Especially, it must have seemed, in the circumstances of a parliament.

On 21 October, three days after Elizabeth's departure, Mary opened her fourth parliament. Her circumstances had weakened immeasurably in the year which had passed since the last meeting of the assembly. Then Mary, newly married and apparently pregnant, had carried all before her. Now she was exposed as barren and deserted. The political initiative deserted her with her husband.

There was, perhaps, a change of personnel also. 'Whether by accident or design,' the Venetian ambassador reported, the House of Commons 'is quite full of gentry and nobility (for the most part suspected in the matter of religion), and therefore more daring and licentious than former houses, which consisted of burgesses and plebeians.' The judgement is snobbish, as we would expect from a representative of the aristocratic republic of *La Serenissima*. And historians of parliament have been unable to prove it statistically. But there are some clear signs of electioneering, including the almost unheard-of return of two of the Verney brothers for both county seats in Buckinghamshire. (Francis Verney was Elizabeth's bold and brazen servant whose behaviour had so outraged Bedingfield the previous summer, and Buckinghamshire was an area where Elizabeth was a major landowner.)[3]

Taken by themselves, these facts do not necessarily amount to much, though later events were to give them an added significance. But, whatever its cause, the change of mood in the House was indisputable.

At first, though, it was masked by the formidable personality of Bishop Gardiner. As was then usual, Gardiner as Lord Chancellor made the opening speech and carried parliament through its first, essential, task of voting taxation. But this was Gardiner's last service for his sovereign. He was already mortally sick and he died on 12 November. Gardiner's trenchant wit and forceful opinions turned him into Protestant apologists' favourite whipping-boy for the régime, and he was blamed for many things that he did not do. But the charges are a sort of back-handed compliment to his competence and authority in a council which lacked both and, with his removal, the government's hold on the Lower House weakened and then collapsed.[4]

The crises came over two key pieces of legislation. The first was a bill to allow Mary to return the Crown's remaining ecclesiastical lands and revenues to the Church. The bill was close to Mary's heart: she felt that the moneys were tainted with blasphemy and sacrilege, and she was determined to divest herself of such unhallowed wealth. But her determination to be rid of them was matched only by the reluctance of the Commons to see them go. They were outraged that an already impoverished Crown was wilfully surrendering further income – income which they knew they would have to make up out of taxation on their own property. They also guessed, correctly, that Mary's sacrifice was intended to put moral pressure on them to make a similar surrender of their own gains from the Church. And that even the most pious of them was determined absolutely not to do. The vote on the third reading was postponed, and opponents lobbied. Finally, on 3 December, the bill was carried by 193 to 126.

To us, this sounds like a convincing result. But then divisions were comparatively rare and, in any case, the outcome was only obtained by treating the Commons like a recalcitrant jury and locking them in the House from early in the morning until three in the afternoon without food or drink.

So only someone as politically tone-deaf as Cardinal Pole could talk of the vote as a 'great victory'. And, in any case, it soon proved Pyrrhic.[5]

Three days later, on 6 December, the other great measure of the session came up for its final reading. This was the bill to confiscate the estates of the 'exiles', as those who had fled abroad to escape political or religious persecution at Mary's hands were known. Mary herself hotly denied the validity of the term, with its hint of legitimate opposition. They were not exiles, she vehemently informed the French ambassador, who had dared to use the word in her presence, but 'those wretches, those heretics, those traitorous, execrable villains'. Calling them names, however, would not break their resistance. But confiscation of their property might. Existing law allowed only for the seizure of movable goods, most of which were already out of the government's reach. The proposed bill provided instead for the confiscation of landed wealth. That was how most of the rebellious nobility of King Philip's territories in the southern Netherlands were to be brought to heel. No doubt the same measure would have been just as effective in England. But it was not to be.[6]

For almost every member of the political élite, starting with Elizabeth herself, had friends and relatives among the exiles. Among the first to leave was Sir Francis Knollys, who had married Catherine, daughter of Mary Boleyn, Anne's sister. This made Catherine Elizabeth's first cousin. The two were close and when Catherine departed abroad Elizabeth wrote her an impassioned letter, signing herself 'cor rotto' – 'broken heart'. The letter is the first example of Elizabeth's high style and is densely interwoven with alliterations, antitheses and apothegms. Later on, the style became a mere affectation, like a bad habit the Queen could not shake off. But the circumstances of this letter show that it was born out of necessity. Mary's government had become an intrusive tyranny: if you had to write something down (you had much better not), it was sensible to make sure that it had two meanings or perhaps none at all. Elizabeth's high style fulfils this requirement admirably: it is obscure because, at first, it had to be.[7]

One thing, however, stands out plainly in this cloud of squid's ink. 'Think this pilgrimage,' Elizabeth told her cousin, 'rather a proof of your friends, than a leaving of your country.' With the failure of Wyatt's Revolt on the one hand, and the beginning of the burnings in early 1555 after the restoration of Catholicism on the other, the stream of those 'leaving their country' broadened from trickle to flood. But, as she had promised, Elizabeth and the rest of her class stood to the 'proof' of their friendship.

The test came with the vote on the 'Exiles Bill'. Opinion in the House was already inflamed by the MPs' treatment three days before over the bill on Church revenues, and the government's enemies were out for revenge. There was also a man of the moment in Sir Anthony Kingston, the knight of the shire for Gloucestershire. He was bold enough to turn the government's earlier tactics against it. With the connivance of the sergeant-at-arms, the doors were locked, this time from within. As MPs milled around, unsure of what to do, Kingston stood with his back to the door and thundered that this measure, at least, must not be passed in defiance of many consciences as the other had been. Despite the procedural outrage, he carried the House and the bill was defeated. Three days later parliament was dissolved and Kingston himself briefly imprisoned in the Tower.[8]

Two very different lessons were drawn from the affair. The first was by Elizabeth herself. She, it seems clear, was kept informed of proceedings in parliament by her many friends and supporters there. And she recognized fully the danger to royal authority of the breakdown in trust between the Crown and the Commons. Whenever such a breakdown threatened to occur in her own reign, as in the 1590s, she had the wit to back off. Not till the 1620s and the reign of Charles I, who, like Mary, combined political obtuseness with a sense of his own absolute rightness, were the scenes of 1555 to be repeated.

The other lesson also seems to belong to the Stuart, rather than the Tudor, century, as the leaders of the parliamentary opposition decided that events had shown that they were now powerful enough to carry their resistance from the Commons to the country.

Kingston was released from the Tower on the 23rd, just in time for Christmas. Outwardly, he made a humble submission; privately he was burning for revenge. 'They have put me in the Tower for their pleasures,' he said, 'but so shall they never do more.' There was only one way a subject in Tudor England could thus bind his sovereign and that was by successful rebellion.[9]

Kingston and other parliamentary leaders had already gone far down this road. During the parliament, they had met to plot tactics at Arundel's, a popular tavern in St Lawrence Poulteney in the City. Their conversation had ranged widely and they had observed none of Elizabeth's caution. 'I have perceived of their talk,' one observer later reported, 'that with great wilfulness they intended to resist such matters as should be spoken of in the parliament other than liked, and that they did very sore mislike such Catholic proceedings as they perceived the Queen and all Catholic men went about, as indeed they did everywhere declare themselves to be right Protestants.'[10]

The question, however, was how to make this resistance good once the special forum of parliament was removed. For the position of the gentry was a paradox. They owned most of the land and they ran local administration on behalf of the Crown. But they were not used to taking either the political or the military initiative. These still remained with the titled nobility. The nobility numbered only a few dozen. Individually, they were the richest people in the kingdom. But collectively, their wealth was only a fraction of the gentry's. The issue loomed large in the 1550s because a political gulf had opened between gentry and nobility. Most of the leading gentlemen of the southern shires sympathized, more or less openly, with the stand taken by Kingston and his friends. But the peerage did not. Instead, 'the chief nobility and principal personages', as the Venetian ambassador noted, 'show themselves well disposed' to Mary's government. Partly, their attitude was the result of the usual caution of the very rich, who have much to lose. But many of the nobility were religious as well as political conservatives. And, in the purifying fires of Mary's reign, religious conservatism was turning into a convinced and confident Catholicism.[11]

Without their traditional noble leaders, the radicalized gentry had to find new ones. Wyatt had tried to go it alone and make up for the smallness of his forces with a direct appeal to the people of London. That had failed, though only just. In 1555 Kingston and his friends turned instead to the new brand of military adventurer who had arisen in the wars and ideological struggles of the last twenty years. They would supply leadership and, with luck, semi-professional troops drawn from the ranks of English exiles in France.

In late November, even before the dissolution of parliament, Kingston had met with one of the most important of these adventurers, Sir Henry Dudley. Henry Dudley came from the senior branch of the family, which had been displaced and dispossessed by the rise of the younger branch in the person of John Dudley, Duke of Northumberland. Despite this, Henry Dudley had attached himself to Northumberland, and in 1553 had been sent to France to win backing for Queen Jane at the price of the surrender of Calais. Mary, as part of her task of undoing Northumberland's legacy, restored Henry Dudley's elder brother to his peerage. But Henry Dudley himself remained loyal to his 'Protestant' allegiance and now came up with a comprehensive scheme to overthrow Mary's government. Kingston would raise the Welsh Marches. Dudley would lead an invasion from France bringing troops and arms. Richard Uvedale, the captain of the key fortress of

Yarmouth on the Isle of Wight, would open Portsmouth to Dudley's forces. Dudley and Kingston would then march on London, exile Mary, marry Elizabeth to Edward Courtenay, the Earl of Devon, who had been exiled to Venice after his release from custody, and make the pair Queen and King.[12]

The ultimate objective of crowning Elizabeth and Courtenay was the same as in Wyatt's Revolt in 1554. But the tone is different from the earlier rebellion. In 1555 the leaders talked of Elizabeth more freely and with apparently fuller knowledge. And the involvement of her household servants was more extensive and better documented. There is of course nothing which directly implicates her. But the circumstantial evidence for Elizabeth's complicity is much stronger than in 1554. Why then, after making so much of a rather slight case in 1554, did Mary's government ignore a much stronger one in 1555? And why have modern historians turned a blind eye as well?

Part of the answer lies in the name conventionally given to the revolt. Wyatt was indisputably leader in 1554 (though under interrogation he hinted darkly at greater men behind him). But Dudley's role in his eponymous revolt is more debatable. The nature of the revolt required that he should have been a charismatic leader; but little trace of his personality emerges. This could be the result of caution, though there are few other signs of it in his behaviour. More likely, it is because the real, inspirational leader and energizer was not Dudley but his father-in-law, Christopher Ashton. Ashton was a swashbuckling adventurer of the authentic, larger-than-life Elizabethan type. And, like his Elizabethan descendants, he idolized Elizabeth.

CHAPTER 30

Elizabeth's First Adventurers

At Christmas 1555, as the comings and goings of the plotters were concealed by the festivities, Elizabeth, the ostensible focus of all this activity, was twenty-two. She was, or should have been, an unknown entity. She had ridden in a few processions and had her portrait painted a handful of times. But no poems or pamphlets had been written about her and her name appeared in only one document in the public arena: her father's will. Yet somehow, whether by word of mouth or by wishful thinking, she had acquired the persona that was the rebels' best recruiting sergeant.

Christopher Ashton, perhaps half in love with her himself, made particularly free use of her name. When the rebels were victorious, he assured Henry Peckham, another of the conspirators, 'your father shall be made a duke; for I tell you true, that the Lady Elizabeth is a jolly liberal dame, and nothing so unthankful as her sister is; and she taketh this liberality of her mother, who was one of the bountifullest women in her time or since; and then shall men of good service and gentlemen be esteemed'.[1]

The image of Elizabeth as 'a jolly liberal dame' is half fantasy at least. But it is no more (and no less) false than the stories about modern celebrities which fill the pages of our tabloid press. Such people live by and in the imaginations of their worshippers as much as in reality, and their actions are mythic, too, beyond any ordinary test of truth or fiction. Elizabeth already belonged to this select band.

Christopher Ashton, one of her true believers, was a Berkshire gentleman and JP who held the part-time court office of Gentleman

Usher of the Chamber.[2] Like many of his type, he had taken his first step up the ladder of fortune by marrying a widow. But she was no ordinary widow. She was born Lady Catherine Gordon, the daughter of the second Earl of Huntly, and cousin of James IV of Scotland. James had married her to Perkin Warbeck, the pretender to the English throne who claimed to be the younger of the Princes in the Tower. Catherine had accompanied Warbeck on his ill-fated invasion of England in 1497 and, after his defeat, capture and eventual execution, was given honourable refuge by Henry VII. Later, Henry VIII took a shine to her, giving her denizenship, or naturalization, and a clutch of lands round Fyfield near Abingdon in Berkshire. Her manor-house still survives and her tomb-niche still stands in the church.[3]

Catherine retained Fyfield through two more marriages. And when, three decades later, once more a widow, she married Ashton, her new husband was given a thirty-one-year lease of the lands after her death. It is one of the curious twists of Tudor history that the lands of the widow of Perkin Warbeck, Henry VII's 'great rebel', were used to help finance rebellion under Henry VII's granddaughter Mary, in the interest of his other granddaughter, Elizabeth. Catherine's royal connections probably reinforced Ashton's taste for high politics and gave him some of his insights, real or imaginary, into royal personalities.[4]

The other source of Ashton's wealth was monastic plunder. He quickly spotted the way the wind was blowing and, with Thomas Cromwell's connivance, dunned the Abbot of Abingdon for the choicest bits of the soon-to-be-dissolved abbey estates. In so doing, he clashed with another existing lessee, John Audelett. While Audelett was alive Ashton harassed him through his position as JP. And after his death, Ashton's servants seriously assaulted his widow's cousin, from whom she had sought advice. Mrs Audelett complained, and Ashton's status as trouble-maker was eloquently confirmed by Lord Sandys of the Vyne, a neighbouring magnate. 'Ashton', Sandys wrote, 'is a man of marvellous perverse and evil conditions. Many honest persons, in town and country, stand in great dread of him.'[5] He may have had his come-uppance in 1544, when the Privy Council detected him in extortion in running the county musters. But it seems unlikely. Instead, he helped his son, also called Christopher and clearly a chip off the old block, to make a successful career as a captain in the wars of the 1540s.[6]

Ashton, in short, was a natural malcontent and, whether at Fyfield or in London, he lay at the centre of the plots of the winter of 1555–6 like a spider in its web. Sir Anthony Kingston hastened to Fyfield after his release from the Tower and Ashton gave him half a split coin. It was agreed that when Ashton sent him the matching half from France, Kingston would raise the standard of revolt. Sir Nicholas Arnold, one of the former leaders of Wyatt's Revolt, also visited Ashton, though, according to Arnold, his own part in the conversation was a model of political propriety. And it was John Bedell, Ashton's servant and factotum, who acted as organizer-in-chief, sacrificing his Christmas to the cause: 'all the holidays and after [he] did nothing else but ride from one to another, [for] his mind could not be in quiet till he had brought this matter to some purpose.'[7]

So far, the plot had gone smoothly ahead. Now it encountered its first obstacle. The conspiracy had been put together under the aegis of the French ambassador and the assumption was that the French king, Henri II, would finance it. But at the beginning of February 1556 Henri II signed the truce of Vaucelles with Philip II and so was, temporarily, uninterested in destabilizing Philip's position in England.

Henri II's chief minister wrote to the French ambassador in England to inform him of the change of circumstances. Be circumspect, he was told.

> And above all, make sure that Madame Elizabeth does not begin, for anything in the world, to undertake what you have written to me. For that would spoil everything and lose the benefit which they can hope for from their schemes, which must be played out in the long game, waiting meanwhile until time gives them the opportunity.[8]

The French, then, took Elizabeth's part in the plot for granted. Perhaps we should too. Perhaps indeed Ashton was not such a fantasist as he seems.

The conspirators now had to find alternative sources of finance. They hit upon the idea of robbing the Exchequer. Once again, events began by smiling on them. They got access to the treasure and even managed to weigh some of the bullion. Confident now of success, Dudley and Ashton sailed for France to organize the invading force. Ashton travelled, *Godfather* style, with a retinue of deferential servants and side-kicks, including one Reeve, 'his chamberlain'. But disaster struck. One of the Exchequer plotters, who may have been a double agent all along, got cold feet and informed Cardinal Pole. On 18 March the London plotters were arrested and Kingston himself sent for. The investigation was entrusted to a select group of Mary's most reliable Catholic councillors and they were unpleasant in their methods. Even Bedingfield, who now exercised his talents as gaoler on a larger scale as Lieutenant of the Tower, was 'ready to cast his gorge' (that is, vomit) at the befouled cell into which one of the prisoners was flung.[9]

Nevertheless the plot was unravelled only slowly. Two of the leaders, Ashton and Dudley, were out of the government's clutches in France, while the third, Kingston, died, perhaps by his own hand, on his way to the Tower after only a preliminary examination in the country.[10] So it was only at the end of April that the leading country gentlemen plotters were rounded up and only in the course of May that the arrests spread to Elizabeth's own household. But they penetrated deep.

First to be run in was Elizabeth's young gentleman-servant, Francis Verney. He had lived up to the judgement reported two years previously by Bedingfield, that 'if there be any practice of ill in England, this Verney is privy to it'. He had also involved his elder brother, Edmund, and his brother-in-law Henry Peckham, who had been the recipient of some of Ashton's more extravagant confidences. But the biggest catch of all was Verney's uncle on his mother's side, John, Lord Bray. Bray was not Elizabeth's servant; but Elizabeth was, as he put it, 'my neighbour of Hatfield'. Actually, she was his still-nearer neighbour at Ashridge. But, wherever or however it had been established, by acquaintance or imagination, Elizabeth had left Bray with the same sense of her open, generous character as Ashton: 'if [my lady Elizabeth] might once reign,'

Bray had said in early January, 'he should have his lands and debts given him again, which he both wished for and trusted once to see'. The hope was not merely treasonable; it was a personal insult to Mary, and Bray probably paid for the latter as much as the former. He was harshly imprisoned for one of his status, and for some weeks the betting was that he would be executed.[11]

The round-up of the Verney group of conspirators seems to have rested on new information screwed out of the prisoners in the Tower. John Daniel was seen as a particularly good source and stick-and-carrot methods were employed on him, with considerable success. On 23 April, one of the inquisitors gave instructions for 'Daniel to be gently used and given some freedom'. But on 7 May he was tried and condemned and on the 12th he was plunged back into the horror of the stinking dungeon which had previously turned Lieutenant Bedingfield's stomach. It worked. The following day the inquisitors noted that 'Daniel, being yesterday removed to a worse lodging, beginneth this day to be more open and plain than he hath been, whereby we think he knoweth all . . . and will utter the same'. The indictments suggest that it was Daniel who supplied the principal evidence against the Verneys; he also probably led to higher game.[12]

For, a few days after Daniel had started to sing to the satisfaction of his interrogators, two events of major importance happened. On 25 May Antoine de Noailles, the French ambassador, took his formal leave. The interrogations, he knew, had uncovered his deep involvement in the conspiracies against Mary's throne and he feared for his life if he remained. Immediately after Noailles' flight, the arrests struck at the heart of Elizabeth's circle. On about 26 May, two royal agents, Sir Henry Jerningham and John Norris, arrived at Elizabeth's house with a posse of troops. Their rank showed the importance of the mission. Jerningham was a privy councillor, vice-chamberlain of the household and a member of the inner circle of councillors charged with investigating the Dudley–Ashton conspiracy; Norris was the senior gentleman usher of the court and a passionate Catholic in a family which had otherwise embraced the cause of the Boleyns and the Reformation. Both knew

Elizabeth well and were able to cope with her. First they arrested (yet again) her lady mistress, Catherine Ashley, her Italian teacher, Battista Castiglione, and three other women of her household. Then they put an armed guard on the house and confronted the princess. She said 'she thought they would fetch all away at the end'. Elizabeth was putting a brave face on a situation which, she must have known, was at least as dangerous as in 1554.

Jerningham and Norris returned with their prisoners to London. According to the usually well-informed Venetian ambassador, they all confessed to knowing about the plot. Still worse, a search of Catherine Ashley's chambers at Somerset House, Elizabeth's London residence, uncovered a cache of illicit books and pamphlets, some attacking Catholicism, others satirizing the King and Queen.[13]

Short of wringing a confession out of Elizabeth herself, the case against her was complete. But what should be done? Should Elizabeth be summoned to court? Or sent to the Tower? Or carried off and safely exiled to Spain? Everything depended on Philip's advice.

As soon as the incriminating evidence from Ashley and the rest was assembled, the Venetian ambassador learned that Francesco Piamontese, Mary's confidential courier, was immediately sent back to Brussels to report and get instructions, 'since nothing is done, nor does anything take place, without having the King's opinion about it, and hearing his will'.[14]

We know of Philip's response thanks to *The Life of Jane Dormer*. Jane Dormer, along with Elizabeth herself, is one of the most interesting women of the age. Indeed, the two resembled each other closely – so closely that, as their lives intertwined, their relations were characterized by friction as much as affection. Jane, like Elizabeth, was handsome, intelligent, courageous and assertive. Her career, too, was filled with surprises. She was born the daughter of an English country gentleman yet died the Duchess of Feria, a grandee of Spain, and a candidate for the governor-generalship of the Netherlands. Finally, like Elizabeth, she is a testimony to the way the Reformation divided families and friends.

Jane was daughter of Sir William Dormer of West Wycombe in Buckinghamshire and Mary Sidney. The Dormers, successful

gentlemen-woolmerchants who had married into the Woodvilles and
the royal blood of England, remained Catholic. But the Sidneys were one
of the families which led and benefited from the Reformation. Sir
William Sidney, Jane's maternal grandfather, was chamberlain to Prince
Edward; while Sir Henry Sidney, Jane's maternal uncle, married the
daughter of the Duke of Northumberland. Jane, whose own mother had
died young, spent much time with her grandfather in Prince Edward's
household. There she got to know Elizabeth, who was only five years her
senior. They met again in 1554, when Jane greeted the captive princess
at the Dormer mansion of West Wycombe, where Elizabeth
overnighted on her way to her imprisonment at Woodstock. But it was
to Edward's other half-sister, Mary, that Jane gave her personal service
and religious commitment. She became one of Mary's closest attendants
in the last years of her reign and witnessed the relationship, changing and
always fraught, between the Queen and her sister as power ebbed from
the one and flowed to the other.

And it was Philip's decision about Elizabeth's involvement in the
Dudley–Ashton conspiracy that, as Jane recognized, marked the turning
of the tide. In arriving at his decision Philip had to weigh two very
different considerations. On the one hand, there was Habsburg dynastic
self-interest: if Elizabeth were removed by whatever means from the
succession, the English throne would go to Mary, Queen of Scots and
future Dauphiness of France, and so strengthen immeasurably the
Habsburgs' greatest enemy. On the other hand, there was the plain
evidence both of Elizabeth's dabbling in treason and of her support for
everything which undermined Mary's life-work of restoring England to
the faith. One might have expected the two sets of considerations –
equally weighty and utterly opposed – to have given Philip some pause.
But there was none. For in his mind the latter counted for nothing
against the former: justice, Elizabeth's evident guilt and his wife's
feelings, all were nugatory when balanced against his family policy.

There was therefore neither ambiguity nor delay in the message that
Piamontese brought back from Brussels. Mary was instructed not only to
drop any inquiry into Elizabeth's guilt, but also to give out that she

believed that those of her servants who were implicated in the plot had used their mistress's name without authority. Mary obeyed promptly and to the letter. On 8 June two of the inner circle of her council were despatched to Elizabeth to undo the work of their colleagues a fortnight previously. They were instructed to inform Elizabeth of the confessions by which her servants, then imprisoned in the Tower, had accused her of complicity in their conspiracy. However, following Philip's instructions, they were also to assure her that the Queen believed nothing of the sort. Instead, Mary thought her too wise and prudent ever to undertake anything against her sister and sovereign. And, in testimony of the Queen's good opinion, they brought Elizabeth the gift of a valuable diamond, a symbol of purity. They also removed the guard which had been put on her, a step that the princess probably valued more.[15]

Clearly, Mary had not been able to resist tweaking Philip's instructions by informing Elizabeth in detail of the extent of her servants' admissions. According to the French ambassador (Giles de Noailles, who had stepped into the hastily vacated shoes of his brother Antoine), four were mentioned by name: 'Werne', 'Eschelle', Peckham and Daniel. 'Werne', as has long been realized, is Francis Verney and 'Eschelle', almost certainly, Catherine Ashley (say it quickly and transliterate into French!). Both are well known as Elizabeth's servants. But Peckham and Daniel are not otherwise so identified. There seems no reason, however, to doubt the accuracy of the ambassador's information. Or the conclusions we should draw from it. For, with four of her servants among the plotters, Elizabeth's guilt becomes double-dyed – the more so as both Peckham and Daniel had been involved in the conspiracy from the beginning and in its every aspect.[16]

Mary understood all this. But, because her husband had insisted, she lied through her teeth about her belief in Elizabeth's innocence. We should not underestimate what this lie cost her. Mary, unlike Elizabeth, had a rather simple attitude in such matters. Right was right and the truth was the truth. Now, to please Philip, she had solemnly to declare that black was white.

The least Mary expected was that Elizabeth would show a proper gratitude and volunteer to come to court. Once again, Mary showed how little she understood her half-sister, who was so close in blood but so different in character. For Elizabeth had inherited from her father the ruthless instinct of the beast of prey. He was the lion; she was the lion's daughter. When either scented weakness, they moved in for the kill. So it was now. Elizabeth brushed aside the suggestion of a visit to court with transparent excuses. Such boldness, Mary sadly reflected, could come only from the covert support of the English magnates or of a foreign prince. Mary's tragedy was that the name of the foreign prince who was now Elizabeth's most powerful supporter was Philip, King of England and Spain, and Mary's own husband.[17]

CHAPTER 31

Honourable Imprisonment

Mary made her real feelings about Elizabeth plain by a single gesture: on 8 June Sir Thomas Pope, loyalist privy councillor, and Robert Gage, son of Elizabeth's old enemy Sir John Gage, were put in charge of both her person and her household. But this was far from a return to the days of her imprisonment at Woodstock. Pope himself, 'a rich and grave gentleman of good name', was a more sympathetic character than the stiff-necked Bedingfield; while the terms of his appointment were specifically designed to keep up appearances: according to the Venetian ambassador, Elizabeth was 'in ward and custody' but in 'decorous and honourable form'. After all, Philip had decreed no less.[1]

With Elizabeth, the chief suspect, barely rapped over the knuckles, the proscription of the remaining Dudley–Ashton conspirators slowly petered out. Peckham and Daniel were executed. Francis Verney was tried and condemned but eventually pardoned, as was his uncle, Lord Bray. Mary was particularly touched by the efforts made for Bray by his estranged wife, remarking sadly that 'God sent oft-times to good women evil husbands'.

Christopher Ashton, adventurer-extraordinary, disappears from the record, probably dying in France. Ashton's end, like his career, was oddly prophetic, since so many later adventurers, who were to risk all in Elizabeth's name, vanished like him into oblivion on some foreign shore.[2]

There was a renewed flurry of activity in July, when a young Suffolk schoolmaster called Cleobury impersonated the Earl of Devonshire and had himself and his 'beloved bedfellow', Elizabeth, proclaimed King and

Queen at Yaxley in Suffolk. The rebellion never got off the ground and the pretender was promptly executed. But the affair gave both Mary and Elizabeth the opportunity to re-enact the new roles, written for them by Philip, of trusting sovereign, on the one hand, and devoted subject, on the other.[3]

On 30 July the council wrote on the Queen's behalf to Pope to inform him of the wicked behaviour of the conspirators, 'and how for that intent they had abused her grace's name' by proclaiming Elizabeth Queen.[4] The gesture called for a corresponding one from Elizabeth. The moment the news arrived she got to work to write to Mary to assure her of her loyalty. The result was one of her most extravagant effusions. She began by contrasting 'the old love of paynims [pagans] to their prince' with 'the rebellious hearts and devilish intents of Christians in name, but Jews indeed, towards their anointed king'. She invoked St Paul to confirm that rebels were indeed devilish. She wished for some maker of 'anatomies of hearts' so that the purity of her own could be laid bare. She was confident that 'whatsoever other should suggest by malice, yet your majesty should be sure by knowledge' – and here she soared into the empyrean – 'so that the more such misty clouds obfuscate the clear light of my truth, the more my tried thoughts should glister to the dimming of their hidden malice'.[5]

Elizabeth's letter was dated 2 August. The council's original letter, sent from Eltham, cannot have arrived at Hatfield before 31 July. So Elizabeth took no more than a day to draft her reply and write it out fair. Ascham would have been proud of his pupil's facility in composition and penmanship – though not, one hopes, of her high style and low deviousness.

Elizabeth's highly coloured letter was, as it happens, probably the most exciting event of Pope's governorship. That is not how his period of office is usually presented, of course. Instead Pope is supposed to have laid on for Elizabeth an extravagant series of masques and hunting parties that rivalled the future entertainments offered to Gloriana by her fawning favourites. Why Pope, vastly rich but wholly self-made, should have suddenly become spendthrift in his old age is never explained. Nor is it made clear why he should have incurred the certain wrath of the Queen in return only for amusing her as yet unrecognized heir.

Happily, these conundrums can remain unexplored since Pope's Elizabethan entertainments turn out to be wholly imaginary. How they were invented has its own interest. Thomas Warton was an eighteenth-century Oxford don and dilettante. Like some of his successors, he found teaching a bit of a bore and devoted himself instead to *belles lettres* glorifying his *alma mater*. He published a heroic poem on Oxford, *The Triumph of Isis*, a collection of Oxonian anecdotes and wit entitled *The Oxford Sausage*, and, most substantially, a life of Thomas Pope, the founder of his own college of Trinity. The *Life* claimed to be founded on 'original evidences with an appendix of papers never before printed'. The most interesting of these give the descriptions of Pope's entertainments of Elizabeth. The descriptions, with their supposed manuscript references, were picked up by Elizabeth's nineteenth-century biographers and have been repeated, with a few honourable exceptions, ever since. The first to smell a rat was the Frenchman Louis Wiesener, in his admirable *La jeunesse d'Elizabeth d'Angleterre*, published in 1878. Wiesener showed that one incident could not have happened and that the remainder did not appear in their supposed source, now known as *Machyn's Diary*. A decade later, the disputed passages were conclusively proved to be forgeries. Warton, who was an amiable man, is probably innocent of the fraud. But he made the mistake of relying on others to do his research for him.[6]

The worst consequence of the forgeries lay in the dates Warton assigned to Pope's governorship. This has distorted the whole chronology of Elizabeth's life at a crucial stage. According to Warton, Pope remained Elizabeth's governor for the remainder of Mary's reign; in fact, his tenure lasted less than six months and he was discharged from office in the second week in October 1556. At the same time, Catherine Ashley, imprisoned since her arrest in May, was released. But she was 'deprived not only of her office as governess, but forbidden ever again to go to her ladyship'.[7]

The loss of Catherine, her mentor since childhood, was sad. But now, for the first time since her own arrest after the Wyatt Revolt, Elizabeth was free. Some of her exaltation comes through the cautious phrases of the letter she wrote a fortnight later on 29 October to her old friend and protector on the council, the Marquess of Winchester. The letter has been misdated and misunderstood. Yet it certainly belongs to 1556 and its meaning is transparent. Elizabeth begins by reminding Winchester of his role – unwilling, she clearly hints – in escorting her on that terrible journey by boat from Whitehall to the Tower. He was 'constrained to come the first unto me in the entry of my troubles'. But, she hopes, 'I wish yourself to be now the last that should freely end the same'. And the way to draw a line under the past, she insists, is for the council to believe only her testimony about her own actions – hers and no one else's. 'For in the earth, my lord,' she continued, 'none of my state hath been and yet is, more misused with them of mine own family than myself.'[8]

This phrase has been described as 'a *cri de coeur*, which rings eloquently down the centuries'. In it, Elizabeth, goaded beyond endurance, is supposed finally to give vent to her feelings about her harsh usage by her family – that is, by the Queen herself. Alas, the very notion shows tone-deafness to both Tudor politics and Tudor English. It makes Elizabeth commit the inconceivable solecism of complaining about the Queen's behaviour to the Queen's second-ranking minister. It also makes her write modern English, which is the last thing she did. For the word 'family' did not have its present sense. When the Tudors referred to the biological family, they talked of 'kindred'. 'Family', instead, was employed in the latinate sense of 'household servants'. And that is how Elizabeth is using it here: she is following the official line and blaming everything on the unauthorized actions of her servants, who have 'misused' her by their involvement in so many treasonable plots.

Whether anybody believed her is another matter. It is unlikely that she even believed herself. And, in any case, her hopes that 'her troubles' were at an end proved optimistic.

CHAPTER 32

Marriage with Menaces

For a long time Elizabeth, who was no country girl, had been anxious for permission to spend some time in London. She is probably referring to this request when she ended her letter to Mary of 2 August 1556 'with the new remembrance of my old suit more for that it should not be forgotten than for that I think it not remembered'.[1]

The principal reason for refusing Elizabeth permission to visit London was, of course, that she was a security risk. The capital was volatile, and Elizabeth was fuel to any flame. But there was also the more human element of jealousy. Mary was the older, uglier sister, with the ugly sister's taste for parading in rich fabrics and flashy jewels. Elizabeth, with the conscious simplicity of dress and ornament she affected at this time, was the rural Cinderella of Hatfield.

These problems never entirely disappeared. But during the autumn of 1556 relations between the two sisters continued to improve. The sudden death in Padua on 18 September of the Earl of Devon, whose name had so often been linked with Elizabeth's in rebel manifestos, removed an important ground of contention. In October, Mary freed Elizabeth from Sir Thomas Pope's supervision. Finally, in November, she decided that Elizabeth should be allowed to have her wish and go to the ball of the London season. She was not to be wholly unchaperoned, however, and Mary couched her permission in the form of an invitation to spend Christmas with her at court. Elizabeth was delighted. But it all ended suddenly and in tears. As with Cinderella, the problem was a prince who proposed marriage and would not take no for an answer.[2]

Elizabeth set out from Hatfield in high spirits and fine style. She was escorted by two hundred gentlemen on horseback, all wearing her livery. She entered the City in similar order, processing from Smithfield, through the Old Bailey and along Fleet Street to her residence at Somerset House. Her followers' chains and velvet coats cut and trimmed with black excited particular attention – as much, that is, as could be spared from the princess herself, who, as usual, was the centre of all eyes. But the courtiers were more cautious than the crowds. The crowds cheered anyway; the courtiers waited for a signal first. But the signal came and, three days later, Elizabeth was received by the Queen with every sign of friendliness and honour. An even greater portent was the behaviour of Cardinal Pole. A year previously, when Elizabeth was at Greenwich to support her sister through the pain of her leave-taking of Philip, he had cold-shouldered her. Now he paid her court with the rest.[3]

It was like a transformation scene in a play. And it proved just as insubstantial. Suddenly her invitation for Christmas was withdrawn and on 3 December Elizabeth retraced her steps through the City on her way back to Hatfield. Her departure had been so unexpected that the Venetian ambassador, who made it his business to meet everybody, had no time to arrange an audience.

Elizabeth was in disgrace again. She did her best to disguise the fact by her proud bearing on her return journey. But her pale complexion, tinged yellow with jaundice, and her short, rapid breaths betrayed her strain.[4]

What had happened? The best guess of contemporaries is that sudden, overwhelming pressure had been put on her to marry. When she resisted, Mary's carefully constructed mask of benevolence fell away. For Elizabeth was now defying not only Mary but Philip himself.

The intended bridegroom was Emmanuel Philibert, Prince of Piedmont and titular Duke of Savoy. The French had seized the duchy in 1536 and, when Emmanuel Philibert succeeded his father, he inherited little but the limbo in which dispossessed monarchs dwell. His best hope of escaping from it lay in his connections with the Habsburgs. He was Philip II's cousin and friend. Perhaps the Habsburgs could also find him a wife. Elizabeth was the most promising candidate.

The scheme of neutralizing Elizabeth by marrying her to Emmanuel Philibert, who was both a good Catholic and a good Spaniard, was first floated in the aftermath of Wyatt's Revolt. Charles V himself had scotched it. But the Emperor changed his mind when the Prince showed signs of marrying a French princess and reconciling himself to France. At Christmas 1554, therefore, Emmanuel Philibert came to England as the recognized candidate for Elizabeth's hand. Elizabeth, still in her imprisonment at Woodstock, made her unwillingness plain. But Emmanuel Philibert was more disturbed by the Channel weather, and his seasickness was so bad that he did not reach London till 27 December. There he joined a brilliant court, in which the grandees of Spain and the princes of the empire celebrated the acquisition of England as the Habsburgs' most recent accession.[5]

But, amid all the rejoicing, events brought home to Emmanuel Philibert the vulnerability of his position between the upper and nether millstones of France and Spain. On the one hand, news came that the French had captured Ivrea in Piedmont, hitherto one of his most valuable remaining territories. And, on the other, he was probably told that the terms for his marriage to Elizabeth were the surrender to Spain, without compensation, of the rest of his lands to the west of the Alps, including Nice and Villefranche. He left England on 28 January 1555, without either a wife or a kingdom, amid the scarcely suppressed guffaws of the French.[6]

In October 1555, as one of his last acts before he abdicated, Charles V had compensated Emmanuel Philibert by making him Lieutenant of the Netherlands in Philip's absence and giving him a rich pension. Now, in December 1556, Philip was also reviving the plan to marry his friend to Elizabeth. He exerted formidable pressure, though of course, since he was abroad, at one remove. He leaned on Mary, who in turn threatened Elizabeth with a parliamentary declaration of her bastardy and an acknowledgement of Mary, Queen of Scots, as heir. In despair, Elizabeth not only fled the court but also contemplated fleeing from England.[7]

As she had done previously in similar extremities, she pushed at the ever open door of the French embassy. The ambassador himself had tried to make contact with her on her brief visit to London, but had found her too well guarded. Instead, it was left to Elizabeth and her inventive and intriguing household to find the means. The chosen instrument was the Countess of Sussex, who visited the ambassador twice in disguise. Her mission was to discover whether it were possible to smuggle Elizabeth out to safety in France.[8]

The message was extraordinary. And so, for that matter, was the messenger. Anne Calthorp, the second countess of Henry Radcliffe, Earl of Sussex, was one of the many remarkable sixteenth-century women who found themselves attracted, like planets to the sun, to Elizabeth's orbit. Anne's marriage had soon become impossible: she was a deeply unconventional woman married to a husband who was conventional in every way – religiously, socially and politically.

Religion provided the first grounds for disagreement between the couple. The Countess of Sussex was one of the small coterie of high-ranking court ladies who shared Catherine Parr's advanced religious opinions. This made her vulnerable herself and a means to injure others, including Catherine Parr. When, for instance, the martyr Ann Askew was put to the torture, her tormentors tried to make her incriminate the countess. Elizabeth, who had spent much of 1544 in Catherine Parr's household and shared its religious enthusiasms, would certainly have met the Countess of Sussex at this dangerous time. Under Edward VI, Protestantism became conventional. So the unconventional countess moved on to sorcery, which in 1552 led to her imprisonment in the Tower. But a letter from the Duke of Northumberland makes clear that she had dabbled in treason as well, by claiming that a son of Edward IV was still alive. Mary's accession gave her fresh opportunities for dissent. The Earl of Sussex was one of Queen Mary's earliest backers; the countess fled abroad within eighteen months. This proved the last straw and in 1555 Sussex divorced his 'unnatural' wife.

Free at last from the restrictions of an unhappy marriage, the countess returned to England and the marriage-free zone of Elizabeth's service. There she was in her element. Not content with her undercover visits to the French ambassador, she also seems to have gone to France to spy out the land directly. On her return in April she was promptly imprisoned.[9]

For once, in contrast, the French ambassador counselled caution. At least that is what he claimed over twenty years later. He had reminded Elizabeth, he said, of the parallels between her situation and Mary's in *her* times of trouble. Then Mary had a choice between flight to the protection of the emperor and sticking it out under Edward VI. If she had chosen the former, she would have left the way open to the usurpation of Jane Grey. And if Elizabeth chose flight, she too, the ambassador warned, would lose the throne.[10]

Elizabeth heeded his advice. Indeed it is uncertain how serious the project of exile had been. But that she should even have toyed with the idea, apparently for the first time, gives some idea of the corner she felt she had been backed into.

And the pressure remained unrelenting. In January 1557, the French king, Henri II, broke the truce of Vaucelles and resumed the war with the Habsburgs. Two months later, Philip II landed in England. Previously, he had been deaf to Mary's pleas to return. But now he had his own business to transact. His first concern was to bring England into the war with France; his second finally to conclude the marriage between Elizabeth and Emmanuel Philibert, who was now his chief general in the north.

In this latter war of Venus Philip brought with him two potent allies, his cousins the Duchesses of Parma and Lorraine. Their exact purpose was unclear. Maybe the duchesses were intended to demonstrate to Elizabeth that life as a Habsburg spouse need not be so bad after all. Or perhaps they were to offer a cover of respectability for a plot to kidnap Elizabeth, smuggle her abroad and marry her willy-nilly to Emmanuel Philibert. The French ambassador warned Elizabeth of the existence of the plot by means of the Marchioness of Northampton.[11]

Like the Countess of Sussex on an earlier occasion, the messenger was well chosen to warn Elizabeth of the perils of marriage. Elizabeth Brooke had married the Marquess of Northampton as his second wife after he had divorced his first for adultery. But Mary had refused to recognize the second marriage and had forced the couple to separate. This left Lady Northampton in limbo: she was a wife who was not a wife; she was also a marchioness who was not a marchioness, as Mary had stripped her husband of his titles as well as his spouse. She became one of Elizabeth's closest and most confidential servants.[12]

When Lady Northampton returned to her mistress with the French ambassador's message Elizabeth exclaimed that she would rather die first than suffer a kidnap and forced marriage.[13]

Strikingly, Mary suffered almost as much as Elizabeth from Philip's determination to push the marriage to a conclusion. The evidence consists of a draft reply to Philip, written in French in Mary's own hand and heavily and revealingly corrected by her. The draft is undated but Wiesener has shown that it certainly belongs to the early months of 1557 – in other words to the period just after the resumption of the war with France in January and just before Philip's visit to England in March. The draft is thus a window into a continuing debate. It was clearly a brutal one.[14]

Philip, who was a considerable strategist, had attacked Mary on her two most vulnerable points: her religious faith and her marriage. First, he and the friars he had left in England had argued that Mary was bound to bring about Elizabeth's marriage as a matter of faith and conscience. In the final state of the draft Mary limited herself to saying that she did not understand the argument. Originally, though, she was far blunter, and explained that the conscientious scruple she had against the marriage went back twenty-four years – that is to Elizabeth's birth in 1533. But she crossed the passage out.

Philip's other point of attack was even crueller. Mary (who had finally learned about the powers of parliament, though too late to make much use of the lesson) had explained that the marriage could take place only with parliamentary consent. Philip had retorted that if parliament refused its agreement, he 'would impute the blame to me'. It was a body-blow and Mary was frank about the pain it caused. Do not do that, she begged him, otherwise 'I shall become jealous and uneasy about you, which will be worse to me than death'. Then she permits herself a touch of bitterness: 'for I have already begun', she continues, 'to taste [of such jealousy and uneasiness] too much to my great regret'.

But, despite Mary's protestations of humble, wifely duty, she never quite forgot that she was one sovereign negotiating with another, or that she was a woman trying to get her way with a man. So she concluded with her trump card: circumstances, she insisted, would not enable the marriage to take place in Philip's absence. But let him come to England, let them kneel together and pray to the Lord, 'who has the direction of the hearts of kings in his hand', and all would be well.

Mary had hit the spot. To get what he wanted, both with Elizabeth and the war, Philip came.

It is easy to see why Philip staked so much on the scheme to marry Elizabeth to Emmanuel Philibert. His own marriage to Mary and the attempts to absorb England directly into the Habsburg empire had begun well, but were clearly ending in failure. Characteristically, Philip looked for the next best option. As such, the Savoy marriage promised much: there was something in it for everybody, starting with the bride and groom. In return for agreeing to the marriage, Elizabeth would be formally recognized as heiress, while Emmanuel Philibert would get a kingdom in return for surrendering what was left of his ancestral lands. In the broader perspective, the English would get a man who had the makings of a more acceptable consort than Philip. Most important was what Emmanuel Philibert was not: he had no discernible national identity to arouse English xenophobia, and (thanks to French conquests and Spanish greed) few foreign territories to keep him abroad. But his qualities were not simply negative. He was personable, competent, pious, and pro-Habsburg. With luck, therefore, the Habsburgs would get most of all, as King Emmanuel Philibert delivered them an England that remained both Catholic and a client state.

The advantages, then, were obvious. Why, Philip must have wondered, could neither Elizabeth nor Mary see them? Viewed from their perspective, however, things looked very different. The French ambassador had advised Elizabeth to take the long view. She now applied this lesson to the Savoy match. If she accepted it, she would have an immediate assurance of the succession. But it would come to her as a gift of the King of Spain. If she waited, however, it would come to her anyway, with no strings attached. She decided to wait. But she did not take the decision coolly. Instead, the *form* of the proposal roused her to fury. For this was marriage with menaces. If she did not marry Emmanuel Philibert, she was threatened with imprisonment, disinheritance and even death. This bullying revolted her, both as a woman and as a princess. As a woman, she was determined to marry – if she married at all – only the man she wanted to and no one else. And as a princess, she would never yield to compulsion. She had inherited this trait from her father, and it was the thing that made her a natural sovereign. It also made her resist the Savoy marriage with all the force of her being.

Mary, who had the usual high Tudor view of her own absolute authority, normally found Elizabeth's intransigence infuriating. This is why she reacted so violently when Elizabeth had rebuffed the proposed marriage face to face at their interview in early December 1556. But gradually deeper considerations came into play. These were the long-standing conscientious scruples about Elizabeth to which she referred in her draft letter to Philip. They were invoked because of the payoff between the marriage and the succession which lay at the heart of the scheme. Elizabeth rejected the payoff because she refused to receive the succession at Philip's hands. Mary rejected it because she thought that Elizabeth should not receive the succession at anyone's hands at all, least of all her own. For, far from being heiress of the true blood-royal of England, Elizabeth, Mary's conscience told her, was an open bastard and a secret heretic and so unworthy to inherit anything, much less a throne.

As usual, the two sisters were at loggerheads. But, though they came to it from completely opposite starting points, they finally agreed on one thing: absolute opposition to the Savoy marriage. Faced with this alliance of opposites, Philip's support counted for nothing and the scheme collapsed.

But did this tactical alliance between the sisters blossom into an *entente cordiale*? Elizabeth's romantic biographers have thought so.

Relations between the sisters were warmer than ever before, they gush. They also adduce some apparently convincing detail. In April 1558 Mary visited Elizabeth at Hatfield, and was entertained with singing and bear-baiting. Not to be outdone, when Elizabeth paid a return visit to Richmond in the summer, Mary sent for her a barge wreathed in flowers and gave her a sylvan banquet. Alas, for these floral fictions! They too are among the forgeries of Warton's *Life of Sir Thomas Pope* and have to be rejected with the rest.

Instead, 1558 was a harsh and brutal year for both sisters. In it, life and power drained away from Mary, while Elizabeth moved to secure the throne by force if it did not come to her by law.

CHAPTER 33

Two Portraits: Mary and Elizabeth

S hortly after his mission ended in 1557, Michieli, the Venetian ambassador who had been such an engaging observer of the English scene, wrote his *relazione*, or report. It is a tour through the geography, history, and constitution of England, and a discussion of the character of its government and people. It is clear, accurate, and, like all the best travel-writing, packed with insight into the foibles and idiosyncrasies of us strange islanders. Best, perhaps, is the penultimate section in which Michieli writes successive character-studies of the two women who dominated the English political scene: Queen Mary and Princess Elizabeth.[1]

The two portraits are conscious pendants, in which the women are scored against each other. It is as though Michieli recognized that they were at the beginning of the last act of their mutual struggle. At the end of his account, there is not much doubt about who will win.

Mary, according to Michieli, was short and 'moderately pretty' (rare indeed is the monarch who is admitted to be ugly). She was well educated and a good linguist, and had all the usual feminine accomplishments, such as music and needlework. 'She was devout and staunch in the defence of her religion.' She also had a piercing eye, which could 'inspire reverence and respect, and even fear', and a deep voice, 'almost like that of a man'. These last are the traits of the Mary who, with eyes flashing and voice thundering, had taken the City fathers by storm at the crisis of the Wyatt Revolt. But, Michieli implies, such command of the political scene was the exception rather than the rule for her. Instead, his Mary was a victim – of her history, ill health and unpopularity, of the poverty of the Crown and conspiracy against her, even of 'her passion for King Philip, from whom she is doomed to live separate'. It is a long list, and, he says frankly, it overwhelmed her.

But one last pain was more bitter than the rest. 'She is, moreover, a prey to the hatred she bears my Lady Elizabeth.' There were two grounds for this. One looked to the past 'and has its source in the recollection of the wrongs she experienced on account of her mother'. The other, sharper one, looked to the future, and to 'the fact that all eyes and hearts are turned towards my Lady Elizabeth as successor to the throne'.

Some historians have questioned Michieli's assessment of Mary's feelings towards Elizabeth as uniformly icy. No doubt there were thaws between them. These were frequent when they were young, for then they were in essentially the same situation. But once Mary was Queen and childless, with Elizabeth as her likely successor, who would undo everything that Mary had done, the logic of their positions worked in the opposite direction and drove them apart. Michieli is clear that Mary would have liked to take this logic to its natural conclusion and disbar Elizabeth from the throne. He is equally clear that it was largely Elizabeth's cleverness that stopped her.

For his Elizabeth is a victor, not a victim; she is not overwhelmed by circumstances, but their mistress. She even looks the part, as she is tall, well-built and handsome. Her complexion is pale, but she has fine eyes and beautiful hands, 'which she takes care not to conceal'. Michieli does not bother to list her accomplishments; instead, he notes that, as she has Greek and more fluent Italian, she is a better linguist than the Queen. But it is her superiority in 'spirits and understanding' that really impresses him. The 'proof', according to Michieli, is that Elizabeth has repeatedly confronted adverse circumstances and overcome them. She has been surrounded by suspicion and danger, but (wisely, he thinks, for after all he is the countryman of Machiavelli) 'she concealed her religion and comported herself like a good Catholic'. She knows 'what sort of woman her mother was', but she insists on her legitimacy and is 'proud and dignified in her manner'. She glories in the fact that she and Mary were treated equally in their father's will. Though she knows Mary hates her, Elizabeth has extorted a formal respect, in that the Queen treats her 'with every outward sign of affection and regard'. Above all, 'she has contrived to ingratiate herself with the King of Spain', who stops his wife from punishing Elizabeth. Elizabeth indeed deserves punishment – 'For whatever plots against the Queen are discovered, my lady Elizabeth, or some of her people, may always be sure to be mentioned among the persons concerned in them.'

Michieli's bias is evident. But his picture is essentially true: Mary is a loser, and Elizabeth a survivor and almost a winner – but not quite yet.

First, there was the little matter of making good her claim to the throne. For she was not without competitors. Michieli's *relazione* ended with a list of half a dozen of them, each with a different basis for their claim. Elizabeth headed the list and her title – grounded on the will of Henry VIII and the act of succession – was the strongest. But the other candidates were not inconsiderable. Next was Mary, Queen of Scots, who claimed (as all her Stuart descendants were to do) an absolute hereditary right which was indefeasible and could not be set aside by statute or other human ordinance. Then there were the surviving sisters of Lady Jane Grey, who claimed precedence of Elizabeth by virtue of the will of Edward VI. Finally, there was Lady Strange, the descendant of Eleanor Brandon, the second daughter of Henry VIII's sister, Mary. She insisted that, as the Greys had been disbarred by Jane's attainder, she was the only native, legitimate issue of the royal house.

Elizabeth, in short, could take nothing for granted. But it was not in her nature to make such a mistake.

CHAPTER 34

Power Ebbs

The year 1558 began disastrously for Mary. Philip had failed in one objective of his visit to England the previous year: the Savoy marriage. But he had succeeded in the other: he had brought England into the war against France. This was not only the result of his persuasive skills. Indeed, at first the Privy Council stuck by the letter of the marriage treaties which allowed England to remain neutral even if Philip's other dominions were at war. But then the French played into Philip's hands by supporting yet another invasion plot, this time by Sir Thomas Stafford. Stafford sailed with a couple of ships and took Scarborough with a handful of men, some English, some French. He was quickly captured and executed. But the insult to English honour was one too many and the council agreed to declare war. When the English herald conveyed the English Queen's defiance to Henri II, he laughed, saying she would do anything to please her husband.[1]

He was, however, soon laughing on the other side of his face as Emmanuel Philibert, Elizabeth's rejected suitor, proved his merit, on the battlefield at least, by inflicting a catastrophic defeat on the French army trying to relieve St Quintin. The English army under the Earl of Pembroke did not arrive in time for the battle. But the English played an impressive part in the capture of the city which followed. Included among the officers were such former rebels and plotters as Lord Bray, Sir Peter Carew, Sir Nicholas Throckmorton and the surviving sons of the Duke of Northumberland, including Lord Robert Dudley. Their presence showed that the old tactic, used at least since the days of Henry V, of turning ne'er-do-wells into heroes by sending them off to fight the French could still work under King Philip. But, like so much else in Mary's reign, this success proved a flash in the pan.[2]

Instead, England's intervention in the war turned into catastrophe. The English had received several warnings that another French army under the Duke of Guise was likely to turn against Calais. But nothing was done. After all, had not even the shrewd Michieli commented that Calais was impregnable? 'Although', he added cautiously, 'there are persons skilled in the art of fortification, who doubt that it would prove so if it were put to the test.' In the event, the doubters were only too justified. The garrison was under-strength and poorly equipped and the maintenance of the fortifications had been scrimped. And Guise took the defenders by surprise by attacking in the middle of winter. The actual moment of the assault on New Year's Eve was chosen because Guise's astrologers said it would be propitious; shrewder prophets would simply have forecast that the English were likely to be drunk.

The great complex of fortresses fell in slow motion. First, Calais itself was taken. But Guisnes, the main satellite castle, held out. The Earl of Rutland was commissioned to head a relieving force. But then Guisnes too surrendered. At this point the English seem to have lost their nerve. Rutland was countermanded. Philip's offer to launch a two-pronged counter-attack was rejected by the council, which decided that England could not afford to take part. Parliament began by considering a huge emergency grant but a more modest levy was quickly substituted. Feria, Philip's enjoy to England, was openly contemptuous of the indecisiveness and disarray. The rats were not yet leaving the sinking ship, but they were sitting on their hands.[3]

Calais was more than a place; it was a symbol. It had been in English hands since its capture by Edward III in another famous siege in 1347. And it was the sole remnant of the Anglo-French empires, which, one way or another, had endured from the Normans to the Wars of the Roses. Now it was gone. It was like Saratoga, which led to the loss of the American colonies, or Suez, which heralded the end of the British Empire. The English nation – for the term is not anachronistic – was humiliated and a stamp of failure was put on the whole Marian régime.

How far Mary herself was aware of this is unclear. Legend claims that she died saying that the word 'Calais' would be found engraved on her heart. But there is no contemporary evidence for the remark, and it seems at odds with Mary's customary indifference to mere worldly policy. Usually, her mind was on higher things.

Philip's visit in 1557 had lasted for just over three months. As before, Mary quickly convinced herself that her husband had left her pregnant. But, more cautious this time, she resolved not to make it public until she was sure. She herself entertained no doubts. She knew her mission was divinely ordained. And she knew that all would be lost if she did not leave a Catholic heir. God could not, would not, desert her.

But He was taking His time. Philip had left England on 5 July 1557. Seven months, eight months, came and went, and still there was no sign. But on 30 March 1558, as the ninth month approached, Mary made her will.

The will was founded on Mary's absolute assumption that she was pregnant. True, she rather took for granted her own death in childbirth – 'foreseeing the great danger which by God's ordinance remains to all women in their travail of children', she began. But she was confident that her child would survive. So the Crown was left to 'the heirs, issue and fruit of my body'; while Philip, 'according to the laws of this my said realm for the same provided', was appointed guardian and regent for the future prince or princess. Armed with this certainty of issue, she was similarly confident about the survival of the rest of her work. Money was left for the reburial of her mother in Westminster Abbey and to the religious houses she had refounded. Her 'most reverend father in God and my right entirely beloved cousin', Cardinal Pole, was enjoined to continue the task of restoring ecclesiastical revenues from the Crown to the Church, which Mary had begun but had not time to complete. And, finally, all her subjects were commanded, on their allegiance, to obey Philip, 'whose endeavour, care and study hath been, and chiefly is, to reduce this realm unto the unity of Christ's Church and true religion, and to the ancient and honourable fame and honour that it hath been of, and to conserve the same therein'.

The will was signed by four witnesses. Heading the list is

Bedingfield. He had continued to rise without trace and had recently been made vice-chamberlain. It is hard to think of any less natural courtier.[4]

Most royal wills contain a strong element of wishful thinking, as the testators assume, almost always wrongly as it turns out, that they can continue to command from the grave as they have done in life. Even so, Mary's is unique in its unreality. It is not so much a disposition of the Crown as a testament to delusion and obsession. Mary was deluded about her physical condition; deluded about the state of the body politic; deluded about her husband's aims and objectives. No doubt she had to deceive herself to make her life tolerable. And no doubt, at this safe distance of time, her delusions appear pathetic and arouse sympathy. But then they were dangerous. And they were dangerous, above all, to Elizabeth. For, in her ecstatic detachment from reality, what might Mary not do?

CHAPTER 35

Power Flows

It is difficult to glimpse Elizabeth in these earlier months of 1558. On 25 February, she came to London and stayed in Somerset House. Whatever hesitations the nobility previously had about paying her court were now thrown to the winds and she was escorted 'with a great company of lords and noblemen and noblewomen'. But she stayed little more than a week before returning to the country on 4 March. Had she been summoned, as two years previously, to be with her sister during her lying in? Had she left when it appeared that it was false? Or had Mary dismissed her as a thing of ill omen? We do not know. At any rate the sisters seem never to have met again.[1]

The only other solid information about Elizabeth is that she was worried about money. Balancing the books was to be her life-long preoccupation as Queen. But at first it is difficult to see why she should have had a problem as princess since her father had left her an apparently generous settlement of £3,000 a year. Under Edward VI, this had been enough to enable her to accumulate a substantial cash surplus. But by 1557 Michieli reports that she was 'always in debt'. Part of her problem, no doubt, was the difficulty caused for everybody on fixed incomes in this period by the rapid rate of inflation. Worst affected was the Queen herself, which is one reason why Mary did not show herself sympathetic to her sister's financial plight.[2]

But the real reason for the Queen's lack of generosity to her sister is that Elizabeth's financial difficulties were essentially political in origin. Elizabeth was not spending recklessly on clothes or jewels; instead, she was buying, or at any rate consolidating, support. This also is made clear by Michieli's account. Elizabeth's debts, he says, would have been much worse, 'did she not steadily restrain herself to avoid any increase of the

Queen's hatred and anger . . . by increasing the number of gentlemen servants of her household'. All the same, the pressure on her to enlarge her following was continuous: 'there is not a lord or gentleman in the kingdom', he continues, who did not endeavour 'to enter her service himself or to place one of his sons or brothers in it'. Michieli attributes this to 'the love and affection borne her'. No doubt. But equally at work was the desire of the ruling élite to hitch their wagons to the rising star. This is why Elizabeth was able to enter London with scores or even hundreds of gentlemen in livery; it is also why Mary adamantly refused to increase her allowance. Otherwise Mary would have been subsidizing an alternative government, or perhaps even paying for the overthrow of her own.

This to-ing and fro-ing between the sisters could have continued indefinitely. But in late August Mary seems to have fallen victim to the pandemic of influenza, which hit England repeatedly in the 1550s. It did not kill her. But it exacerbated her underlying medical condition. Her health fluctuated in September. Then, in early October, she became dangerously ill. Her doctors started sending regular bulletins to Philip, and the news obviously reached Elizabeth. Both reacted in their own way.[3]

The decline in Mary's health raised the issue of the succession in an acute form. Would Mary now try to divert the succession from Elizabeth? Even if she did not, would one of the other candidates try to step into the breach? Elizabeth had to assume the worst and seek to prevent it.

And she had the means to hand. For most of Mary's reign her household had been a hotbed of conspiracy. Servants of Elizabeth's had plotted the raising of troops and the stockpiling of weapons. They had debated strategy and sounded out allies. Latterly their numbers had grown, despite her half-hearted attempts to keep them in check. Women were involved, too, and not a few waverers seem to have been won over by their wives. The result was that her supporters were planted throughout the length and breadth of England: in country houses and castles, in town and villages, in ships and garrisons, even in the court and the council. Now it was only a matter of setting the machinery in motion.

And it was done with a smoothness and speed that gives the lie to all Elizabeth's previous disclaimers and disavowals about her involvement in conspiracy. In the thick of things, as usual, was the organizing genius of Thomas Parry, Elizabeth's Cofferer and chief of staff. In October and November Parry sent three letters to Sir John Thynne at Longleat. Thynne was Protector Somerset's former factotum. He had survived the wreck of his master's fortunes and had turned himself into a great power in Wiltshire. Now he was placing this power at Elizabeth's command. Parry transmitted her thanks in the biblical language to which Elizabeth often resorted at moments of crisis. 'Blessed is the servant to whom the master, when he comes home, may say, "I have found thee a faithful and good servant",' Parry was told to write. Also prepared to help in Elizabeth's 'coming home' was the bulk of the garrison of Berwick-on-Tweed, which was the largest concentration of troops on mainland England. Forty years later Thomas Markham reminded the Queen that Parry had summoned him to Brocket Hall on her behalf. He had duly appeared with written undertakings from his fellow captains to serve Elizabeth with ten thousand men 'for the maintenance of her royal state, title and dignity'.[4]

Brocket Hall seems to have been Elizabeth's operational head-quarters in these crucial weeks. It lay only two and a half miles to the north of Hatfield. It was presumably chosen because it was more defensible than Elizabeth's own sprawling, ex-episcopal mansion, and also because, being on the River Lea, it had better communications. Three and a half years earlier, during the Wyatt Revolt, Elizabeth was accused of planning to use Donnington Castle in a similar fashion. But there was an all-important difference. During the Wyatt Revolt, the Dudley–Ashton conspiracy and the rest, Elizabeth had carefully covered up her role. Now there was no need for concealment. On 28 October she wrote a letter of thanks from Brocket Hall to one of the many who had offered support: 'you may well assure yourself', the addressee was told, 'that we neither do, nor can, forget the same'. At the top left is written in

Elizabeth's own hand, 'Your very loving friend, Elizabeth'. The die was cast. She must have been very sure of her grounds.[5]

Elizabeth drew strength, of course, as Mary became physically weaker. On the 28th, the day Elizabeth signed the letter to her unknown supporter, Mary added a codicil to her will. She acknowledged she was 'presently sick and weak in body'. Then came a greater admission: 'God', she wrote, 'hath hitherto sent me no fruit nor heir of my body, and it is only in his most divine providence whether I shall have any or no'. God alone knows what it cost her to confront the fact. And still she clung to a hope. But it was a faint one. Instead the likelihood was that, in the absence of her own 'issue and heir', she would be succeeded by 'my next heir and successor by the laws of this realm'. It was a step towards reality. But it was a very short one. She could not bring herself to mention Elizabeth by name. She did not even acknowledge Elizabeth's gender. Instead, she exhorted 'them', 'by the bowels of the mercy of God', to respect her bequests to the Church.

Finally, Mary turned her thoughts to her husband, whom she would never see again. She acknowledged that, for lack of children, he would have no further part in the government of England. Instead, she besought Philip, for love of her and for the ancient amity between the Low Countries and England, to continue to take England under his wing: 'to show himself as a father in his care, as a brother . . . in his love and favour, and as a most assured and undoubted friend in his power to my said heir and successor, and to this my country and the subjects of the same'.[6]

Whatever the criticisms, and there are many, of Mary's exercise of sovereignty, there can be none about this farewell to power. It is dignified, and, in a situation where bitterness might have seemed the only thing left to her, utterly without rancour.

Philip is unlikely to have been moved by the appeal to sentiment (when he heard the news of his wife's death he wrote that he felt a 'reasonable regret'). But he had his own, self-interested motives for benevolence towards his soon-to-be former kingdom.

Back in March, Simon Renard, the former ambassador who had been so hostile to Elizabeth, reviewed the English succession and advised

Philip that there was no alternative to Elizabeth. She would succeed anyway; and the only hope of limiting her freedom of action, Renard thought, was for Philip to insist on the Savoy marriage. Philip, however, had long known that the marriage was a non-starter. So he decided to make the best of a bad job by acquiescing in the inevitable. Indeed, he would go further: instead of opposing Elizabeth's succession, Spain would do everything possible to facilitate it. This was not altruism. Rather, the hope was that if Philip and his agents made enough noise about assisting her to the throne, Elizabeth might be deceived into thinking that Philip had brought about her succession. It was worth a shot – though Philip, bearing in mind his good opinion of Elizabeth, can hardly have thought that it would be easy to pull wool over her eyes: she knew that her succession would happen anyway, with or without his support.

But the attempt *was* taken seriously. At the end of October, the Count of Feria was once more despatched to England as Philip's special envoy. He was already betrothed to Jane Dormer and, unlike most Spaniards, spoke excellent English. If anyone could pull wool, it was Feria.[7]

Meanwhile, the Catholic circles closest to Mary, and even the Queen herself, seem to have arrived at a similar conclusion: that the best hope of saving something from the wreck was to co-operate with Elizabeth rather than to resist her. On 6 November, therefore (three days before Feria's arrival), Mary bowed to the inevitable and accepted Elizabeth as her heir. According to the Spanish agent, the Queen 'showed herself much content'. If so, the reaction was play-acting. In fact, she was so anxious that she was still trying to impose conditions. Her agent, curiously, was Jane Dormer.[8]

Jane was sent to Hatfield, to inform Elizabeth that Mary had nominated her as her heir. As a token, she brought the jewels that Mary normally kept at court. In return, Mary asked for Elizabeth's assurance on two points: that she would pay Mary's debts and keep the Catholic religion as it had been re-established. There is a vivid account of Elizabeth's response in Jane's *Life*. 'She prayed God', Elizabeth is said to have exclaimed, 'that the earth might open and swallow her up alive, if she were not a true Roman Catholic'.[9]

There is another, equally vivid account of a similar interview, given in a letter from Edwin Sandys, then in exile but soon to return to become one of the leading Anglican bishops. According to Sandys, the Privy Council also sent two of their number to Elizabeth to inform her of her recognition as heir. This message added another condition: as well as requiring Elizabeth to settle Mary's debts and maintain Catholicism, she was also enjoined to keep the composition of Mary's council unchanged. So far, the drift is recognizably the same. But Elizabeth's reply, according to Sandys, was different in content and still more different in tone. Spiritedly, she insisted that she was as free to choose her councillors as Mary had been to select hers. As for religion, 'I promise this much', she is said to have stated, 'that I will not change it, provided only that it can be proved by the word of God, which shall be the only foundation and rule of my religion.'[10]

Who is right, the exile-reporter, or the direct participant in affairs? Evidently, it is Jane Dormer's word that carries weight. It would also have been out of character for Elizabeth, offered the chance of a smooth takeover of power, to have risked upsetting it by a few histrionics about her faith. Equally, I fear, it was also *in* character for her to make sure that the opposite impression was conveyed to her hardline Protestant supporters – hence the story that reached Sandys.

On 9 November, Feria himself arrived in London. He first had an audience with the Queen. She received him with joy but she was too ill to read the letter he had brought her from Philip. The count then convoked an emergency meeting of the council. He reported on the progress of peace negotiations with the French; welcomed the decision to acknowledge Elizabeth as heir, and informed them of Philip's warm endorsement of her succession. They listened to him without enthusiasm, almost indeed without interest. For power was ebbing from them with Mary's life; as for Feria, they treated him, he commented wryly, as though he had been the bringer of 'bulls from a dead pope'.[11]

The following day Feria, as ordered, went to Elizabeth to offer her

Philip's congratulations and assistance. He found her at Brocket Hall, attended on by Lady Clinton. Her husband, Lord Clinton, was Lord Admiral and one of Elizabeth's staunchest supporters on the council, while Lady Clinton, born Elizabeth Fitzgerald, was one of the beauties of the age. The youngest daughter of the Earl of Kildare, she had been immortalized by the poet-Earl of Surrey as his 'fair Geraldine'. The Archbishop of Canterbury, on the other hand, thought that she should have been whipped in Bridewell as a strumpet. As a girl, she had been one of Mary's attendants at Hunsdon. But, in later life, she became close to Elizabeth and ended as one of the leading ladies of her bedchamber. Feria, clearly, had interrupted a house-party in which the future shape of England was being determined.[12]

He was invited to dinner, which he ate with Elizabeth and Lady Clinton. During the meal, they 'laughed and enjoyed [them]selves a great deal'. But after dinner, Elizabeth turned abruptly to business, informing Feria that he could speak freely as her women could only understand English. Feria replied expansively that he would be happy if the entire world understood what he was saying. Elizabeth was equally gracious in turn, acknowledging that she owed Philip a heavy personal debt for his protection and good offices while she had been imprisoned by her sister.

All seemed to be going to plan. Emboldened by Elizabeth's charm, Feria determined to clinch matters. She owed, he now explained at length, the acknowledgement of her right to the throne neither to the dying Queen nor to the council, but solely to the King of Spain. It was the worst thing he could have said. It was fear of seeming to 'owe' the English throne to Spain which had made Elizabeth reject the Savoy marriage so vigorously. She was equally vehement in repudiating Philip's patronage now.

Feria reported the outburst in detail. It was the people, Elizabeth insisted, 'who had put her in her present position, and she will not acknowledge that your majesty or the nobility of the realm had any part in it'. To put her attitude to foreign interference beyond any doubt she commented explicitly on the Savoy marriage. Philip, she said, had tried hard to get her to marry the duke. But she had refused because she knew 'that the Queen had lost the affection of the people . . . because she had married a foreigner'.[13]

The tables were turned. Feria had failed to pull the wool over Elizabeth's eyes; rather, she had unveiled his. He had looked at her hard, without sympathy but without much prejudice either. And he wrote to Philip what he had seen. The result is a remarkable portrait of a young woman who knew what she wanted and knew moreover, as Feria sadly realized, that she was on the point of getting it.

> "She is a very vain and clever woman. She must have been thoroughly schooled in the manner in which her father conducted his affairs, and I am very much afraid that she will not be well-disposed in matters of religion, for I see her inclined to govern through men who are believed to be heretics and I am told that all the women around her definitely are. Apart from this, it is evident that she is highly indignant about what has been done to her during the queen's life-time."[14]

Nor had Elizabeth limited herself to generalities. Instead, she had bluntly divided the council into sheep and goats. The sheep, in the main, were Mary's 'old' councillors, whom she had inherited from her father and brother; the goats, on the other hand, were the Queen's new, Catholic councillors. Elizabeth also had hard words for the Earl of Pembroke. He was the man who, in the divisive parliament of 1555, had taken Mary's side against Elizabeth's partisans, headed by Sir Anthony Kingston. Clearly Elizabeth, as those on both sides of her list would discover, had a long memory. But she had two especial *bêtes noires*: John Boxall, Mary's Secretary of State, who was both a pious Catholic clergyman and a close personal friend of Cardinal Pole, and the Cardinal himself. 'I fear', Feria wrote of Pole, 'she will cause his downfall.' Elizabeth's invective was wasted; Pole, too, was on his deathbed.[15]

But the princess's tirade had shocked Feria. His relationship with Jane Dormer took him to the heart of Mary's inner circle. Its members were his friends; they were also, it was now plain, regarded by Elizabeth as her personal enemies. Feria feared for them. He also feared for Elizabeth, if her actions matched her words. So he warned her against showing a desire for vengeance or anger against anyone. Nothing could be worse than to display such feelings, he said, since everybody hoped to find her a woman filled with kindness and a good Catholic princess.

Elizabeth had an obvious retort to hand. She could have reminded Feria that Mary herself had come to the throne with just such a reputation for gentle, feminine moderation. She had sedulously cultivated it as long as she needed Protestant support to secure her throne. Then, when she did not, she had thrown off the mask and begun her policy of sincere but savage persecution.

But, instead of raking up old embers, Elizabeth's reply was moderation itself. She only wanted, she said, certain members of the council to realize that they had behaved badly towards her; for the rest, she would pardon them. It was the same formula she had used after her release from imprisonment at Woodstock, when Gardiner had tried to make peace between her and the council.[16]

Were Elizabeth's protestations of moderation real? Or were they as false and disingenuous as Mary's had been? The moment of proof was at hand.

Feria extracted one last pointer to the future from his visit. He obviously viewed Elizabeth's 'heretical' entourage with distaste. But he managed to make himself pretty well informed about it. Elizabeth, he thought, would favour the Earl of Bedford, Lord Robert Dudley, Sir Nicholas Throckmorton, Sir Peter Carew, Thomas Parry and John Harington. He also thought it a foregone conclusion that Sir William Cecil would be Secretary of State. The list of names was a familiar one. It included exiles and plotters. Some were condemned traitors; others were merely disaffected with the Marian régime. And, to a man, they were Protestant. True, Cecil stood somewhat apart by his studied moderation and conformity. But even he had breathed the intoxicating fumes of open opposition in the parliament of 1555. If Feria's prediction of the future shape of Elizabeth's government were accurate (as indeed it proved to be) it pointed to a régime as partisan and narrowly based as Mary's – though drawn, of course, from the opposite end of the politico-religious spectrum.[17]

The only difference lay in the character of the future Queen, which Feria noticed almost in spite of himself. One moment he was wringing his hands about the nature of Elizabeth's following. 'There is not a heretic or traitor in all the kingdom, who has not joyfully raised himself from the grave in order to come to her side,' he lamented. In the next sentence, he conceded that 'she is determined to be governed by no one'. In that determination, inherited from her father, lay the hope that she would be able to offer something better than the extremist, faction-ridden governments of her brother and sister.[18]

CHAPTER 36

The Enemy: Cardinal Pole

When Feria got back to London after his interview with Elizabeth, he had two urgent pieces of business to transact. First he got in touch with Lord Paget to press on him the need to fix Elizabeth up with a good Catholic husband as soon as possible. Paget's response was world-weary. He had arranged Mary's marriage to Philip, he said, and that had turned out badly. He would have nothing to do with any more royal matchmaking. Feria then informed Cardinal Pole of Elizabeth's feelings towards him. On 14 November Pole sent his chaplain, Seth Holland, the Dean of Worcester, to Elizabeth, with a letter and an oral message. In the letter, Pole said that when he died he wished to leave everybody satisfied with him, and especially Elizabeth. We know of the contents of the message only from the report of the French ambassador. Pole apparently told Elizabeth that there was no hope of the Queen's life, and that he prayed that whoever succeeded her would maintain the true faith. Elizabeth replied to him as she had done to Mary: protesting that if God ever brought her to that dignity, she would keep to the religion of her sister. She also murmured some expression of regret for Pole – 'showing herself very happy all the same', adds the ambassador. The remainder of her message was just as sincere.[1]

Elizabeth's antipathy to Pole was partly personal: the shrewd, utterly unillusioned princess was (in an inevitable pun) poles apart from the devout cardinal, with his lack of 'mundane experience'.[2] But she objected to what Pole stood for as well as to what he was. As legate, he was the symbol and the agent of the restoration of papal supremacy in England. As the Queen's leading councillor, he was the supreme example of

Mary's reintroduction of churchmen into public life. Finally, as both legate and chief councillor, he bore the ultimate responsibility, along with the Queen herself, for the policy of religious persecution which led to 300 burnings, to 100 other deaths in custody and to 800 fleeing into exile. The statistics are familiar. But familiarity does not lessen their horror. Each unit of those hundreds represents unspeakable suffering, which is not diminished because it was both inflicted and borne for the highest motives.

Pole's involvement in all these policies was not without its ironies and ambiguities. He was the agent of the papal restoration in England. But in 1557 England quarrelled with Rome and Pole was the principal victim. The Pope, Paul IV, was passionately anti-Spanish. But Mary had allied herself with Spain in the war against France. So Pope Paul revoked Pole's legacy and summoned him back to Rome on – the final insult – charges of heresy! It was strange, too, that Pole finished up as Mary's political confidant, when he had spent most of his earlier life bitterly denouncing Mary's father, Henry VIII, for introducing politics into religion. To the outsider, it is not clear why Erastianism (the domination of the Church by the State), which Henry VIII practised, is worse than theocracy (the domination of the State by the Church), which was the upshot of Mary's single-minded concentration on religion. Both are contaminating and lead to hopelessly mixed motives. And both resulted in horrible persecution.

And it is on the question of religious persecution that Pole's position was most awkward and inconsistent. Pole was an evangelical. That is to say, he shared the same passionate, Bible-based Christianity as most of the Reformers (which is why, of course, the charges of heresy against him had a certain awful plausibility). But not only was his religiosity similar to the Protestants', he was also committed to an ideal of Christian charity and love. Personally, he practised this with extraordinary success and in his circle of intimates and friends was venerated as a near-saint. But his intensely personalized Christianity did not translate well onto the public stage where he was condemned to play his final part. There, this most sensitive of men could appear triumphalist and rigid.

And in his handling of the martyrs he was at his worst. He did not want to burn people, and showed a fastidious disinclination to get involved in the sordid practicalities. But he did not dissent from the principle of persecution. For him, as for Sir Thomas More and Mary herself, heresy was the worst of offences. Other crimes, such as theft or murder, threatened only men's goods and their bodies. Heresy imperilled their immortal souls. And as the soul is so much more precious than the body, so much greater therefore must be the penalty.

Neither Mary nor Pole expected to have to burn many. They thought that the English had been led astray by a few bad shepherds. When the bad shepherds were eliminated and punished the flock would return, rejoicing, to the true faith. Most did indeed return. But more than expected persisted in their errors.

At this point, prudence, as well as charity, would have suggested a reconsideration of the policy. And indeed on 10 February 1555, after the first batch of burnings, Alphonso a Castro, Philip's chaplain, preached at court before the King. He attacked the bishops for the burnings, 'saying plainly that they learned it not in scripture, to burn any for his conscience'. Whether the sermon was sincere, or whether it was 'a very gross artifice' to deflect blame for the executions from Philip and the Spaniards, who were popularly accused of responsibility, is unclear.[3] And, in any case, it made no difference. The burnings continued: out of zeal, out of legalism and, increasingly, out of desperation.

So long as Mary sat on the throne, Elizabeth uttered not a word of condemnation of any of this. On the contrary, she professed a passionate Catholicism. But the religious complexion of her household, and her bitter personal hostility to Pole, hinted at her true feelings. Soon, she would no longer have to conceal them.

CHAPTER 37

Two Deaths

By the time Elizabeth received Pole's messenger, her half-sister was at death's door. In her final days, Mary frequently lapsed into a trance-like state. When she awakened, she told her attendants that she had been surrounded by little children singing and playing 'like angels'. On 16 November, towards midnight, she received the last rites. A few hours later, she was able to make the responses at the early-morning mass, which, as usual, was said in her bedchamber. Then, between 5 and 6 a.m., she slipped away. No one noticed the precise time, so gently had body and soul parted. She whose life had been a catalogue of trouble had the most peaceful of deaths.[1]

Across the river at Lambeth, Pole's attendants tried to keep the Queen's death from the Cardinal. But one of his chaplains let the news slip. Pole 'remained silent for about a quarter of an hour, but though his spirit was great, the blow nevertheless, having entered into his flesh, brought on the paroxysm earlier, and with more intense cold'. In the late afternoon, he heard vespers and compline. Then he too drifted quietly away. He had said 'that in the course of the Queen's life and of his own he had ever remarked a great conformity'. The resemblance extended to their deaths, Pole dying twelve hours after his cousin.[2]

Meanwhile, 'all London sung and said *Te deum laudamus* in every church'. They were rejoicing for Elizabeth's accession; some, no doubt, were thanking God for the deaths of Pole and Mary as well.

CHAPTER 38

Accession: A New Government

Parliament, which had reconvened on 5 November, was in session on the day of Mary's death. In the late morning the Speaker and the MPs were summoned to the House of Lords. There, Nicholas Heath, Archbishop of York and Lord Chancellor, announced Mary's death and Elizabeth's accession. The words we all imagine to have been used on such occasions – 'The Queen is dead. Long live the Queen!' – were not employed. But Heath's brief speech contrived a similar perfect balance between grief and rejoicing. Mary's death, he said, is 'most heavy and grievous unto us'; on the other hand, 'we have no less cause another way to rejoice with praise to Almighty God for that he hath left unto us a true, lawful and right inheritrix to the crown of this realm, which is the lady Elizabeth, second daughter to our late sovereign lord of noble memory, King Henry VIII, and sister to our said late Queen'. Heath then added, with heavy emphasis, these words: 'of whose most lawful right and title to the crown, thanks be to God, we need not doubt'.

'God save Queen Elizabeth!' the assembled lords and commons cried. It was not quite a parliamentary confirmation of Elizabeth's title, as there was no legislative act. But it amounted to much the same thing. Both houses of parliament then went in a body to Whitehall for the Queen's formal proclamation, which took place around noon.[1]

It was an extraordinarily smooth accession for one who had had such a rocky road to the throne. From the death of Mary to Elizabeth's proclamation took only six hours. Between these two events, the transition of power was seamless: swift, open, legal and unchallenged. Not since 1422 when the infant Henry VI had succeeded his all-conquering father, Henry V, had England seen anything like it.

It had not of course happened by accident.

Elizabeth, almost certainly, knew of her sister's death before she was proclaimed in London. The verse *Life* of Sir Nicholas Throckmorton perhaps tells us how. Fearful of being tricked by a premature rumour of Mary's demise, Elizabeth had sent Throckmorton to London. As soon as the Queen died he was to ride post-haste to Hatfield with the black enamelled betrothal or engagement ring which Philip had given her. Throckmorton duly set off at a fine gallop on the morning of the 17th. But he passed the information on to an acquaintance *en route* and when he arrived 'my news was stale'.[2]

Perhaps. But whenever the news arrived, Elizabeth and her entourage were fully prepared. Feria had already sensed what was afoot, and he had named the key figure, Sir William Cecil, who, he reported, was certain to be Secretary. He was right. For when the 17th dawned, Cecil was already at Hatfield and at his desk. As he would for the next forty years, Cecil kept careful notes. These enable us to reconstruct the events of the following few days, when, with clinical precision, the Marian régime was dismantled and Elizabeth's government slotted into its place. The operation was quiet, unassuming and fiercesomely efficient – like Cecil himself.

Cecil and Elizabeth had known each other since at least the reign of Edward VI. For in 1548 Catherine Ashley had written to Cecil, who was then Protector Somerset's secretary, about the exchange of an English prisoner in Scotland and Elizabeth had added a postscript in her own hand: 'I pray you further this good man's suit. Your friend, Elizabeth.' They had probably been introduced by Thomas Parry, who was Cecil's remote kinsman. Both Parry and Cecil were members of the 'Tudor Taffia': the much-intermarried group of anglicised Welsh gentry, who had flourished under the greatest Anglo-Welsh family of all, the Tudors. Another link lay through the pattern of local landholding. Cecil's father and grandfather had settled the family at Stamford in Lincolnshire. This lay just to the north of one of the most important but most remote clusters of Elizabeth's estates, centred round her great-grandmother's house at Collyweston. Lady Margaret helped set the Cecils on the first rungs of the ladder of promotion; Elizabeth took them to its summit.

But, in late 1549, this was far in the future. Then, Elizabeth was struggling to rehabilitate herself from the Seymour affair. She used Cecil as her intermediary to make sure that her letters reached Somerset. The following year, probably as a reward for his services, he was made surveyor of Elizabeth's estates. The fee of £20 p.a. was relatively paltry. But there were richer pickings in the form of land-grants.[3]

By this time Cecil was a figure of independent weight. He had survived the disgrace and partial rehabilitation of his former master, Somerset, and had been appointed royal Secretary. But the relationship between Elizabeth and Cecil was already more than mutual back-scratching; instead, there was also mutual sympathy. For both Elizabeth and Cecil had a common intellectual formation: both were products – Cecil directly and Elizabeth indirectly – of St John's College, Cambridge, during its brightest epoch. Cecil had gone up in 1535 at the age of fifteen. His elder contemporary was Roger Ascham, who became Elizabeth's most important tutor. And Cecil and Ascham shared the same teacher in the great John Cheke. Cecil's own academic career was seen as distinguished – at least retrospectively. And it is rather surprising that he went down in 1541 without a degree. The explanation, probably, is that his relations with Cheke had ceased to be merely professional. Instead, he had fallen in love with Margaret, Cheke's sister. They were married shortly after his leaving Cambridge and a son, Thomas, was born almost nine months to the day after the wedding. Margaret herself was dead within two years but the closeness between Cecil and Cheke remained. More importantly, Cecil, like Elizabeth, retained his intellectual interests for the rest of his life. He always carried a copy of Cicero's *De officiis* (*On Duties*), and it was at a dinner hosted by Cecil at court that Ascham decided to write *The Schoolmaster*, which supplies so many anecdotes of Elizabeth's own education and scholarly achievements.[4]

So Cecil was learned. But, unlike Cheke or that other contemporary scholar-politician Sir Thomas Smith, he proved fully able to adapt himself to the cut-and-thrust world of the court and council. Most academics, then as now, fail to make the transition to politics: they are too rigid intellectually and (often) too uncouth personally. Cecil, on the other hand, might have boasted with his colleague, William Pawlet that he was sprung of the willow rather than the oak. This flexibility, this reluctance – literally – to go to the stake for anything, established yet another bond of sympathy between him and the supremely pliable princess.

But, above all, Cecil could work, and thanks to his experience under Edward VI, he knew the business of government inside out. This alone got the Elizabethan régime off to a flying start.

In almost all his earlier contacts with Elizabeth, Parry had been Cecil's intermediary. And it was Parry, as we have seen, who had mobilized Elizabeth's following for a possible *coup d'état*. If Elizabeth had had to seize the throne by force, Parry would probably have emerged as the strong-man of the régime. But, with the peaceful transfer of power, it was Cecil's administrative talents that were required.

Nevertheless, Parry remained Cecil's coadjutor and co-equal until his sudden death in 1560. And they worked together in the immediate aftermath of Elizabeth's accession.

There was plenty to do. First, Cecil's notes suggest, there was the drawing up of the accession proclamation. This is a remarkable document in itself. With effect from 'the beginning of the 17th day of this month of November, at which time our said dearest sister departed from this mortal life', it informs all the Queen's subjects that 'they be discharged of all bonds and duties of subjection towards our said sister, and be from the same time in nature and law bound only to us as their only sovereign lady and Queen'. So the proclamation not only announces Mary's death and Elizabeth's accession; it also buries Mary politically. The copy in the State Papers is endorsed '17 November'. If this really was the text proclaimed at Whitehall at noon that day, communications between London and Hatfield had been extraordinarily fast.[5]

On this same first day of the reign, Cecil also drew up a new councillor's oath. It seems to have been designed to deal with the leaks and factioneering of Mary's council. Heavy emphasis was placed on secrecy; and, if councillors were aware of any incriminating matter against fellow councillors, they were not to communicate it either to the suspect 'or to any other allied . . . to [him]'.[6]

By this time, nicely on cue, a delegation of Mary's councillors had arrived from London. Just how many is not clear. According to Feria, immediately after proclaiming Elizabeth at Whitehall at noon, Mary's council decided that six of their number should 'go to the new Queen and perform the usual ceremonies' while the rest stayed in London. But the remainder took their relegation badly, since, as Feria contemptuously noted, 'everybody wanted to be first to get out'. One of those not in the official party, Lord Montagu, certainly went and later reported his reactions. And in all likelihood some if not most of his colleagues joined him on the road to Hatfield.[7]

There are two accounts of their reception. According to the first, the straggling party of councillors found Elizabeth walking in Hatfield Park, under an immemorial oak. As they knelt to her, she in turn fell upon her knees, uttering the words of Psalm 118: 'A Domini factum est; Et mirabile in oculis nostris' ('This is the Lord's doing; it is marvellous in our eyes'). It is a pretty story. But it is unlikely to be true. It has no contemporary authority, being reported only in Sir Robert Naunton's confection, *Fragmenta Regalia*, written seventy years after the event and as pretentious as its title. And it is also psychologically wrong. Naunton describes Elizabeth as speaking only 'after a good time of respiration'. Now the untutored Victoria may have received the news of her accession with such female flutterings; Elizabeth surely did not. And in any case, of course, she knew already.[8]

Fortunately, there is a more convincing alternative. In the collection of materials largely assembled by Elizabeth's godson Sir John Harington, *Nugae antiquae* (*Ancient Trifles*), there is a speech headed 'Words spoken by the Queen to the lords at her accession'. The speech is normally assigned to 20 November. But it makes no sense then. Spoken, however, three days earlier, in the fading light of her accession day, it becomes a thing of purpose. It addresses Elizabeth's own immediate concerns in the hours following her sister's death. And it addresses, still more pointedly, the concerns of her audience of ex-Marian councillors, most of whom were 'extremely frightened of what Madame Elizabeth will do with them'.[9]

As an individual, she explained, she was sorrowful for her sister's death and 'amazed' at the great burden which had fallen to her. But she was 'God's creature', and, as it was His design, she would do her best in the office to which He had called her. This royal office made her a 'body politic to govern', as well as a human 'body natural'. In this task of reigning, she asked for the assistance of 'chiefly you of the nobility', 'that I with my ruling and you with your service may make a good account to Almighty God'. 'I mean', she added emphatically, 'to direct all mine actions by good advice and council.'

It was this passage which Montagu recalled when, a few months later, he ventured to offer Elizabeth very blunt advice in the House of Lords on the question of religion. It was, he said, 'my chance . . . to be present when . . . her highness declared the great confidence she reposed in her nobility', and the value she placed on their duty 'to advise her highness as they in conscience thought'. As so often, Montagu had heard what he wanted to. For Elizabeth went on to offer a piercing analysis of just what she meant, both by nobility and by 'good advice and council'.[10]

She divided the councillors before her into three groups. First were those of the 'ancient', by which she meant the pre-Tudor, nobility; then there were the office-holders who had recently been ennobled by her father, brother and sister; finally, there were Mary's personal followers, who had been 'called to her service only and trust'. Elizabeth thanked them all, both for their past and future services and their 'faithful hearts'. But she made clear that, 'for council and advice', she would choose only 'such . . . as in consultation I shall think meet and shortly appoint'. And to those chosen, 'with their advice', 'I will join to their aid, and for ease of their burden, others meet for my service'.

Elizabeth was signalling, unmistakably, a heavy cull. But she had words of consolation: those who were left out, she concluded, were not to think that they were excluded 'for any disability in them, but for that I consider a multitude doth make rather disorder and confusion than good council'. It was a light but effective sugaring of the pill.[11]

After the speech, it was dark. Courtesy demanded that the (ex-) councillors were invited to supper. Then they went their separate ways. A few (they already knew who they were and had smugly listened to Elizabeth's clever choice of words) stayed to be part of the process of 'consultation' about the new council. But most dispersed to overnight lodgings in the town, before returning to London and – for the bulk of them – political oblivion or outright opposition.

Elizabeth had promised not only consultation, but that it would take place 'shortly'. She was as good as her word and it began the next day. Early in the morning (the usual starting time for council meetings was 8 o'clock), the small group of ex-Marian councillors met with a rather larger number of Elizabeth's own designated advisers. Just how many is not clear. But we should imagine no more than eight or ten round the table. The agenda was prepared by Cecil and has survived. He listed the items to be discussed down the left-hand side of the page; then, as the meeting proceeded, he made jottings against each item on the right-hand side. These noted what was decided and who was responsible for executing the decision.

The first item on the agenda was 'to admit the late Queen's councillors'. A bit of detective work establishes that three of them were present: the Earl of Pembroke, Lord Clinton, the Admiral, and William, Lord Howard of Effingham. They swore Cecil's new conciliar oath and then, with their new colleagues, proceeded to the remainder of the crowded agenda.[12]

High on the list was 'the admittance of the council in London'. Cecil wrote seven names against this. Six – Archbishop Heath, the Marquess of Winchester, the Earls of Shrewsbury and Derby, Sir Thomas Cheyney and Sir William Petre – were leading members of Mary's council whom Elizabeth had decided to retain. But the seventh was the Earl of Bedford. He was a convinced Protestant and was the only member of the high nobility who had gone into exile under Mary. While he was abroad, Cecil had helped manage his affairs and the two had remained close. Bedford's appearance as a councillor shows that the ex-Marians would have to get used to some fresh priorities.

Second only in importance to the council for the shape of the new régime was the court. This was the subject of two more items of the agenda. The first was 'to put the Chamber in order'. The 'Chamber' was the name given to the household 'above stairs'. It dealt with the public ceremony of the court, as opposed to the personal body service of the privy lodgings. And it was particularly important in the circumstances of 1558. Normally, posts at court were valued according to the closeness of access they gave to the monarch. But, with a female ruler, almost all the body servants of the privy and bedchambers had, by definition, to be women also. So posts in the chamber became the next-best thing for men, which in turn made competition for them intense.

Cecil's notes here were laconic. The first read: 'Lord William Howard, Lord Chamberlain'. This is the first indication that Howard, more properly Lord Howard of Effingham, was to replace Lord Hastings of Loughborough as departmental head of the Chamber. Howard had been one of the members of Mary's council who was most consistently friendly to Elizabeth in her frequent troubles under her sister; Loughborough, on the other hand, was an ultra-Marian and, it was thought, a diehard Catholic. An even more extreme exchange of personnel was effected in the next ranking post of vice-chamberlain. 'Mr Bedingfield discharged', Cecil noted. And replacing Elizabeth's despised gaoler was Sir Edward Rogers, who had been a co-conspirator with Wyatt.

A briefer agenda item referred simply to 'the stable'. This was the transport department of the court and was headed by the Master of the Horse. Opposite this entry Cecil wrote simply: 'Lord Robert Dudley'. Dudley had been imprisoned in the Tower with Rogers (and indeed with Elizabeth herself). Now, it appeared, this gaolbird, too, was to make good. The mastership of the horse was another position that carried enhanced prestige under a female sovereign, since its holder was, *ex officio*, closest to her outside the confines of the palace: when she hunted, went on progress or rode in a procession, he was beside or behind her. Dudley was not slow to exploit the possibilities.

A few more items of business – which included the decision to use 'King Henry VIII's funeral book' as the precedent for Mary's interment – concluded the meeting. When it was over, the essential shape of both Elizabeth's council and court was clear: key appointments had been agreed and the essential process of fusing the 'old' and 'new' councillors had begun. The meeting was not an official session of the council, since it does not appear in the Privy Council register. But this scarcely matters. Much important business was kept off the record, by accident or design. And the business of this meeting could scarcely have been more important since it effectively constituted the new council: it set it up and it agreed the main compromises – personal and political – that were necessary for its functioning.

On the 19th, the frantic pace of activity seems to have diminished. For some reason, Elizabeth was proclaimed again in Hatfield town; more importantly, the clerks of the Privy Council arrived from London with their travelling chest of books and records. This meant that the first formal meeting of the council could take place on the 20th. The attendance was probably much the same as on the 18th, but with one all-important addition. Elizabeth, who was to be present in the council only on the rarest occasions, put in an appearance to inaugurate the governing body of her reign. She also made three official appointments. 'By the Queen's majesty's commandment and in her presence', Parry was made Controller of the Household, Rogers Vice-chamberlain and Cecil Secretary. All three were also appointed members of the Privy Council. All this, of course, was only rubber-stamping decisions which had been agreed at the crucial meeting on the 18th or even earlier. But it has a certain symbolic importance.

Most symbolic of all were the few words Elizabeth said to Cecil on this occasion. It is one of her shortest speeches and one of her best:

> I give you this charge, that you shall be of my Privy Council, and content yourself to take pains for me and my realm. This judgement I have of you, that you will not be corrupted with any manner of gift, and

that you will be favourable to the state, and that, without respect of my private will, you will give me that counsel that you think best. And if you shall know anything necessary to be declared to me of secrecy, you shall show it to myself only, and assure yourself I will not fail to keep taciturnity therein.[13]

Though neither Elizabeth nor Cecil can have known it, it was a compact that was to last forty years.

The next day, the 21st, the news that Elizabeth had chosen her council and filled almost all important posts, outside the council as well as within it, was all over London and was swiftly reported by Feria back to his government.[14]

And still the reign was only four days old.

We, of course, can be blasé about this. Two hundred years of parliamentary government, followed by another hundred years of parliamentary democracy, have made us used to changes of administration that are smoother and more dramatic. Famously, the new prime minister enters the front door of Downing Street in triumph, while, at the back door, the goods and chattels of his defeated predecessor are loaded ignominiously into a removal van. But succession in monarchies was normally a tamer affair. The difference in 1558 was that Elizabeth had not only been Mary's successor, she was also leader of her (dis)loyal opposition. It was a role she deprecated, in public at least. But, like all successful leaders of the opposition, she had put her time in the wilderness to good use by preparing for her takeover of power. And she reaped the fruits of her preparation.

It is, however, another commonplace of parliamentary politics that the politicians who are the most effective in opposition do not necessarily make the transition to government. So it proved in 1558.

Much the most impressive member of the opposition to Mary was Sir Nicholas Throckmorton. Politics – assertive, dramatic politics – were in his blood. His father, Sir George, built the magnificent gatehouse of Coughton Court in Warwickshire. This combines the arms of the

Throckmortons with the royal arms and badges of Henry VIII in a statement of flamboyant loyalty. But Sir George became the most outspoken parliamentary critic of Henry VIII's divorce. When the King asked him in person why he opposed his remarriage to Anne Boleyn, Throckmorton replied by accusing the King of sleeping with both Anne's sister and her mother. The double charge of incest stunned Henry, into blurting out 'Never with the mother'. It was left to Cromwell, who was present, to try to rescue the situation by insisting 'Nor never with the daughter neither'.[15]

Nicholas Throckmorton's cousin, John Throckmorton, took the opposite side. He was one of the leaders of the Dudley–Ashton conspiracy which had aimed to dethrone Mary in favour of Elizabeth. After his arrest, he resolved to die like a Roman by remaining silent under torture. Many made the vow. Throckmorton was one of the few with the courage to maintain it. A generation later, Francis Throckmorton, Nicholas's nephew, who came of the main, Catholic line of the family, also died in agony, when he suffered the full horrors of treason for his leadership of the Throckmorton Plot against Elizabeth.[16]

But even in this family gallery of heroes and traitors (which is which, of course, depends on your point of view), Nicholas was outstanding. His portrait shows a handsome and fashionable man. His beard and hair were auburn – the same colour as Elizabeth's. He had piercing eyes, which disconcert with their intense, sideways glance. And there is a suggestion of fastidiousness in the spotless handkerchief whose folds protrude from his purse. His mental qualities were of a piece. His brilliant defence when he was tried for his part in the Wyatt Revolt had persuaded the jury to break all the conventions of a Tudor treason trial and acquit him. But he did not push his luck twice and, after the collapse of the Dudley–Ashton conspiracy in which he was also involved, he decided to trust his safety to his legs rather than his tongue and fled abroad. There he kept in touch with Elizabeth, who, he claimed, showed a 'gracious acceptance' of his communications.[17]

Now he decided to bid for power in the new régime by sending Elizabeth at least two letters which together formed a fully worked-out blueprint for her accession. They were written sometime before the event, in response to a false rumour of Mary's death. And, since in the first he referred to the fact that he was 'so far distant from your presence', he may have still been abroad at the time.[18]

Throckmorton's composite paper was a superb performance, envisaging a sequence of events very similar to what occurred between the 17th and the 20th. It began with a draft of an accession speech for Elizabeth addressed to Mary's ex-councillors. It gave an annotated list of candidates for each of the major posts in the administration. And it supplied a complete list of Mary's existing council so that Elizabeth could begin the process of weeding. No one, in short, could fault his knowledge of men or of measures. And he wrote with his usual, cocksure confidence that his advice would be listened to and acted upon.

Indeed, Throckmorton was deficient in only one thing: his understanding of Elizabeth. He took her for granted, even patronized her. And he paid the price. For she followed his scheme only incidentally, if at all. She differed from him both in the particular and in the general. She appointed few of his specific recommendations (apart, of course, from Cecil, who was his choice as secretary, but then Cecil had been universally touted). And, more importantly, she pursued an overall strategy that was the antithesis of the one he had suggested.

No doubt Elizabeth paid much more attention to Cecil's advice. Nothing survives from him on paper. Instead, he seems to have taken soundings in person. We can glimpse these in the case of the lord chancellorship, which, in the hands of Wolsey under Henry VIII or Gardiner under Mary, had been the dominant post in the government.

'Three or four days' before Mary's death, Cecil had gone to the house of Archbishop Heath, Mary's Lord Chancellor. Archbishop Heath had asked for the meeting because he was anxious to talk over the likely direction of policy under Elizabeth, particularly in matters of religion. He also had a personal matter to broach. He knew there was talk that Elizabeth wished him to continue as Chancellor: Throckmorton, for instance, put him at the head of his list for the post, and Feria had heard the same rumour. But Heath would have none of it. Instead, he begged

Cecil that he would be 'a mean' (that is, an intermediary) to Elizabeth that 'I might be utterly disburdened of mine office'.

Cecil agreed and then asked for Heath's views on his possible successors. Heath replied that 'there was choice enough' but diplomatically didn't name names. Not to be put off, Cecil ran a list of suggestions past him. Later, Heath remembered only that Dr Wotton was one of the names canvassed. 'No word', however, was spoken of the eventual appointee, Nicholas Bacon, Cecil's brother-in-law.[19]

The discussions with Heath were unusually harmonious. But we should envisage a similar process in the case of most other major appointments, with Cecil as Elizabeth's acknowledged broker.

Finally, however, it seems clear that the choice of Elizabeth's government rested primarily with Elizabeth herself. Those whom Feria heard she liked were in; those whom she told him she disliked were out, almost to a man. There were a few exceptions, including Pembroke. Elizabeth distrusted him but finally felt that he was too powerful militarily to leave off the council. Mary, curiously, had experienced the same dilemma and had resolved it in the same way. And Pembroke rewarded Elizabeth, as he had Mary, with his loyal service.

What is most striking, however, is not whom Elizabeth decided to appoint or dismiss but the speed and decisiveness of her decisions.

Here we must return to Throckmorton's scheme. Throckmorton, like most of the men of action of the 1550s, was soon to join the following of Elizabeth's first and greatest favourite, Robert Dudley, Earl of Leicester. As their previous history in the armed opposition to Mary would suggest, the Dudley faction are supposed to have been in favour of bold and decisive action. Elizabeth and Cecil, on the other hand, are characterized as being cautious, dilatory or even congenitally incapable of taking decisions at all. The events of the first three days of the reign tell a different story.

For it was Throckmorton who had counselled caution. Elizabeth was, he told her at the beginning of his paper, 'to succeed happily through a discrete beginning'. All his other recommendations were subordinate to this overriding concern. At first sight, he seems to be doing little more than suggesting delay: 'defer the swearing and nominating' of counsellors; 'suspend your grants to all persons with good words for a time'; be especially cautious about replacing the officers of the Tower, 'wherein there is to be used cunning dealing for avoiding alarm'. And so on.

But towards the end of his paper it becomes clear that he had a much more ambitious strategy. 'Until [Mary's] funeral be past', he wanted the realm to believe that Elizabeth intended the Broadest-Church administration, with 'the advice of many and the wisest'. His idea therefore was to keep Mary's existing council in being, on the one hand, and for Elizabeth to do nothing formally to constitute her own council, on the other. Of course Elizabeth would need fresh blood and different advice. But let there be no official appointments: 'that no oath be ministered to any nor no commission be had or used for a time of privy councillors'.

Not only would this bamboozle popular opinion. With luck, it would also take in the political élite as well. 'It shall not be meet', Throckmorton concluded, 'that either the old or new [councillors] should wholly understand what you mean, but to use them as instruments to serve yourself with: for some be mete to countenance your service and some mete to give advice and serve indeed'. Throckmorton *was* ambitious. He thought Elizabeth (guided by his master-hand) could not only deceive all the people all the time, but most of the politicians too. Throckmorton's belief in deception is why Richard Moryson called him a 'Machiavellian'.[20]

Now, there were many times when Elizabeth was prepared to deceive and delay, and to prove herself a mistress of both arts. But the beginning of her reign was not such an occasion. Nor were her councillors an appropriate subject. The relationship between sovereign and councillor, as Elizabeth understood it and set it out in her speech to Cecil, was one of mutual trust and openness: deceit, on either side, would destroy it; delay would render it nugatory.

This is one reason why Elizabeth's behaviour at her accession was a kind of point-by-point refutation of Throckmorton's scheme. Far from concealing what she meant from friends and enemies, she broadcast it from the roof-tops – or, at least, told Feria, which amounted to much the same thing. Far from delaying appointments to office or the council, she made both as rapidly as possible. And far from pretending that Mary's council was still in existence, she announced its dissolution in the most public fashion possible. There is an important general lesson here. When Elizabeth thought it was time for clarity, no one could be blunter. And when she was persuaded that she should be decisive, no one could act faster or more ruthlessly.

But Elizabeth had a deeper motive as well. She was reacting not only to Throckmorton's advice, but also to Mary's behaviour at *her* accession. For Mary had picked up her councillors as a rolling stone gathers dross – as automatically, and with as little thought. From the moment she had raised her standard against Northumberland, each day had brought fresh reinforcements. And as each group of East Anglian bigwigs had joined her swelling forces, so their leaders had been sworn of her council. When she had reached London and Edward's council had given their allegiance to her, they had been taken on board as well. So were backwoods peers and prisoners in the Tower, including Norfolk and Gardiner. The result had been an exceptionally large council of over thirty, and an exceptionally fractious one.

Elizabeth's accession speech, with its blunt reference to the fact that 'a multitude doth make rather disorder and confusion than good council', shows that she recognized the problem. Her actions show that she was prepared to do what was necessary to overcome it. She dismissed almost thirty of Mary's councillors and appointed only ten of her own. The result was a council that shrank back in size to her father's and contracted still further as old favourites died and Elizabeth hesitated to fill their places with the young and untried.[21]

This conscious reaction against Mary's behaviour – which was also a determination to trust to her head rather than her heart, as Mary had done, and to act with advice rather than on impulse – is the key to Elizabeth's reign and to her triumph.

CHAPTER 39

Between Old and New

B ut Elizabeth was not all head. She had a heart and she knew, as
well, how to appeal to other people's hearts. Nor was she simply
queen-in-council. Monarchy, then as now, lived in ceremony
and by ceremony. And it died in it, too. Before Elizabeth was fully
Queen, Mary had to be buried, while Elizabeth had to take possession of
her capital and be crowned. This process, it turned out, was as protracted
as the setting up of her government had been brief. It also seems to have
owed much more to Sir Nicholas Throckmorton's advice.

Throckmorton's scheme for Elizabeth 'to succeed happily by a
discrete beginning' had focused on Mary's funeral. Here he showed a
shrewd understanding of contemporary ceremonial – and a
characteristically cynical appreciation of how it could be exploited for
political ends. For until Mary was buried she was not fully dead. Her
physical body, which was also the motor of practical government, had of
course died in the early hours of 17 November. But the monarch was
more than a physical body. Elizabeth alluded to this in her accession
speech when she distinguished between her 'body naturally considered'
and her 'body politic to govern'. It was this body politic, created by the
investing rituals of coronation, that survived the death of Mary's natural
body. And it would continue to survive until it was exorcised by the
divesting rituals of a royal funeral, in which the image of monarchy was
finally separated from the all-too-human carcass of the defunct
sovereign.[1]

In the interval between Mary's death and her funeral, which could
be as long as Elizabeth wished, not only would the image of Mary's
monarchy survive, so too would the shadow of her government. And it
was Throckmorton's suggestion that Elizabeth should conceal herself

behind this shadow as her own government and policies were matured. As we have seen, Elizabeth had rejected this tactic over personnel. But she seems to have followed it over policy, especially religious policy.

The crucial choice about Mary's funeral had been made at the proto-council meeting at Hatfield on 18 November. By deciding to use 'King Henry VIII's funeral book' as the precedent for Mary's interment, it was decided also, in effect, that her burial should be traditional, Catholic and lavish. The council gave the Marquess of Winchester overall charge of the arrangements. It was an excellent appointment. He was the most senior of Mary's surviving councillors: rich, ceremonious and, despite his soft spot for Elizabeth, a Catholic at heart. He went to work with a will. On the 21st the council considered his letter about progress. Some of the nominated mourners had refused to take part. Winchester was told that Elizabeth would be prepared to order their participation. He also asked for an immediate cash advance of £3,000. This was an enormous sum, especially for a cash-strapped government, and a decision was postponed for a few days until the council arrived in London.[2]

But the first, practical steps had already been taken. Mary had died at St James's. As soon as she was cold, her grieving ladies withdrew and the grim process of preparing the body for its long, above-ground limbo began. The belly was laid open and the viscera removed. These were buried separately in the Chapel Royal at St James's. The wounds were cauterized and the cavity filled with preservative herbs and spices and closed up. Finally, the royal plumber soldered the body into a lead coffin, which in turn was placed in a wooden chest.

By the 21st, Feria reported, Mary's body had been removed to lie in state 'in the chamber outside the one she slept in'. This was probably the privy chamber. There, in life, Queen Mary had received privileged visitors under the cloth of estate. Now, in death, mourners paid homage to the body in the same place. Meanwhile 'the house', Feria noted, 'is served exactly as it was before'. Food was placed on the royal table; gentlemen ushers officiated with their white wands; guards stood at doors. The royal hive continued to hum though the Queen was (un)dead.[3]

Meanwhile, what of the other, living Queen?

Also on the agenda at the meeting on the 18th was an item 'concerning the meeting of the Queen's majesty at Barnet'. Barnet, just to the north of London, had been chosen as the place where Elizabeth would be greeted by her loyal subjects and escorted towards London.

But Elizabeth, supposedly so eager for the embrace of her people, was taking her time. She did not leave Hatfield, scene of so many joys and sorrows, till the 23rd. Her procession was a thousand strong. But she had often mustered more than that as princess. At Barnet, as arranged, the sheriffs of London met her and rode before her to the gates of her next lodging, the Charterhouse.[4]

The gates of the Charterhouse still survive. It was here, in the Carthusian monastery, that the young Thomas More had gone into retreat to consider taking monastic vows. And it was to these gates, thirty years later, that one of the mangled quarters of the prior of the monastery had been nailed. The prior had been hanged, drawn and quartered because he had joined More in resisting Henry's marriage to Anne Boleyn. Now Elizabeth, the daughter of that marriage, was coming to stay in the smart new house which Lord North, one of Henry's councillors, had built on (and out of) the ruins of the monastery.

Even now, however, the new Queen was near London but not yet in it. For the Charterhouse lies two hundred yards to the north of Smithfield, which roughly marks the City boundary at this point. There Elizabeth stayed for another five days. The council met frequently and the wheels of the new government began to turn smoothly.[5]

Not till the 28th, a full ten days after her accession, did she make her entry into the City. Her route avoided Smithfield, scene of the Protestant martyrdoms of her sister's reign (though, in contrast, the Catholic ghosts of the Charterhouse do not seem to have troubled her slumbers). Instead, she rode along the Barbican, entering the walled city through the Cripplegate. Once inside, she immediately turned east, by the little church of St Alphage, and went along London Wall to Bishopsgate, where she turned south towards the heart of the City. She went through Leadenhall, Gracechurch Street and Fenchurch Street. All the way the streets were newly gravelled. Bishopsgate was richly hung with tapestries; the City waits (that is, musicians) played; children made speeches and there was singing and playing on organs.

For the first time, Elizabeth was showing herself as Queen. At the head of the procession rode her escort of lords and gentlemen, 'all the trumpets blowing' and the heralds 'in array' (that is, wearing their brilliantly coloured tabards of arms). Then came the royal party. Elizabeth was preceded by the Earl of Pembroke, bearing the upright sword that was the symbol of sovereign power. The Queen wore another symbol: she was mounted on horseback and clad from head to foot in velvet of royal purple, with (incongruous among all this majesty) a scarf around her neck to protect her from the cold to which her 'body naturally considered' was all too susceptible. The sergeants-at-arms rode on either side and, immediately behind, pranced Lord Robert Dudley, the Master of the Horse, in his place of honourable proximity. The yeomen of the guard with their halberds brought up the rear. 'There was such a shooting of guns as was never heard before.'[6]

London had its Elizabeth.

But it quickly yielded her to the Tower. She had last entered it four years before, under suspicion of treason and in danger of her life. Did she remember those days now, in the moment of her triumph? We do not know. Nor do we have any but the most general indications of her behaviour to her people in the course of the procession.

For despite the trumpets and guns and the purple, it was a little low-key – and, in comparison with what came later, downright amateurish. Shrewd propaganda management would turn Elizabeth's future *entrées* and progresses, whether in the capital or in the country, into fulsome love-ins between the Queen and her subjects. Of this there was little trace in 1558. Was it just a question of early days? Or was Elizabeth, who was rarely backward at coming forward when it suited her, deliberately pulling back? I suspect the latter.

Elizabeth stayed for six days in the Tower. Then, on 5 December, St Nicholas's Eve, she left by water and, 'with trumpets playing and melody and joy and comfort', was rowed upstream, towards Westminster, the seat of government. But, once again, she stopped short. On her way to London, she had halted at the Charterhouse, outside the City boundaries. Now she paused again and, instead of entering the royal palace-complex that made up much of the present City of Westminster, she dropped off at Somerset House, her own town residence as princess. Here she stayed till the very eve of the Christmas celebrations.[7]

There were good, practical reasons for Elizabeth's decision to use Somerset House for this extended stay. As her own residence, it was ready to receive her whereas Whitehall was not. But symbolic considerations were at least as important. By staying in the house which she had used as princess, Elizabeth left Mary's body in possession of Mary's palace; she also left Mary, residually, in possession of the sovereignty.

But now this extraordinary six-week dual monarchy of the quick and the dead was drawing to its end as Mary's body began its leisurely journey to its final resting place. On 10 December a solemn procession of heralds, lords, ladies and officers of the household all in black entered the privy chamber at St James's, took up Mary's coffin and carried it to the Chapel Royal of the palace. Above, in the panelled plaster ceiling modelled after Serlio were the interlaced emblems of Henry VIII and Anne of Cleves. Even in death Mary could not escape the ironies of her parenthood.[8]

On the 10th the body of Cardinal Pole also began its journey from Lambeth to its burial place at Canterbury. Even after death, the strange parallelism between his life and Mary's, which Pole had noted, continued as they departed the political stage together.[9]

Mary's body lay three more days before the altar at St James's; then, on the 13th, the funeral proper began. The coffin, draped in cloth of gold, was placed in a wheeled 'chariot'. On top of the coffin lay a life-sized and life-like image of the Queen. Though heavily mutilated, it still survives. The image was robed in crimson, with the crown on its head, the sceptre in one hand and the orb in the other, and 'many goodly rings' on its fingers. Mary had always been fond of jewellery.[10]

The route of the funeral procession lay along the present Pall Mall to Charing Cross and then south, through King Street which bisected Whitehall Palace, to Westminster Abbey. It was the same ground over, which Wyatt's rebels had fought and lost at the climacteric of Mary's reign.

At the head of the procession were great banners representing the different branches of the English royal house to which Mary was heir; then her household servants, in black, on foot and marching two by two in order of rank; more banners; and then gentlemen mourners, also on foot and in black. The procession now became mounted and squires, carrying more banners, preceded four figures bearing the regalia: the Marquess of Winchester with the banner of the English royal arms embroidered with gold; Chester herald with the royal helmet; Norroy king of arms, with the royal shield; Clarencieux king of arms with the royal sword; and Garter king of arms with the Queen's coat armour. These were the personal symbols of English sovereignty, which were, of course, exclusively knightly and male. After their bearers came the late sovereign: the body below; the image above. And still in attendance on their mistress were her ladies-in-waiting. At each corner of the funeral chariot a herald on horseback carried the banner of one of the four English royal saints; in front went the clergy of the Queen's Chapel Royal, while behind followed the monks whom she had restored to their places and the bishops who, even then, were wrestling with their conflicting loyalties to the Queen whom they were about to bury and to the Queen whom they were shortly expected to crown.

Once at the Abbey the coffin with its image was carried to the great hearse, or catafalque, which had been built specially to receive it. There it lay all night, watched over with burning torches.

The next day came the mass of requiem and the burial, which finally separated the Queen's two bodies. At the offertory of the mass the regalia were offered up on the altar: one by one the Queen's coat armour, sword, shield and banner of arms were returned, symbolically, to the God who had bestowed them. Then, the mass ended, the Queen's image and all other tokens of royalty were removed from the coffin. The image was taken to St Edward's Chapel, where Mary had retired (robed, crowned and sceptred like the image) after her coronation. Meanwhile the board coffin, stripped of majesty and now a merely human receptacle, was carried further east to the great chapel built by Mary's grandfather, Henry VII, the founder of the house of Tudor. A vault had been opened in the north aisle of the chapel and into this was lowered the body. Earth was cast on top. 'Ashes to ashes and dust to dust': both Mary the woman and Mary the Queen had departed and the last trace of power was gone.

Next, each of Mary's officers of state and household broke his wand of office and threw the pieces into the grave on top of the earth: their power, too, had gone with their mistress's and its fragments were laid to rest with her.

The final disvestiture was performed by the heralds. They tore off their tabards and hung them on the hearse.

Then, at long last, a month after Mary's physical death, the heralds raised their cry: 'The Queen is dead; long live the Queen!'

Less reverentially, the people cut up the banners and cloth hangings of the church for souvenirs.

But the power had not gone without a struggle. And its memory still had the capacity to stir. All this was made clear by the explosive sermon preached at the funeral by John White, Bishop of Winchester. The sermon dealt, as was proper, with those great dichotomies – life and death, and the past and present queens – that filled the minds of all involved as they crossed from one side to the other of the strange middle ground between two reigns. But it did so in sensational language.[11]

White chose as his text two verses from the book of Ecclesiastes: 'I praised the dead which are already dead more than the living, which are yet alive'; and, in contrast, 'for a living dog is better than a dead lion'. Just to make sure everybody understood, he gave his own down-to-earth English translations of the Latin of the Vulgate. It was probably a mistake. Before him was a congregation that consisted of all of the old

English establishment and much of the new. As one, their jaws must have hit the floor. The bishop was saying, wasn't he, that 'a dead Mary was better than a living Elizabeth'? He was even saying, wasn't he, that 'Elizabeth was a living dog and Mary a dead lion'?

Actually, White was saying neither of these things. Instead, with all the ingenuity and learning one would expect from a star pupil at Oxford and a former warden of Winchester College, he glossed his texts into harmless, even improving, analogies.

But his English translations and the heat of the moment had done the damage.

That is one reading of the situation. The other is that White was not a fool and knew full well what he was doing when he chose his provocative texts.

When he came to his actual parallel between Mary and Elizabeth, he of course used more cautious language. Mary, he said, was royal herself and royal in every degree of relationship, as the daughter and wife of kings and the sister of kings and queens. And she had suffered in each capacity. She had all the public and private virtues. She was pious, merciful, and generous. She had wedded herself to her realm with her coronation ring, which she had never removed. Instead, she had discharged her duty by purging the realm of heresy. She had died as well as she had lived. After this encomium, the best that White could say of Elizabeth was that she was royal like Mary and held the realm 'by the like title and right'. He ended by 'wishing her a prosperous reign in peace and tranquillity' – 'if it be God's will', which he clearly rather doubted.[12]

When the Venetian Michieli had drawn up his contrast between Elizabeth and Mary he had judged them by the qualities that make for worldly success and he had unhesitatingly awarded the palm to Elizabeth. White used different criteria. He had judged them by spiritual standards and had arrived at the opposite conclusion: 'Mary hath chosen that good part' (Luke 10.42).

After these fireworks, the last word of the service belonged to the emollient Heath, Archbishop of York. Then the trumpets blew and the mourners and peers, the officiating clergy and the officers of Mary's household, all went to dinner in the Abbot of Westminster's lodging. It was the last supper of a régime.

Soon after, the council commanded White to keep to his house.

For White had committed a worse sin than calling the Queen a living dog – if indeed he had. Instead, his real offence was to wake the sleeping dogs of religious controversy which it had been Elizabeth's intention to let lie as long as possible. 'The wolves be coming out of Geneva,' White had cried. And this time he and his fellow-Catholic bishops and their allies in the lay nobility would not, he made clear, roll over in front of the attackers, as they had done at the time of the last invasion of European Protestant divines under Edward VI.[13]

It proved a remarkably accurate prophecy. But its author, Cassandra-like, was neither listened to nor thanked. For the last thing Elizabeth wanted was a fight between irreconcilable, principled religious positions. This is what White offered and it was what Elizabeth was determined to avoid.

Instead, she had begun, as she meant to continue as long as possible, with studied vagueness. Her royal style, as set out in her accession proclamation, ended in an 'et cetera'. This might mean that she intended to claim to be Supreme Head of the Church. Or it might not. The proclamation also concluded by giving 'straight . . . charge and command' to all the Queen's subjects to keep the Queen's peace 'and not to attempt upon any pretence the breach, alteration, or change of any order or usage presently established'. The intention, clearly, was to prevent any do-it-yourself religious reformation. But it was felt safer not even to use the word 'religion' lest the word alone arouse the passions that Elizabeth feared.[14]

Four days later, on 20 November, the first Sunday of the new reign, Dr Bill, Elizabeth's almoner, preached at St Paul's Cross. This was the most important pulpit in the kingdom, and was regularly used as the sounding board for government religious policy. On this occasion it emitted a note of exquisite harmony with Bill, hand-picked for the job, so neatly side-stepping controversy that the diarist Machyn, who was a good, old-fashioned Catholic, called it 'a godly sermon'.

What had changed? – people wondered, as they were intended to. The Protestants, Feria noted, had at first thought that the tables would be turned completely and that they would be able to persecute the Catholics. Now they were not so sure. The Catholics, on the other hand, had thought that all was lost. Now they were not so sure either, what with Elizabeth continuing to go to mass privately at Somerset House and Bill, if the rather garbled version of Feria's despatch in the *Spanish Calendar* is to be believed, having told the people to pray for the Pope.[15]

What indeed was going on? Was Elizabeth merely engaged in a game of cynical double-cross, as the Machiavellian Throckmorton had recommended? There was an element of that. But it is not the whole story. A cynical explanation fails to reckon with the inwardness of Elizabeth's religious experiences under Mary. The conventional understanding of these is that Elizabeth had conformed only outwardly, remaining at heart a convinced Protestant all along. This would be an accurate reading in the case of that other Marian conformist, Elizabeth's confidant, William Cecil. But Elizabeth was more complex and her years of Catholic practice did not simply run off her, like holy water off a duck's back.

Instead, her experience of Catholic worship left two permanent marks: one positive, the other negative. First, they imbued her with a taste for religious ceremonial – if indeed she had ever lost it, since the days when her father had left her with such an opulently equipped chapel. Not that she seems to have attached much liturgical significance to it. Rather, like that other dressy monarch, King Edward VII, who asked that his clergy should wear their 'best clothes' – by which he meant their copes – for his coronation, Elizabeth wanted her clergy at court to be as well turned out as she was.

Hers was, that is to say, an attitude that treated the *forms* of religious worship as pure externals. Or, in the fashionable phraseology of the time, they were 'adiaphora' or 'things in different'. But, in Elizabeth's case, this indifference to externals was bred out of a powerful internal conviction.

Once again this was a product of her Marian experiences. These had required her to enact belief in things in which she did not believe. Her refuge had been to decide that what she was doing did not matter. Most of her contemporaries, of course, thought it mattered a great deal. They thought that the forms of a religious service were 'the outward and visible sign[s] of an inward and spiritual grace'. This meant that for a Protestant to take part in a Catholic service or vice versa was hypocrisy at best and apostasy at worst. Elizabeth would have none of this. True religion, *her* religion, lay between Man and his Maker. The outward forms, on the other hand, were the work of human hands. Still worse, they were primarily the work of clerical hands. And Elizabeth had an abiding contempt for the clergy.

Elizabeth's contempt for the clergy, like so much else, may have been imbibed directly from her father. It was reinforced when, under Edward VI, the leading Prostestant clergy had attacked her alleged lover Seymour for his wantonness and ambition and herself for her bastardy. And it had been further exacerbated under Mary when the Catholic clergy, led by Bishop Gardiner, had seemed her most inveterate enemies at court and on the council. Elizabeth took a quick revenge. She made plain to Feria her especial dislike for Mary's clerical councillors. And, with the short-lived exception of Archbishop Heath, she eliminated any clerical representatives of whatever colour – Catholic, Protestant or Anglican – from her council for thirty years. She also made a practice of humiliating her clergy in public, by browbeating them like naughty schoolboys.

Now here was White, one of the worst of the breed, stirring up a hornets' nest. Still worse, the damn bishop (Elizabeth swore like a man) had put his finger on the two most sensitive spots – *her* two most sensitive spots: the nature of the sacrament and the royal supremacy.

He did so, appropriately, by using the deceased Mary as an exemplar to the living Elizabeth. Mary, he claimed, had expired in ecstasy at the moment of the elevation of the mass. This, if more reliable accounts of Mary's end are to be believed, was a pious fiction. But it was true to Mary's life. True also was his claim that Mary's own learning had encouraged her to repudiate the royal supremacy as a thing accursed. She knew as a fact, White asserted, that no prince had exercised such a supremacy in all the fifteen hundred years since Christ's death. She also knew that the royal supremacy, unheard of in a man, was impossible in a woman because of her sex.[16]

These two issues – the elevation (which betokened the real presence in the sacrament) and the royal supremacy – were Elizabeth's two sticking points. White, an intelligent and able controversialist, knew this, and, worst of all, he knew that she knew that she could not continue to sit on the fence indefinitely.

And so it proved. As Throckmorton had mapped out, Mary's funeral provided the turning point. With Mary buried, Elizabeth no longer had any excuse for not occupying the royal palace. Characteristically, she still procrastinated a little and it was not till 23 December that she moved from Somerset House to Whitehall.[17]

Here, a sudden spotlight shone on Elizabeth's religious practice. On Sundays at least she had to worship in all the publicity of the Chapel Royal. What form would the services take? And how, above all, would she cope with the forthcoming feast of Christmas? This was the high point of the court calendar, when the celebrations, in the chamber and the chapel, were at their most magnificent. It was also the time when the court was at its fullest. Usually, one imagines, the goings-on in the chapel were the last thing on most people's minds as they got ready for the fun and games; this year, however, all eyes were on the young Queen as, on Christmas Day, she processed with her ladies to hear Mass in the Holyday Closet of the Chapel Royal.

The Holyday Closet was a sort of dress circle. It was situated at the west end of the chapel at first-floor level and was linked to the royal apartments by a series of galleries. The closet was comfortably furnished and heated and had glazed windows with casements opening into the chapel itself. Through these casements Elizabeth was later to lean to make several shattering interventions when she was discontented with the preacher. On this occasion, however, she limited herself to dumb show. There was also a staircase linking the closet to the body of the chapel. On days of high religious ceremony the monarch would descend this stair and process to the high altar. Christmas Day was one of these days when, traditionally, the monarch made an offering of £1 in gold. At the offertory of the mass he prostrated himself (on a carpet and cushion deftly slipped beneath the royal knees by a gentleman usher) before the King of Kings; kissed the paten of the chalice, took the coin from an assistant lord, and offered it up.

On this Christmas morn Elizabeth watched from her closet as the officiating bishop, Oglethorpe of Carlisle, vested himself for the mass 'all in the old form'; waited until the gospel was said; then, when she should have descended to the altar to make her offering, instead turned round, and, accompanied by her lords and ladies, withdrew from the closet and the mass and returned to the privy chamber.[18]

The action spoke louder than words.

The words followed in the form of a royal proclamation issued at court on 27 December and proclaimed in London three days later on the 30th. Ostensibly, the proclamation was intended to prohibit unlicensed preaching and prevent innovations in ceremonies. But in fact it slipped in crucial changes, as Wriothesley, the herald-diarist and another conservative, instantly understood when the proclamation was enforced in the City: 'the parson or curate', he reported, 'should read the Epistle and Gospel of the day in the English tongue in the Mass time and [employ] the English procession now used in the Queen's chapel'.[19]

'The English procession' was in fact the Litany. This text was written by Elizabeth's godfather Cranmer for her father, Henry VIII. Elizabeth had used it during her imprisonment in the Tower and she clung to it as long as she dared thereafter. Now it was restored to the services in her chapel and throughout the land, where it dwelt alongside the Latin mass, as it had done in the last years of Henry VIII's reign.

Was this the beginning of change or its end? Whatever it was, it went too far for her bishops (that is to say, *Mary's* bishops) and nearly wrecked the coronation service.

CHAPTER 40

Coronation

Visible preparations for the coronation got underway during this same Christmas week, as workmen began to build 'scaffolds' or platforms for the pageants and the conduit in Cheapside (which would flow wine) was regilded. The day chosen for the ceremony was 15 January. This had been selected as propitious by the astrologer, Dr John Dee, in a horoscope which he had specially cast. The suggestion for the horoscope had been Lord Robert Dudley's, but the acquaintance between Dee and Elizabeth was an old one.[1]

Dee, who was a serious mathematician and Greek scholar, as well as a necromancer, had been educated at St John's College, Cambridge (where else?). At the beginning of Edward VI's reign he had gone to Louvain for further study. There his pupils had included Sir William Pickering, who was to be a conspirator and exile under Mary and, under Elizabeth, allegedly a candidate for the Queen's hand. On Dee's return to England, Cheke had introduced him to King Edward and Secretary Cecil. For Dee, therefore, as for so many of his ilk, Mary's accession was a disaster. And, with his background, he almost inevitably cast his lot with the oppositionist circles round Elizabeth. He also cast nativities for her in June 1555, as well as horoscopes for Philip and Mary. This was dangerous work in the immediate aftermath of Mary's phantom pregnancy. Even more dangerous were Dee's close contacts with members of Elizabeth's household, such as Parry, and with some of the most outspoken of Mary's opponents, such as Sylvestra, Lady Butler, who wished that 'the King and the Queen were at sea in a bottomless vessel'. Dee was arrested and interrogated but released on sureties. Lady Butler, however, continued her involvement in armed opposition by making her late husband's houses in

Gloucestershire and London available for meetings of the Dudley–
Ashton conspirators.[2]

We have already encountered the close links between astrology and
treason in the case of another of the eccentric ladies in Elizabeth's circle,
the divorced Countess of Sussex. And, under Mary, Elizabeth herself was
evidently up to the hilt in both. This interest in the occult is one of her
few points of divergence from her father and it is an instructive one.
Henry VIII was indifferent to astrology, if not actively contemptuous of
it. He had had one devastating experience of its fallibility when, two
months before his mother's death at thirty-seven, William Parron,
Henry VII's court astrologer, had predicted that Elizabeth of York would
live to the age of eighty! This incident was probably decisive in itself. But
there is a broader question of experience and the psychological reaction
to it. Henry VIII, his mother's death apart, had a singularly happy and
secure childhood and an untroubled succession to the throne. Secure
indeed, why should he bother with the dubious speculations of
necromancers and conjurers? Elizabeth's youthful experience was the
opposite: insecurity was her daily bread and astrology (as it is for the
rootless and the buffeted and the famous-for-nothing-but-being-famous
of today) was one of her comforts and hopes.

She clung to it and to 'Dr' Dee (his doctorate, naturally, seems to
have been bogus) now in the hour of her prosperity and throughout her
reign.

And she was determined to appear prosperous indeed on the great
day of her coronation, whatever the state of the treasury.

On 30 November the council ordered the London importer to hold
all incoming consignments of crimson silks, so that the Queen could
have the pick of them 'towards the furniture of her coronation'. As the
date of the event had not been published yet, he was to keep the order
secret. Over the next few weeks, huge quantities of precious fabrics were
bought: cloths of gold and silver, crimson and purple velvets, damasks
and taffetas, and plain cloth for the smaller fry. As well, there were furs
and feathers and gloves and stockings and even cotton wool to dab up the
anointing oils. Everybody who took part was given one set of clothes; the

Queen herself had at least three, though two, curiously, were hand-me-downs from Mary. The first of these was the coronation mantle itself of cloth of gold and silver, furred with ermine and closed with a tasselled lace of silk and gold, and the matching dress. Elizabeth wore this outfit for the eve-of-coronation procession, and she sat in it for her coronation portrait. A new bodice was made for the dress to show off her elegant figure in contrast to Mary's pudgy frame. Also her sister's were the parliament robes of crimson velvet trimmed with ermine, with the matching cap of estate, that she wore for the short journey from Westminster Hall to the Abbey. The third set – the purple velvet robes of estate, later known as the mourning robes – were, however, specially made for Elizabeth and part of them may have had their first outing when she made her entrée into London on 28 November. All three sets were retained in the royal wardrobe and – constantly aired, repaired and remade as fashion demanded – survived to the end of Elizabeth's reign and beyond.

Despite these characteristic little economies in her own dress, the cost of materials for the coronation exceeded £16,000. No doubt the loans immediately raised for Elizabeth on the Antwerp money market by her 'agent', Thomas Gresham, came in useful. The mayor and aldermen of the City were required to provide their bonds as security, but, as their citizens were the prime beneficiaries of the show, they probably did not complain too much.[3]

But one thing could not be bought, not even by Gresham and all his gold. And that was the usual complement of bishops to carry out the ceremony. The *Liber Regalis*, the authoritative text of the coronation, laid down firm rules. 'The right of anointing the kings and queens of England belongs above all by ancient custom, hitherto followed, to the Archbishop of Canterbury.' In addition, the *Liber* stipulated that the Bishop of Durham should support the king's right hand and the Bishop of Bath his left.[4] But in 1558 Pole, the Archbishop of Canterbury, was dead, Tunstal, the Bishop of Durham, was eighty-four and was licensed to be absent because of age, and Gilbert Bourne, the Bishop of Bath, was cousin of Mary's Secretary of State, John Bourne, and wholly identified with the previous régime. Nor was it easy to find substitutes. One of the senior bishops, White of Winchester, was still under house arrest because of his inflammatory funeral sermon for Mary; and another, Bonner of London – 'bloody Bonner' to the Reformers – was personally unacceptable to Elizabeth because of his record as a persecutor. Nor, in any case, would he have been prepared to act.

In these circumstances, Heath, the Archbishop of York, would have been the obvious choice to crown the Queen. He was well disposed to Elizabeth and she had a high regard for him. But he found the religious innovations she had introduced at Christmas unacceptable and declined. And so did all the other leading members of the bench of bishops, which, admittedly, had been much depleted by the influenza epidemic. Finally, one of the most junior of the episcopate, Oglethorpe of Carlisle, was prevailed upon to officiate. He had already experienced Elizabeth's heterodoxy, when she had withdrawn from his celebration of the Christmas mass. So he was forewarned. Probably, indeed, a deal was done. And at least he looked the part, since Elizabeth, who (as we have seen) liked her clergy to be nicely turned out, ordered Bonner to lend him his richest vestments for the day. The remaining bishops were present in their copes and mitres; they processed with their crosses and sang the anthem *Salva festa dies* (Hail festive day!) and swore the oath of allegiance. But they took no further part in the ceremonies.

What would have happened if Oglethorpe had not agreed is anybody's guess. But at least the defection of so many of the bishops reminded Elizabeth of the need to consolidate such support as she had. Nor did she hesitate about where to turn. 'She puts great store by the people and is very confident that they are all on her side,' Feria had told her brother-in-law, King Philip, a few days before her accession. Now she would demonstrate her popularity to the world – and perhaps, above all, to herself.[5]

On 12 January Elizabeth went back to the Tower to prepare herself for the traditional eve-of-coronation procession. The route was to go through the City, along the Strand, and so to Westminster. The monarch, the *Liber Regalis* required, should 'ride bareheaded . . . in suitable apparel offering himself to be seen by the people who meet him'. In short, the procession was – and was designed to be – a test of the sovereign's popularity. Anne Boleyn, Elizabeth's mother, barely passed. Henry VIII had spared no expense and everywhere there were his and Anne's initials or cipher, 'HA', laced by a lover's knot. But the women, especially, hated the flashy mistress made good and as she passed mocked her by crying out, 'Ha, ha!', in parody of her royal cipher. That, at least, is what Chapuys, Anne's inveterate enemy, claimed. But Elizabeth, Anne Boleyn's daughter, passed her test with flying colours.[6]

She left the Tower at about 3 o'clock in the afternoon of 14 January. She was carried in a litter, borne by two powerful mules. It was covered in yellow cloth of gold lined with white satin and it was open on every side – just as Elizabeth, back in 1554, had faced down rumours that she was pregnant by having the curtains of her litter pulled open and braving the stares of the crowd. The litter, where she sat on a draped and cushioned chair with a white damask quilt for warmth, was also vast, so that she would not be overwhelmed by the number of presents thrust into it, as had happened on her journey to her imprisonment at Woodstock. And she, above all, was ready herself. Elizabeth's quick tongue and mind made her a natural at all public events. This time, nature was improved on by making sure that a reporter was at hand to record the Queen's impromptu remarks (which, like all the best impromptu remarks, had no doubt been practised a little beforehand).[7]

As she left the Tower, she had to hand a carefully turned prayer, in which she recalled the earlier days of her imprisonment and thanked God for her deliverance, like Daniel from the lions' den – all this, theatrically, as she passed the lions in the Tower menagerie. When she saw in the crowd an old man weep and turn his face away, she quipped that she was sure that the tears were tears of joy. She kept a sprig of rosemary given her by a poor woman in Fleet Street all the way to Westminster. And when she heard the cry 'Remember old King Henry VIII', her smile became even wider. Everywhere, she was prepared to stop, to listen to suits, and to reply to good wishes with kind words.[8]

But it was her replies to the formal pageants that were the most important. Pageants, or dumb-shows with accompanying texts, spoken or written, were a long-standing element in coronation processions and other *entrées*. Holbein had designed a famous example for Anne Boleyn's procession and a particularly elaborate series had greeted the *entrée* into London of Catherine of Aragon, Mary's mother, when she arrived for her first marriage to Prince Arthur. But it was the pageants which London had mounted for Mary's own marriage to Philip which were in everyone's mind in 1559. These had been a bold exercise in revisionist history. The half-Spanish Mary and the wholly-Spanish Philip were presented as authentically English. And Roman Catholicism, which for twenty years had been reviled as foreign and traitorous, was reinstated as the true, patriotic faith of England. But though the revisionism was bold, in the event it failed. Mary's barrenness, the burnings, and, above all, the loss of Calais, killed it off. The pageant series which greeted Elizabeth's procession was a deliberate dancing on the grave. And Elizabeth, who had previously maintained a public face of loyalty to her sister, danced with the best.[9]

There were five pageants, each one a stage in a coherent argument, which followed Elizabeth's route across the city from east to west. The first pageant at Gracechurch Street dealt with Elizabeth's genealogy. Out went the newfangled double descent of Philip and Mary from John of Gaunt; back came the union of York and Lancaster – but with a special, complimentary twist to Elizabeth. Just as Elizabeth of York had brought the Wars of the Roses to an end by her marriage to Henry VII, so would Elizabeth, her namesake and granddaughter, bring peace once more to strife-torn England. Elizabeth, who had been unable to hear the message of the pageant because of the roaring of the crowds, turned back to make sure that she had understood it. And she ever after showed a partiality to her Yorkist ancestry, restoring the tombs of the Dukes of York at Fotheringhay but suffering those of the Lancastrians at Leicester to be destroyed.[10]

In the second pageant, at Cornhill, Elizabeth's government was shown as upheld by four virtues: True Religion, Love of Subjects, Wisdom, and Justice, each of which trampled its opposite vice. In the third scene, at Soper's Lane, the Beatitudes ('Blessed are the poor; Blessed are the meek; Blessed are they which are persecuted for righteousness' sake') were applied to Elizabeth's sufferings under Mary.[11]

The fourth pageant, at the Little Conduit in Cheapside, was the weightiest of the series. It also made the most explicit attack on the previous régime. When Elizabeth glimpsed it from the other end of Cheapside, she asked what it represented. 'Time', she was told. 'Time,' she replied, 'and Time hath brought me hither.' On either side of Time was a tableau: one depicting 'a decayed commonwealth' (Mary's government), and the other 'a flourishing commonwealth' (Elizabeth's government). Time himself had a daughter, Truth, who carried a Bible in English, prominently labelled the 'Word of Truth'. And it was the possession of this, the argument of the pageant stated, that explained the difference between the two commonwealths. When the 'Word of God', as it was then called, had appeared in the 1554 series, Gardiner had made a scene about it. Elizabeth, who certainly knew the story, was so keen to demonstrate the opposite attitude that she almost spoiled the pageant by sending to ask for the Bible in advance! Politely she was told to wait. She then made up for her impatience by the raptures with which she received the book, in proper season, from the child who had interpreted the pageant. 'She . . . kissed it, and with both her hands held up the same, and so laid it on her breast, with great thanks to the City therefore.'[12]

The fifth and last pageant at Fleet Street, just before she left the City, depicted Elizabeth as Deborah, the prophetess who rescued Israel from Jabin the king of Canaan and then ruled over the Jews for forty years. Elizabeth/Deborah was shown enthroned under a date palm (carefully labelled 'A Palm Tree' – like Bottom's 'Wall' in *A Midsummer Night's Dream*). She wore parliament robes and beneath her were figures representing the three estates – the nobility, the clergy and the commons – of England/Israel. At the time of course it looked back to Mary's disastrous later parliaments. But it proved oddly prophetic too: parliaments (though she disliked them) provided the setting for some of Elizabeth's greatest triumphs. And she even exceeded Deborah's four decades of power by ruling for forty-five years.[13]

Throughout, the reporter was close by the Queen, observing, listening and scribbling in his notebook. The results of his labours were rushed into print in a pamphlet, *The Queen Majesty's Passing through the City of London to Westminster*. This created a new literary genre. Normally, pageants were a one-sided dialogue, in which the people addressed their ruler. But Elizabeth's behaviour in 1558 turned the procession into a conversation, in which the prince's replies were as important as the people's address. The author of the pamphlet fully recognized the novelty of what he had witnessed. He also understood its theatricality, without in any way diminishing its significance:

> So that if a man should say well, he could not better term the City of London that time, than a stage wherein was showed the wonderful spectacle of a noble-hearted princess toward her most loving people and the people's exceeding comfort beholding so worthy a sovereign and hearing so princelike a voice.[14]

The reign of Elizabeth, we all know, saw the birth of the English stage. The procession of 14 January showed that its first star was Elizabeth herself. And both star and audience were conscious – supremely conscious – of their nationhood. Elizabeth was 'mere English'; while the author of this pamphlet boasted that London 'without any foreign person, of itself beautified itself, and received her grace'. The contrast with 1554, when the princes were foreign or half foreign and the most conspicuous of the pageants were mounted by the foreign merchant communities in London, was deliberate and complete.[15]

England to itself would now be true – it had no choice, if it was, alone among the great kingdoms of Europe, to embark once more on the course of religious reform.

The first taste of isolation came next day at Elizabeth's coronation when the Spanish ambassador, Feria, absolutely refused to be present (even incognito) for fear of the nameless impieties which he had been told Elizabeth would introduce. In the event, the broad outlines of Elizabeth's coronation were the same as Mary's – though with a few key innovations.

And the one foreign observer who was present, Il Schifanoya, the Venetian agent, was more confused than shocked by what he saw.[16]

The ceremonies began in Westminster Hall. This had been hung with some of the finest of Henry VIII's magnificent collection of tapestries. As an Italian, Schifanoya immediately recognized the set of the Acts of the Apostles as being based on designs by Raphael. In one corner stood a cupboard, loaded with massive gold and gilt plate, again from Henry VIII's collection. Elizabeth then processed to the Abbey. She wore her crimson parliament robes and walked on blue cloth, which had been laid all along the route to the Abbey. As the Queen passed, the cloth was cut up by scavengers and souvenir hunters – just as it had been at Mary's funeral.[17]

At the Abbey, the Queen was conducted directly to the crossing. This was the 'theatre' specially created for the coronation of Henry III, the rebuilder of the Abbey. She would still recognize the place today. To the east is the High Altar, and, behind it, St Edward's Chapel and the shrine of the royal saint. To the west is the choir. Two thrones were placed for Elizabeth before the altar: St Edward's Chair with the Stone of Scone beneath the seat was to the north and a chair of estate to the south. In the body of the crossing was a high stage, also called the 'theatre', with another throne on top and steps on either side. Finally, behind the altar in St Edward's Chapel, there was a traverse, or curtained and heated enclosure, to which Elizabeth withdrew, as was customary, to make the frequent changes of dress required by the ceremony. The traverse also figured for a much less orthodox purpose.[18]

Oglethorpe began by leading Elizabeth up to the stage and (the accounts differ) either proclaiming her Queen at each of the corners of the stage or, in the more traditional form, asking the people four times if they would have her for their Queen. The people had spoken the day before; now they shouted. 'Yea, yea!' they cried, as the trumpets, drums, organ and bells all sounded together. Then the Queen and bishop descended towards the altar. The Queen made the traditional offerings at the altar and then sat in the southern throne to hear the sermon. She knelt for the Lord's Prayer and then took the oath, the bishop giving her a copy of the text which she read. Then the service proper began, with the bishop reading from two service books at the altar.[19]

Here some extraordinary series of manoeuvres seem to have taken place. According to the 'English Report', which is accurate but naïve, at this point 'the Queen gave a little book to a lord to deliver unto the bishop'. 'The bishop', the account continues, 'returned the book to the lord, and read other books. And immediately the bishop took the Queen's book and read it before the Queen.' A little later, yet another volume appeared as 'Secretary Cecil delivered a book to the bishop'. Obviously, this cannot be disentangled for certain. But the usual speculation, that these elaborate exchanges of books related to the oath, seems unlikely, since Elizabeth swore the oath in the usual form. There are two other possibilities. Elizabeth, perhaps, was taking characteristically direct action to make sure that Oglethorpe modified the service where and when she wished. His returning of the book to the lord suggests that he resisted at first. But then a direct command from the Queen forced his compliance. So much for the first exchanges. Cecil's offering, however, was probably the coronation pardon, which, in the absence of either a Lord Chancellor or a Lord Privy Seal, it would have been his responsibility to draft. And this was something of a bombshell.[20]

Elizabeth having won her point, the service now continued without interruption. She withdrew to the traverse to change for the anointing. Traditionally, male sovereigns were stripped to their underclothing or even further. Elizabeth seems only to have exchanged her parliament robes for the kirtle or dress of the gold and silver coronation robes. The dress, like all Tudor costume, was in separate sections, held together by points or ties which, presumably, were loosened to enable Oglethorpe to perform the anointing. She 'leaned' on the cloth-of-gold cushions and carpet which had been spread before the altar and a scarlet satin pall was held over her while the anointing proceeded – on the shoulder-blades, the breast, the inside of the arms, the hands and the head. Apparently Elizabeth complained that the chrism, or holy cream, which had been specially obtained by Mary, 'was grease and smelt ill'. Her reaction was a characteristic mixture of fastidiousness, on the one hand, and, on the other, contempt for a ceremony which smelt, not only of grease, but of too much superstition for her moderately reforming tastes. The chrism (except on the head) was wiped off with the cotton wool; white gloves put on her hands; a white coif or cap on her head, and over her dress, the white dalmatic of a deacon. Then she moved to St Edward's Chair itself for the investiture with the ornaments – the sword, the armils, or bracelets, the mantle, the ring, and the sceptre – and the crowning itself. Some concession was made to her gender and the sword was slung over her shoulder, 'baldrick-wise', rather than being belted on. Then, successively, three crowns were placed on her head, with fanfares of trumpets marking each imposition.[21]

Now she was arrayed in gold from head to foot: from her golden shoes to the heavy gold of the imperial crown on her head. She had the golden sceptre in one gloved hand and the golden orb or 'world' in the other. Half-woman, half-bishop and wholly royal, she was led up the steps to the throne on the stage to be hailed by her people, for the first time, as their consecrated sovereign. She was assisted to the throne. Then, led by Oglethorpe, the Lords Spiritual and Temporal paid homage. Against precedent, the temporal lords went first: each knelt and then kissed the Queen's cheek. Old and young, smooth and bearded, prostrated themselves before the woman who was now their liege lord. Finally, the bishops enacted the same ritual, but first removed their mitres before kissing her.[22]

It was at this point, either before or after the homage, that the coronation pardon was proclaimed. I have speculated that this was the 'book' handed by Secretary Cecil to Bishop Oglethorpe. The pardon itself was a long, Latin text, and it was probably summarized in general terms, rather than being read in full. But when, the morning after, the élite and their legal advisers got round to reading the text, they were in for a shock. For Elizabeth had excepted from the pardon the perpetrators of 'any conspiracy, confederation, abetting, or procurement made or had against our person, or for the imprisonment of our person in the time of our dearly beloved sister, Mary, formerly Queen of England'. Those excluded embraced virtually the whole of the establishment of Marian

England: from the council, who had ordered Elizabeth's imprisonment, to Bedingfield, who had administered the orders, and down to the merest yeoman of the guard, who had only done what he was told. No one, in the event, was prosecuted for their role in Elizabeth's ill treatment. But the threat of prosecution, and almost as terrible, the ill will of the Queen, hung over the perpetrators.[23]

It was now time for the coronation mass, which followed, with Elizabeth's personally enforced innovations. The Epistle was read twice, first in Latin and then in English. Then the bishop brought Elizabeth the Gospel. This too was read twice, in the old liturgical language and again in the Tudor vernacular, which has, to us, become almost as remote, beautiful and hieratic as the Latin. Elizabeth now repeated her gesture of the day before and kissed the Bible – and, it is safe to guess, the English one.

The traditional text of the mass resumed, with its climax in the consecration. Here Oglethorpe had reached the limits of his compliance, and, as he had done on Christmas Day, he defied his Queen and served his God by performing the elevation. Elizabeth's response was the same as well: she withdrew to her closet in the traverse for the elevation and, it seems fairly certain, did not communicate. Instead, changed once more and wearing her robes of purple, she processed from the guttering lights and incense clouds of the Abbey into Old Palace Yard and back to Westminster Hall for the coronation banquet.[24]

She still wore the heavy weight of the crown and carried the orb and sceptre. But her heart was light. She was smiling broadly and greeted joyfully the thousands who pressed up to congratulate her. The people loved it. But Schifanoya was disapproving and thought that her behaviour 'exceeded the bounds of gravity and decorum'. She would not have given a rush. She was happy because she had got what she wanted. She had been crowned in her own way and on her own terms. Now she could face a parliament, and, she was perhaps too confident, get that to do what she wanted as well.[25]

CHAPTER 41

Religion Reformed

Elizabeth's first parliament had been due to assemble on 23 January. But the opening was put off because the Queen did 'not feel . . . herself in good disposition of body'. Was it a cold, or a temporary nervous collapse after the excitement and elation of the coronation the previous week? At all events she recovered quickly and the postponed ceremony took place two days later, on 25 January.[1]

Wearing her parliament robes again, Elizabeth processed from Whitehall to the Abbey for the religious service which preceded the opening of parliament proper. Traditionally, the Queen and her lords heard a special mass, that of the Holy Ghost, whose presence was invoked to inspire their deliberations. Once again, Elizabeth used a calculated gesture to signal her religious intentions. As she approached the Abbey doors she was greeted by Feckenham, abbot of the restored monastery, with his monks, censers swinging and tapers burning. 'Away with these torches,' Elizabeth snapped, 'for we see very well.' The sermon was even more disconcerting to the orthodox. The preacher was Dr Richard Cox. He had been Prince Edward's tutor; then, in exile under Mary, he had defended Cranmer's Prayer Book against the Presbyterian criticisms of John Knox. Calvin, who inclined to Knox's side in the dispute, described Cox as being characterized by 'a proud and confident manner in his carriage and language'. This abrasiveness seems to have been fully on display in his sermon – though, on this occasion, the target of his tongue lay on the other side of the religious divide. According to Schifanoya, that colourful but not always reliable reporter, Cox inveighed against the monks as the authors of the Marian persecutions and lauded Elizabeth as one divinely ordained to destroy monasteries and images and purify the Church. And he went on for an hour and a half.[2]

Neither the length nor (if he be correctly reported) the extreme language and content would ordinarily have pleased Elizabeth. But her reactions are unknown. From the Abbey she went in procession once more to the Lords' or Parliament Chamber in the old Palace of Westminster. Once she was seated on the throne, the commons were summoned to the bar of the house and Sir Nicholas Bacon, the Lord Keeper, read (as was customary) the opening speech.

Bacon was the last to be appointed of Elizabeth's major ministers. His name had not even been mentioned when Heath and Cecil had discussed candidates for the lord chancellorship back in mid-November. And in Throckmorton's list he appears only as an also-ran for the post of Master of the Rolls, who was the Chancellor's deputy. But, one by one, the other candidates dropped out or were eliminated (in some cases because of their Catholicism), with the result that Bacon was appointed. Going for him were the facts that he was a serious scholar, a sound lawyer and Cecil's brother-in-law. But his modest social status (he was the son of a sheep reeve who had been in charge of the flocks of the Abbot of Bury St Edmunds) meant that Elizabeth (who was snobbish about such things) only gave him the junior rank of Lord Keeper of the Great Seal, rather than the fully-fledged chancellorship.[3]

Bacon, in short, was on approval and had to tread carefully. And the opening of parliament would be the first, and very public, test of his abilities. Sensibly, he consulted carefully. He probably talked to Cecil; and he certainly talked to Elizabeth. The result was that his opening speech was a triumph. It was measured and authoritative, as such things had to be. But it was also shot through with a series of surprisingly personal statements about the Queen's hopes and fears, her attitude to policy and her understanding of her role. These can only have come from Elizabeth herself. So though she remained silent on her throne, crowned and robed like an image of monarchy, it is often her voice that we hear speak through her lord keeper.[4]

Their prime task, Bacon told the assembled lords and commons, was to consider the 'well-making of laws for the according and uniting of the people of this realm into a uniform order of religion'. This can have surprised none of them. But when he went on to describe the atmosphere in which Elizabeth wished the debate to be conducted, some of the older members may have pricked up their ears. They were, Bacon told them in his best schoolmaster tones, to avoid 'all sophistical, captious and frivolous arguments and quiddities . . . comlier for scholars than for councillors'. And, above all, they were not to insult each other with terms like 'heretic', 'schismatic' and 'Papist'. These are the sentiments, the tricks of speech and even some of the actual words of Henry VIII in his speech on religion, delivered to the same audience in the same place a quarter of a century previously in 1545. Elizabeth herself, aged only twelve, had not been present. But she had made the words and the sentiments her own. Like her father, she wanted religion to be debated in a way that avoided extremes. And she wanted a settlement that avoided extremes as well, uniting instead as many of her subjects as possible under the protective umbrella of the Crown.[5]

'A middle way' such as Elizabeth's is attacked from both right and left. According to Sir John Neale, the doyen of Elizabeth historians of the previous generation but one, the attack came from the left, and returning exiles (including Cox) forced Elizabeth towards a more radically Protestant settlement than she wished. In fact, the opposite seems to be the case. All the evidence of Elizabeth's behaviour in the first months of her reign shows that she feared the power of Catholicism. After all, the England she inherited was a Catholic country; Catholicism was the religion by law established, and the leading Catholics, clerical and lay, were imbued with a confidence and certainty which they had lacked under her father or her brother. Confronting Catholicism was difficult and dangerous, which is why Elizabeth had delayed revealing her hand for as long as possible. And by the time parliament adjourned for the brief Easter recess she must have wondered whether she could overcome it at all.[6]

Sensibly, the government got agreement to a subsidy (that is, direct taxation) first and only then moved on to the vexed question of religion. The first step was a bill to restore the royal supremacy. This was introduced into the Commons on 9 February and committed, after its two-day second reading, on the 15th. The same day, a separate uniformity bill was introduced into the Commons. This disappeared without trace. But, on the 21st, the committee set up on the 15th returned to the House with a hybrid measure: it had joined the supremacy bill, which dealt with questions of jurisdiction, to a uniformity bill, which dealt with the faith and liturgy of the Church. The result was a comprehensive scheme for a religious settlement. It was debated over three days and on the 23rd Cecil himself, who was sitting as senior knight of the shire for Lincolnshire, intervened to secure the passage of the bill without committal. We do not know what he said. But presumably he indicated that the bill represented government policy and was personally acceptable to the Queen. So far so good. The projected religious settlement had cleared the Commons in a fortnight. Nor had it been resisted by the sort of vehement minority which had opposed Mary's restoration of Catholicism. It was to be a very different story in the Lords.[7]

The bill for a religious settlement was sent to the Lords and read on 28 February. The peers delayed considering it, deliberately, till 13 March and then forced it into committee. The debate was vigorous, with the Catholic lords and bishops, who between them had a majority in the house, mounting a formidably confident and coherent attack. Lord Montagu and Archbishop Heath made the most effective speeches. The former spoke as a layman, the latter as a cleric, and between them they executed a perfect pincer movement on the bill. Montagu, availing himself with relish of Elizabeth's injunction in her accession speech to her lords to counsel her honestly, attacked the political risks of another reformation. A clash with the papacy, he pointed out, was the surest way to destabilize a kingdom: a sentence of excommunication had led to both Navarre and Naples changing hands, and it had triggered a revolution against King John in England. Did the lords wish to embark on that route? Heath scored by the careful moderation which, other things being equal, would have made him a natural Elizabethan. He conceded frankly that the present Pope, the vengeful, almost insanely irascible Gian Petro Caraffa, who reigned as Paul IV, was a disaster. He had been 'a very austere and stern father unto us ever since his first entrance into Peter's chair'. But there was, he insisted, a world of difference between bad relations with one particular pope and a breach with the papacy as an institution. The latter, as England's recent experience showed, led inevitably to heresy and schism. But his shrewdest blow was reserved for the supremacy. By what authority could parliament give it to any monarch, he wanted to know. And how, in particular, could they give it to a woman, who was forbidden by St Paul even to speak in church, much less to be its head.

Emboldened by these arguments, the lords' committee massacred the bill: the uniformity clauses were struck out completely, and the supremacy was reduced to a shadow. Instead of conferring the supremacy on Elizabeth (something which Heath argued that parliament had no power to do), the Queen was merely 'allowed' to assume it *if she chose*. The mutilated bill returned to the House on the 18th, was passed and sent down to the Commons.[8]

Both the commons and the government were in a quandary. The lords' bill as it stood was unacceptable and unworkable. But rejecting it was equally unpalatable, since, if the bill failed, England would be left with a Catholic settlement complete with the heresy laws. Biting the bullet, the commons accepted the lords' amendments on 22 March. 'The heretics', Feria noted with satisfaction, 'are very downcast in the last few days.'

Elizabeth and Cecil found themselves in a mess. And they had only themselves to blame. For they had hopelessly underestimated the effectiveness of the lords' resistance. The Reformer John Jewel, soon to be appointed Bishop of Salisbury, was disarmingly frank about the balance of forces in the Upper House: 'The bishops', he wrote, 'reign [there] as sole monarchs in the midst of ignorant and weak men, and easily overreach our little party, either by their numbers or by their reputation for learning.'[9]

Elizabeth's first reaction was to cut her losses. Immediately after the commons' vote on the 22nd a proclamation was rushed into print. The proclamation assumed three things: that parliament had been dissolved; that the Queen had given her royal assent to the amended supremacy bill and that the bill had (at some stage) been further amended to provide for communion in both kinds – that is, to allow the laity the privilege of communicating, not only with the wafer of bread but also with the wine in the chalice, which, in Catholic practice, was reserved to the clergy.

But the proclamation, though printed, was never proclaimed. Nor was parliament dissolved. And the Queen did not give her assent to the butchered supremacy bill. Instead, probably in the small hours of the night of 23–4 March, Elizabeth changed her mind. She did not process to the parliament house on Good Friday, 24 March, to dissolve parliament; instead, she had it prorogued until 3 April.[10]

Often during these night watches, Elizabeth was to regret what she had done because it was too bold. On this occasion, she reconsidered what she had intended to do because it was too weak. Her God-given authority had been challenged and by men whom she despised. God forbid that she should submit! She could not, would not.

So on Easter Sunday she nailed her colours to the mast. She processed in full form to the Chapel Royal, there to witness a religious revolution. Instead of the Latin mass, the English communion service was used, as under Elizabeth's brother, Edward VI. Instead of a stone altar, there was a wooden communion table, covered in a black cloth. And instead of the chalice being reserved to the officiating priest, the laity as well communicated in both kinds – led by the Queen herself who descended from the Holyday Closet to receive communion kneeling. Before doing so, she professed her faith in the saving body and blood of Christ.[11]

At Christmas, Elizabeth's withdrawal from the chapel had made plain her disapproval of the old religion; now her enthusiastic participation conveyed, equally clearly, her endorsement of the new.

But before her subjects could worship as she had done, the resistance of the Catholic bishops had to be overcome. So there could be no surrender, as she had briefly contemplated on the 22nd. Instead, she would regroup and fight and win. She would also be less choosy about the means. Powerful convention forbade interference with members of either House while parliament was in session. But when it was not, for however brief an intermission, it was another matter.

Curiously enough, it was the long-delayed fulfilment of a royal promise that provided the means. Back in November, as Mary lay dying, Elizabeth was supposed to have reassured the council that she would not change the Catholic religion 'provided only that it can be proved by the word of God'. It was now decided to submit the rival religious positions to such proof. A public, set-piece debate was organized in Westminster Abbey in the week of the Easter recess. A team of eight Catholics, four bishops and four doctors, would face as many Protestants of equivalent status. The judges were the Privy Council and the chairman was Lord Keeper Bacon. There were three propositions for debate: first, that it was against the word of God to use a tongue unknown to the people in the worship of God; second, that each Church had the right to change its own ceremonies; and third, that 'it cannot be proved by the word of God that there is in the Mass offered up a sacrifice propitiatory for the quick and the dead'.[12]

The debate purported to be an even-handed exercise, in which 'the best learned of either part' might 'confer together their opinions and reasons and so . . . come to some good and charitable agreement'. In fact, it was nothing of the sort. The questions were framed from the Protestant perspective; and the primarily scriptural method of proof was Protestant too. This already weighted the dice against the Catholics. But it was over procedure that the real gerrymandering took place. The Catholics wanted to debate in Latin, as was customary in university disputations; the government insisted that English was used, with a view to working on the minds of the parliamentary peerage, most of whom were present.

On the first day, a relatively brief statement of the Catholic case on the use of Latin or the vernacular in worship was followed by an immensely long one from the Protestants. When the Catholic bishops rose to reply, Bacon – who had quickly grown into a rather thuggish authority – stopped them, leaving the Protestants with the last word.

The Catholics were not to be caught out twice. At the start of the second day's debate the bishops again demanded a right of reply and, when that was refused, insisted that this time the Protestants spoke first. Bacon stonewalled; the bishops protested, increasingly vehemently, against his evident bias; and the proceedings degenerated into name-calling. Exasperated, Bacon proposed to abandon the debate. Eagerly, White of Winchester accepted: 'Contented, let us be gone; for we will not in this point give over. I pray you, my lords, require not at our hands that we should be any cause of hindrance or let to our religion.'

It was a bold stand. But it gave Bacon his opportunity. He now asked each, personally, whether he would read his prepared paper, as the council required. Led by White, who had already proved his courage in his funeral sermon for Mary, three refused point-blank, one agreed, one fence-sat in squirming discomfort, while Oglethorpe of Carlisle, who had squared his conscience enough to crown Elizabeth, came up with a characteristically ingenious compromise: 'if [the Protestants] should not read theirs this day, so that our writing may be last read, so am I content that ours shall be first read'.

Bacon had manoeuvred the bishops into an explicit act of defiance of the Queen's council and so of the Queen herself, and he ended with words of open menace: 'and for that ye will not that we should hear you, you may perchance shortly hear of us'. That evening White of Winchester and his boldest coadjutor, Watson of Lincoln, were sent to the Tower for contempt and the remainder were bound over.

It was, of course, pure coercion. All that can be said in its defence is that it was much less than had generally been thought necessary. For instance, Richard Goodrich, in his 'Divers Points of Religion', had recommended imprisoning *all* the most important leaders of the Marian régime before any religious reforms were attempted. Elizabeth had not gone anything like so far. Had she indeed gone far enough?[13]

For once again it was touch and go.

Learning from its mistakes in the first session, the government played safe when parliament reconvened on 3 April. First, it made some substantial concessions. In response to the concerns about Elizabeth's gender, which were felt as strongly on the Protestant side as the Catholic, her proposed title was changed from 'Supreme Head' to the less contentious 'Supreme Governor'. There was an important verbal change, too, to the text of the Prayer Book. In the communion service, the formula of the 1549 Prayer Book – 'The Body of our Lord Jesus Christ, which was given for thee, preserve thy body and soul unto everlasting life' – was added to the words of the 1552 text, 'Take and eat this in remembrance that Christ died for thee, and feed on him in thy heart by faith with thanksgiving'. The former allowed for the view that the bread and the wine contained the real presence of the body and blood of Christ; the latter emphasized that they were purely commemorative. The one was Catholic, the other Protestant. And between the two of them it was hoped that most people would be satisfied. Finally, there was a series of small but significant changes to liturgical practice, all in a conservative direction.[14]

A similar caution was shown over strategy. Supremacy and uniformity were divided into separate bills, so that if the one failed the other at least might pass. In the event, the supremacy bill got through the Lords easily. But the uniformity bill, which imposed the revised text of the 1552 Prayer Book as the faith of England, was fought to the last ditch. Bishop Scot of Chester, who had fence-sat in the Westminster debate, made up for it with an impassioned attack on even the compromise wording of the communion service on the nicely Protestant grounds that it was unscriptural! To no avail. The bill passed by three votes. The margin of victory was accounted for by the enforced absence of White and Watson in the Tower and the absence too of Abbot Feckenham of Westminster. Feckenham's absence is normally described as 'unaccountable'. Not so. He had gone out of his way to be conciliatory during the Westminster debate and had decided, I suspect, that, since the government would get its way willy-nilly, it would be better for everyone if the victory came quickly, without the need for further coercion. So, to make sure, he abstained deliberately by his absence.[15]

Elizabeth had won. And the Catholic bishops had been defeated, though honourably. But the Protestant divines were curiously muted in what should have been their triumph. When the clause – the so-called ornaments rubric, which outlined the conservative changes to liturgical practice – was first mooted, Edwin Sandys, shortly to be Bishop of

Worcester, had been blithely complacent: 'our gloss upon this text is that we shall not be forced to use [it]'. Once the whole thing was over, however, the truth dawned. And in that cold light Jewel wrote an almost despairing letter to his friend, Peter Martyr:

> As far as I can perceive at present, there is not the same alacrity among our friends, as there lately was among the Papists . . . The scenic apparatus of divine worship is now under agitation; and those very things which you and I have often laughed at, are now seriously and solemnly entertained by certain persons (for *we* are not consulted) as if the Christian religion could not exist without something tawdry.

'Scenic apparatus', of course, points us in the direction of Elizabeth, with her frequently displayed fondness for the theatrical. Jewel does not go so far as to name her. But Matthew Parker, her choice as Archbishop of Canterbury, was in no doubt. Had it not been for the ornaments rubric, he remembered Elizabeth telling him, she 'would not have agreed to divers orders' of the Prayer Book.[16]

What had Elizabeth achieved? When the full list of changes made under the ornaments rubric came to be totted up over the course of the following year, they amounted to rather a high figure. Images, relics, pilgrimages, candles (mostly) and the 'telling of beads' (that is, saying the rosary) had all gone. But a surprising amount of Catholic practice survived: the congregation was to kneel for prayers to God and to bow and doff their caps at the name of Jesus; altars or communion tables were ordinarily to stand 'altar-wise' at the east end of churches; the traditional special wafers were to be used for communion, rather than the ordinary bread specified in 1552; the clergy were to wear copes when they celebrated communion and the surplice at other times; endowments for choirs and music were to be retained; and, though most processions were to be abolished, the beating the bounds of the parish was to continue, with an injunction to respect property rights! Finally, the prayer – offensive to even the mildest Catholic – to be delivered 'from the tyranny of the Bishop of Rome and all his detestable enormities', was omitted from the Litany as well.[17]

But, despite their number, the Catholic survivals were pick-and-mix and lacked liturgical coherence. No serious, theologically literate Catholic was likely to be taken in by them. But perhaps that was not the point. Ordinary people noticed visible facts rather than theological niceties: if things looked pretty much the same, perhaps they were the same. For throw in the other, non-scriptural rituals retained in the Prayer Book, including the churching of women, the use of the ring in marriage and the sign of the cross in baptism, and there was a framework round which a new ceremonial religion, with its own comforting rituals and much-loved routines, could grow. It would not be Catholicism. But neither would it be a rigorous, stripped-down, Genevan-style Protestantism.

There are, however, two areas where we seem to touch Elizabeth's beliefs rather than her populist reflexes. One is the amendment to the communion service. On the day that the uniformity bill finally cleared the Lords, Elizabeth discussed her faith with Feria. He reported that 'she differed very little from us as she believed that God was in the sacrament of the Eucharist, and only dissented from three or four things in the Mass'. No doubt Elizabeth exaggerated for diplomatic effect. But she does seem, like her father, to have been sensitive on this score. This is why, surely, Bishop Scot concentrated his fire on the unscriptural nature of the 'Holy Communion, which after the order of this book is holy in words and not in deed'. And it is also why, even after the passage of the bill, Boxall, Mary's secretary, continued to press Elizabeth on this same point, using Treasurer Parry as intermediary. It was to meet these criticisms that the 1549 formula had been restored. Finally, after much heart-searching, Elizabeth seems to have felt that it went far enough – as indeed proved to be the case, since it opened the way to the sacramental revival of the Arminians in the seventeenth century and the full-blown Anglo-Catholicism of the Oxford Movement in the nineteenth.[18]

The other 'window into Elizabeth's soul' is the requirement that the clergy should combat 'the vice of damnable despair' by pointing out to their parishioners 'such comfortable places and sentences of scripture as do set forth the mercy, benefits and goodness of Almighty God towards all penitent and believing persons'. To us the passage reads oddly. This is because we have forgotten the fear of hell-fire. But 'damnable despair', that is, the conviction that one was irretrievably damned, was central to the theology of both the extremes – Roman Catholicism on the one hand and Calvinist Protestantism on the other – against which Elizabeth strove. The Catholic combated 'damnable despair' with the magical apparatus of saints and sacraments; the Calvinist, with the bleak doctrine of predestination, which affirmed that they – the elect – were saved, whereas the unregenerate mass of mankind was damned. Elizabeth instead stuck to the Evangelicalism of her youth and insisted that Gospel offered God's grace to all. She also insisted that a copy of an English translation of Erasmus's *Paraphrases*, which expounded the Bible in this sense, should be bought by every parish and set up alongside the Bible in churches. Evangelicalism, to us, seems sweetly reasonable. But already by the 1550s it was old hat, and Elizabeth's determination to stick with it made her more and more out of line with the academic young turks of the clergy who flocked to embrace the fashionable doctrines of Calvin.[19]

And, even for their seniors, the responsible heavyweights who were to make up the first generation of Elizabethan bishops, the peculiar amalgam of the Elizabethan Church was hard to take. Jewel, soon to be Bishop of Salisbury, wrote of the Church as 'a leaden mediocrity', in which nameless 'others' had decided that 'the half is better than the whole'. The 'other', in fact, was Elizabeth. She was convinced that her 'mean' or middle way was 'golden', not 'leaden', and that it promised the precious metal of inclusiveness and unity, rather than the base of clerically inspired extremism. For it was Mary's surrender to that, the Queen was convinced, which had wrecked her sister's Church and would, she feared, destroy her own, if she let her clergy have their head. And that she was determined never to do.[20]

But these struggles were for the future. For the moment, it was with a sense of satisfaction in a job well done that Elizabeth processed once more to parliament on 8 May to hear Lord Keeper Bacon's closing speech and to give her assent to the bills passed. Bacon ingeniously turned the tortuous proceedings over the religious settlement to advantage. The length of the session, he claimed, showed that they had done nothing unadvisedly; instead, they had 'taken such convenient time and leisure as the weightiness of the matters . . . hath required'. Freedom of speech, he continued, had been allowed for every position. And, after exhaustive debate, there had been 'well nigh an universal consent and agreement'. All this made the Queen hope, Bacon reported, that everybody would accept what had been decided in such an exemplary fashion.

Finally, and in sharper language, Bacon reminded his hearers that the settlement occupied a middle ground, and that obedience would be required from 'those that be too swift as those that be too slow', or from 'those . . . that go before the law or beyond the law, as those that will not follow'. In other words, hot Protestants who wore too few vestments would be just as much at risk as obstinate Catholics who wore too many. And so it proved.

'I have said. Read, clerk,' Bacon concluded. It was an impressive end to a remarkable parliament. It was not, of course, quite as remarkable as Bacon alleged: when he claimed near unanimity for the settlement, he stretched a point, just as he glossed over the imprisonment of the two bishops in the Tower. Catholic polemicists returned endlessly to these questions in an endeavour to undercut the legitimacy of the settlement. But, however much the conservatives might cry 'we woz robbed' and complain, like Feria, that 'it is all roguery and injustice', at least they had been allowed to play the parliamentary game. In contrast, under Mary, the leading Protestant bishops were either in exile or in prison on charges of heresy before parliament gave any consideration to a Catholic restoration. It was, moreover, Elizabeth's own personal determination to proceed only by due process of law. The inevitable delays she was prepared for, though the risks probably turned out to be greater than she expected. But she overcame them and rightly reaped the reward. For Bacon's main point was, I think, well taken even by those who had lost: everybody acknowledged that a decision made freely in parliament was bound in law, however much individuals might dissent from it in conscience.[21]

And it was precisely this question of conscientious dissent which Elizabeth had now to face. Once again, her behaviour shows how much she had learned from the débâcle of Mary's reign. Crucial was the issue of penalties. Her sister's religious settlement had been enforced by the terrible penalties of heresy; her father's by the different but equally horrible tortures of treason. Elizabeth's acts of supremacy and uniformity, on the other hand, prescribed a sliding scale of fines, imprisonment and deprivation of office for their breach. Heresy was narrowly defined and rarely employed; while treason was reserved for the aggravated offence of explicitly asserting the papal supremacy on three separate occasions.[22]

Moreover, Elizabeth's hope was that even this relatively tender armoury would not need to be deployed very much. Events were to force her to be harsher than she wished. But still she had tried.

CHAPTER 42

The Limits of Religious Reform: Practice

Legislating a religious settlement was one thing; enforcing it another. Here, too, Elizabeth, personally and decisively, took the lead, pointing the direction of change but also setting limits to it. As with the run-up to the settlement, the liturgical practice of her Chapel Royal acted as Elizabeth's semaphore: first as the signpost of innovation, later as the boundary-post to its radicalism.

On 12 May 1559 the new English service began in the Queen's Chapel. This was only four days after the dissolution of parliament and over a month before the use of the Book of Common Prayer became obligatory on 24 June. By adopting the English service so promptly for her own worship, Elizabeth was giving her powerful and visible endorsement to the changes. The City was equally enthusiastic and many London churches started using the new Prayer Book only two days later than the court, on Whitsunday 14 May. This is because London was one of the relatively few areas in England where there was committed support for Protestantism – at all levels, from the City's governing élite to the unruly ragtag and bobtail of apprentices and servingmen and women. There were individual exceptions, of course, and there were pockets of more organized resistance, like the cathedral clergy of St Paul's who, following the lead given by their bishop, 'bloody Bonner', persisted with the old services and vestments.[1]

But these futile gestures were quickly brushed aside by the 'Visitors', who started sitting in the City in July. The Visitors were Royal Commissioners, charged with overseeing the implementation of the new religious settlement. Head of the London commission was Sir Richard

Sackville. He was a Boleyn cousin through his mother, and Elizabeth had made him Under-Treasurer of the Exchequer and a privy councillor. As a member of her cousinage, Sackville was close to Elizabeth and his appointment shows the sensitivity of his task.

Sackville and his colleagues had to swear the clergy to the royal supremacy and make sure that churches were equipped with the Bibles and Prayer Books that were needed for the new English services. They also had to supervise the destruction of the elaborate apparatus of the old cult. In many areas of the country, the Visitors met with obstruction, evasion and occasionally outright resistance, as they tried to force the surrender of the vestments, images and altar plate that were focuses of local pride as well as piety. But in London conformity was almost embarrassingly eager. It culminated in a deliberate, grand gesture.[2]

23 August was a red-letter day in the City calendar as it marked the opening of the largest of the London fairs, the Bartholomew Fair, held at Smithfield. This year the celebrations were given a special edge. In a series of carefully stage-managed spectacles, confiscated church goods were burned at St Paul's Churchyard, Cheapside, and other strategic sites throughout the City. Great wooden images of the crucified Christ, the Virgin and St John, which had stood over the rood screens which divided the nave from the chancel in almost all the City churches, were tumbled into the flames, along with banners, copes, vestments, altar cloths, cloths for the rood, crucifixes, missals, service-books and anything else that savoured of superstition and Popery.

It was a bonfire of the religious vanities and it prompted a despairing reaction from the conservative chronicler Charles Wriothesley, Windsor Herald. He watched the iconoclastic flames, just as he had watched them before, under Edward VI. He listed what had been destroyed and concluded sadly that the goods burned 'cost above £2,000 renewing again in Queen Mary's time'. After two more entries he abandoned his chronicle and died in January 1562, under the impression, no doubt, that England had gone to the dogs. But perhaps he despaired too soon.[3]

For Elizabeth knew that many of her subjects shared Wriothesley's sympathies. She also knew that it was dangerous to press them too far or too fast. So, from 1560, the signals sent from the court and the Chapel Royal reversed: in 1559, it had been full steam ahead; now, it was right about turn.

The religious observances of Lent provided the opportunity.

Lent, the period of forty weekdays before Easter, was traditionally marked by fasting by abstinence from flesh. This was rigorously enforced by royal proclamation in 1560 and illicit cart-loads of meat were confiscated. To replace more fleshly fare (since man shall not live by bread or beef alone), it was also the custom to offer thrice-weekly sermons, which were delivered at court by selected preachers on the Sunday, Wednesday and Friday of each of the seven weeks in Lent. Some of the sermons took place in the Chapel Royal itself, but most were given in the Preaching Place at Whitehall.

The Preaching Place lay directly to the north of the Privy Lodgings, on the site of the old Privy Garden. It had been laid out by Elizabeth's father, Henry VIII, as a sort of sacred theatre in the new classical style. It consisted of a square, cobbled courtyard, open to the skies, and surrounded by a two-storey loggia. The loggia, also open, was supported by columns below and had a balustrade on the balcony above. In the centre of the courtyard was a pillared and canopied tribune or pulpit. The preacher stood in the pulpit; ordinary folk gathered round to listen in the open-air courtyard, while privileged courtiers occupied the upper storey of the loggia. Elizabeth herself, like her father and brother before her, sat in the window of the first-floor council chamber, which projected into the courtyard like a ready-made holyday closet.

The great advantage of the Preaching Place was its size: its capacity was four times as big as the Chapel Royal. This made it comparable to St Paul's Cross, the great open-air preaching place in the City to the north of St Paul's. Indeed, it became an alternative to it. The sixteenth century had an insatiable appetite for sermons and when the court preachers were stars (like Latimer under Edward VI or the Lenten preachers under Elizabeth), Londoners flocked to hear them. The journey was only a couple of miles on foot or by boat and there were few problems about people gaining entry when they arrived as access to the outer courts of the palace was fairly open.[4]

One of the faces in this crowd of Londoners was Henry Machyn. Machyn seems to have been an undertaker and supplier of the heraldic materials, such as the hatchments or great painted coats of arms which were carried at funerals and hung up in the church afterwards. He lived in the City, in the parish of Holy Trinity the Less, off Thames Street. Close by the parish church in Trinity Lane stood the livery hall of the Painter-Stainers company, whose members painted and decorated many of the objects supplied by Machyn. Finally, Machyn was a diarist, and his diary-cum-chronicle is a unique record of events in the capital in the middle years of the sixteenth century.[5]

As usual, Machyn was assiduous in his attendance at sermons in Lent 1560. Often he heard two sermons a day. In the morning, he walked the few hundred yards north-west to St Paul's Cross. Then, probably, he returned home for dinner, walking along Old Change and Old Fish Street. In the afternoon, he went west to be present at the Preaching Place at court. Perhaps he took a wherry, or light rowing-boat used as a water-taxi, at the landing stage at Queenhithe, which lay close at hand, just to the south of his house. To judge by his frequent references to the crowds, he would have had to queue for a long time for a boat.[6]

Machyn occasionally notes the topic of the sermon. More frequently, there is a general comment, such as 'a notable sermon'. But, in fact, he came to look at what the preachers wore as much as to listen to what they had to say.

And the story, almost always, was the same: on 3 March 1560, Grindal, 'the new bishop of London', preached at St Paul's in 'his rochet and chimere'; the same day, in the afternoon, 'Bishop' John Scory preached at court, similarly dressed. Scory also wore the same attire for his sermon at St Paul's Cross on the 10th, as did Jewel, 'the new bishop of Salisbury', when he preached at court on the 17th. A 'rochet' is a long, loose gown of white linen; a 'chimere', a sleeveless robe of silk worn over the rochet. Together they formed the ordinary attire of a pre-Reformation Roman Catholic bishop.[7]

What a difference a year makes! A year previously, in Lent 1559, the Elizabethan religious settlement was still being fought every inch of the way through the House of Lords. Grindal, Scory, Jewel and the rest were returned exiles, thundering from St Paul's Cross and the Preaching Place against the Catholic Church and the Catholic bishops. Their prayers had been answered: the old Church had been overthrown and a new erected. Then Elizabeth invited the exiles to become the bishops of the new Church. But there was a condition: they had to wear the robes of the old.

This was Elizabeth's personal decision. It was also her personal decision that the 'new bishops', as Machyn called them, were paraded in these old robes in the Lenten pulpits of 1560. In them, they were victims arrayed for the sacrifice – the sacrifice of their convictions on the altar of Elizabeth's compromising religious polity. The hotter sort of Protestants were scandalized at the transformation: this was the dress which Bonner had worn when he had interrogated and burned their brethren. But the ordinary, conservative middle-of-the-roader was reassured: if the new bishops at least looked like bishops, perhaps they would behave like bishops, despite their years of exile in Strasbourg or Geneva. As for Elizabeth, we know nothing of her thoughts as she gazed down from the casement of the Council Chamber on the vestmented figure and the attentive crowd. But we can imagine a smile of satisfaction at a battle won.

The court sermons finished in the late afternoon, at about 5 o'clock. On 24 March, the fourth Sunday in Lent, the preacher was William Barlow. In exile, he had been in the household of Catherine Brandon, the radical Duchess of Suffolk. On her return to England in 1559 she had written to Cecil to denounce the signs of temporizing and compromise in the emerging Elizabethan settlement. 'Christ's plain coat without seam is fairer than all the joggs [that is, elaborately cut or slashed clothes] of Germany,' she insisted. Barlow, as Catherine's former Household Chaplain, was certainly in sympathy with her views. But now, as Elizabeth's Bishop of St David's, he too preached in the rochet and chimere. These were rags of Rome and so more deplorable than the 'joggs of Germany'. But yet worse followed.

Immediately after Barlow's sermon, the diarist Machyn noted, the Queen's Chapel 'went to evensong'. And clearly Machyn had gone with them, crowding into the Chapel and noting what he saw. 'There the cross stood on the altar and two candlesticks and two goodly tapers burning and after done a goodly anthem sung.' The 'altar' is, almost certainly, what Machyn had described more precisely on the previous 6 March as a communion 'table . . . standing altar-wise' – that is, at the east end of the chapel.[8]

The other 'scenic apparatus' can, however, be documented precisely within the inventories of Elizabeth's jewels and plate. In this same year, 1560, the Queen embarked on a revaluation and recoinage of sterling, to restore the value of money in circulation. Though it turned out to be both successful and profitable, it was a risky enterprise and it was decided to melt down a substantial quantity of old and surplus plate from the royal jewel house to underwrite it. One of the principal categories of discarded plate was the elaborate church ornaments of Elizabeth's sister, the pious and orthodoxly ostentatious Mary. Some 8,000 ounces of crosses, monstrances, cruets, pyxes, censers, incense-boats and sacring bells were defaced, stripped of their jewels and delivered to the Mint.

After this holocaust, only some forty articles of 'church plate' were left in the jewel house by the time the next inventory was drawn up in 1574. They included three altar crosses, two pairs of altar candlesticks, a Bible and an English Prayer Book, in massive, matching silver-gilt covers, a silver-gilt font and a miscellany of Catholic liturgical ornaments, including mitres and incense-boats. Other depredations in the course of the reign and a second massive purge in 1600 further reduced these forty items to a handful. But a few key objects survived. A series of notes in Cecil's hand tells us where they were in 1595. The 'two great gilt candlesticks with pricks [that is, spikes to hold the candles] well chased for an altar' and weighing 240 ounces between them were 'in the vestry' at Whitehall; while the smaller pair of similar candlesticks weighing 53 ounces were 'in the closet'. The Bible and Prayer Book with gilt covers were also 'in the vestry at the court'. But the altar crosses were noted as having been 'issued to the mint'.

All this, of course, relates to a period long after Machyn's visit to the Chapel Royal in 1560. But, in view of the extraordinary conservatism of court practice, it is pretty safe to assume that the pair of great gilt altar candlesticks in the vestry in 1595 were the ones that Machyn had seen on the communion-table placed altar-wise thirty-five years before. In the interim, it seems clear that Cecil had managed to get rid of the altar crosses and have them melted down (just to make sure, no doubt, that Elizabeth did not change her mind). Their pride of place on the communion-table had been taken, almost certainly, by the massive gilt-covered English Bible, which Archbishop Laud noted as 'usually standing on the altar at Whitehall'.[9]

In comparison with the lavishly garnished altars of Henry VIII, in which the stone had nearly crumbled under the weight of clustering solid-gold images, crosses and candlesticks, Elizabeth's pair of candlesticks with first a cross and then a gilt Bible were austerity itself. But they still *looked* like an altar. And that was the point – whether for those, like the Puritans, who hated altars as idolatrous, or for those, like Elizabeth and Machyn, who loved them as symbols of the continuity of belief and practice.

Elizabeth, Machyn had noted, had not been present at Barlow's sermon, nor does he comment on her attendance at the following even-song. But a couple of weeks later she personally starred in the Maundy ceremonies which were lovingly described by Machyn.[10]

In the later Middle Ages it was customary for monarchs and other magnates, both male and female, to imitate Christ's example of humility when He washed the feet of His disciples. The ceremony took place on the Thursday before Good Friday and its name came from Christ's address to his disciples which had followed the washing of their feet: 'A new commandment [*mandement* in old French] I give unto you, that ye love one another'. It was the sort of superstitious ritual which rigorous Protestants were keen to abolish; instead, in this most consciously tradi-tional of Lents, Elizabeth went out of her way to follow it to the letter.

On the morning of Thursday, 11 April the Queen entered the hall at Whitehall where some poor women were lined up. She washed their feet and drank to them one by one, using a new silver cup on each occasion. Each woman was then given the cup and a gown (the luckiest getting the Queen's own best gown) and a gift of money. In the afternoon, the Queen made smaller doles to a much larger number of people in St James's Park: there were over a thousand poor men, women and children, both lame and whole of limb, each of whom received twopence. Elizabeth strayed from her text in only one particular. According to Machyn, there were twenty women in the Maundy. If Elizabeth had followed tradition and washed the feet of as many poor women as she was years old, there would have been twenty-five. Had the Queen already got in the habit of lying about her age?

A fortnight later Elizabeth had another opportunity to observe traditional ceremonial – or not. 23 April was St George's Day. St George was the patron saint, not only of England, but of England's ancient order of chivalry, the Garter. To mark the feast of their patron saint, the Garter statutes required the companions of the order to process to Chapel to take part in a special service in his honour. The liturgy of the service also contained prayers for the repose of the soul of founder of the order, Edward III, and the previous knights. Every line of this became contentious during the Reformation. Sincere Protestants objected to saints and saints' days, to praying for the dead, even to processions. Accordingly, under Elizabeth's hotly Protestant brother Edward VI, the Garter statutes had been purged. Under Mary, they had been restored. What would Elizabeth do?

Machyn, as usual, made sure that he was there as an eager witness. On St George's Day 1560, his diary notes, 'the queen's grace and the knights of the Garter went a procession' at Whitehall. They marched round the hall, out to the courtyard and so, presumably, to the chapel. The sovereign and knights were escorted by two key groups of ceremonialists, the Heralds and the Chapel. In the circumstances of 1560, it was the Chapel who attracted Machyn's particular attention.

Like the war reporters of a later day, he counted them in and he counted them out: there were present, he notes excitedly, 'all her chapel in copes of cloth of gold, a twenty-eight copes'.[11]

The cope, a great semicircle of rich fabric slung round the shoulders like a cloak, was, after the mitre, the most triumphalist item of clerical dress. This is why so many had perished in the flames of the St Bartholomew's eve bonfires in 1559. But Elizabeth, on the contrary, had decided that her clergy should wear them. The twenty-eight gold-robed figures in the St George's Day procession were so many gleaming affirmations of her will.

Machyn, though he saw Elizabeth frequently and at close quarters, never seems to have met her. Nevertheless, he is an important character in her life – as important, indeed, as most of her supposed lovers. Because Elizabeth *was* in love with him. She loved the people, and Machyn was the man in the street to whom she played and for whom so much of her policy was designed.

He was not of course the *Protestant* man in the street who had cheered Elizabeth as his Deborah in the eve-of-coronation procession and cheered even louder with the establishment of the new Church. Instead, Machyn belonged to the more conservative, more moderate majority. Wriothesley's *Chronicle* shows how close Elizabeth had come to losing such men in 1559. Machyn's diary shows how and with what success she won them back in 1560.

The key is perception. 'Wolves coming from Geneva' was how Bishop White had described the returning Protestant clergy. It was a nice phrase, turning them into bogymen who were guaranteed to put the fear of the Calvinist devil into good Catholic men and women. On the other hand, Machyn himself knew very well what Genevan meant. Congregational singing, with men and women joining in together, was 'Geneva fashion'; psalms sung in metre were 'Geneva ways'. And wholly Genevan was the service at St Martin's, Ludgate, when Jean Veron, a French pastor, was admitted as minister there on 19 March 1560. Machyn loathed Genevan ways and had a particular animus against

Veron. This got him into serious trouble in 1561 when he had to perform public penance at St Paul's for repeating a rumour that Veron had been 'taken with a wench'. Machyn knelt to Veron and Grindal, as Bishop of London, while all his important friends interceded for him. 'Yet they would not forgive him', Machyn noted bitterly in his customary third person. It was all very unchristian.[12]

So Machyn had sensitized antennae: he could sense 'Geneva ways' at a hundred yards. Yet he encountered nothing of them in Elizabeth's own worship nor in most official acts of her Church. There was nothing of Geneva in the rochet and chimere; nothing in the altar with its cross and candlesticks; and nothing at all in the copes.

It is now clear why Elizabeth herself was so hostile to 'Genevan ways': she had to keep subjects like Machyn happy. And if this hostility to advanced reform involved trampling on Protestant toes, so be it. Protestants had nowhere else to turn.

The most illustrious victim of this approach turned out to be the unforgiving Grindal himself. Elizabeth made him Archbishop of Canterbury in 1575. But he served for only a year before the Queen suspended him for encouraging the Puritan practice of 'prophesying' or clerical self-education groups. With Grindal's fall, Machyn had his posthumous revenge for his humiliation in 1561. He was also vindicated in his judgement that Elizabeth was 'one of us'.[13]

But Elizabeth's attitude to religious ceremony was not mere expediency, nor was it merely to do with religion. Machyn was professionally interested in ceremony as an undertaker and heraldic supplier. Elizabeth was professionally interested in ceremony, too. Her *métier* was queen and queenship had ceremony as its essence.

For some time now historians have understood the structures and organization of the Tudor court. But only recently has its ceremonial been similarly anatomized by Dr Fiona Kisby. She has shown that Tudor royal ceremonial was *intrinsically* religious. The main days of ceremony at court were the great feasts of the Church. On these feast days, the monarch wore special robes; processed to chapel; and performed specific rituals there. To abolish Church ceremony, as the hotter sort of

Protestants wanted, was to eliminate royal ceremony as well – or at least to reduce it to a shadow of its former self. And Elizabeth was nobody's shadow. It was her successor, James I, who wrote: 'No bishop, no king.' But Elizabeth anticipated the doctrine and practised it just as fervently.[14]

CHAPTER 43

The Limits of Religious Reform: Persons

When Elizabeth's sister Mary had changed England's religion, she had burned people rather than church ornaments. In contrast, Elizabeth's handling of the immediate personal consequences of the supremacy is the greatest testament to her moderation and, dare it be said, to her fundamental humanity.

The penal clauses of the act of supremacy came into force thirty days after the end of the parliament, that is, on 8 June 1559. On that day, all holders of office in Church and State (except for peers who were exempt by their rank) were required to swear the oath of supremacy, recognizing Elizabeth as Supreme Governor of the Church. The penalty for refusing to swear was deprivation of office – and potentially worse, if the refusal of the oath were persistent.[1]

In the case of the hated Bishop Bonner of London, enforcing the penalty clauses was a pleasure as well as a duty and the government jumped the gun. He was tendered the oath of supremacy on 29 May and, on his refusal, was immediately deprived of his bishopric and imprisoned in the Marshalsea. He never left it. Nor was the refusal to swear the oath and consequent deprivation of the two other Catholic hard-liners, White of Winchester and Watson of Lincoln, either unexpected or unwelcome.[2]

But, for the rest of Mary's bishops and other leading clergy, Elizabeth had high hopes that her carefully balanced settlement might win converts. In the event, only one, Kitchen of Llandaff, was persuaded. The rest stuck to their faith and patiently submitted to the inevitable deprivation. The real surprise from Elizabeth's point of view was the behaviour of Heath of York and Thirlby of Ely. Both were long-serving royal councillors and administrators and Elizabeth assumed that they would put their loyalty to the Crown above their conscience. When they did the opposite, she deprived them only with profound reluctance.[3]

Even more delicate treatment was accorded to the father of the episcopal bench, Tunstal of Durham. He had been excused attendance at parliament on grounds of his age. But, when the nature of the settlement became clear, he decided to end his isolation himself and asked for permission to come to court to remonstrate with Elizabeth. The permission was granted and on 20 July Tunstal arrived in London accompanied, as befitted his princely status, with sixty horsemen. His audience with Elizabeth took place about a fortnight later. Probably the Queen thought that she could play on memories of his half-century of service to her family to win him over. Instead, Tunstal almost turned the tables.[4]

Tunstal had been Henry VIII's theological adviser during the crucial summer progress of 1538 when the King, who had toyed with Lutheranism in the wake of the Boleyn marriage, had turned decisively against it. We know Tunstal and Henry wrote joint papers and it seems that Tunstal had kept one. For he now confronted Elizabeth with evidence in the King's own hand that her father had rejected the views of the Eucharist which she had just legislated. Dramatically, according to Quadra, the new Spanish ambassador, Tunstal exhorted the Queen 'at least to respect the will of her father, if she did not conform to the decrees of the church'. They also discussed the deprived bishops. 'I am grieved for York and Ely,' Elizabeth remarked. Brutally, Tunstal pointed out the obvious: 'How can you grieve when you have the remedy in your own hands? If you are willing to be a Catholic you can have not only those two among your councillors, but many others beside.'

Once upon a time, Elizabeth had addressed the author of the paper Tunstal thrust in her hands as 'mighty and most merciful father' and his every word had been her law. But that was a long time ago. Henry was dead; the writing was old; she was now Queen. Nevertheless, it was not a happy experience for Elizabeth, and, though Tunstal attended at court in expectation of a second audience, none seems to have been granted.

The open clash, which neither Elizabeth nor Tunstal wanted, was now inevitable. On 19 August, Tunstal wrote a letter in duplicate, sending one copy to Cecil and the other to Parry. Its tone was unmistakable. He could not, he told the two key figures of the new régime, consent to the enforcement of the new religious settlement in his diocese, 'because I cannot myself agree to be a sacramentary [that is, to reject the real presence], nor to have any new doctrine taught in my diocese'. He acted thus, he insisted, not 'for any frowardness, malice or contempt, but only because my *conscience* will not suffer me'. His conscience also prevented him from taking part, as required, in the projected consecration of Parker as Archbishop of Canterbury at the beginning of September and from swearing the oath of supremacy when it was tendered to him late in the month. He was deprived and, pushed beyond the limits of his temper as well as his conscience, he seems to have told the commissioners what he thought of them.

For this 'disordered speech' he was committed to Archbishop Parker's custody at Lambeth. There he was treated with every courtesy. Even now strenuous attempts were made to bring him round. Elizabeth herself was kept personally informed. At first Parker was hopeful, but finally he had to admit defeat. Cecil transmitted Elizabeth's personal regrets: 'for the recovery of such a man would have furthered the common affairs of this realm very much'. The Queen continued to show Tunstal personal kindness and to tempt him with the offer of a pension, scaled according to his conformity. To no avail. He died a few weeks later, on 18 November.[5]

It is possible, though I think wrong, to write off Elizabeth's expressions of regard for Tunstal as self-interested blarney, spun in order to win the old man round with fair words. The same cannot be said for the funeral sermon, preached by Alexander Nowell, Dean of St Paul's. Tunstal was dead and there was nothing more to be gained from him. But that did not stop Nowell, a Protestant, from eulogizing Tunstal, a Catholic, in a sermon which took as its text 'Blessed are they which die in the Lord'. 'Such force hath virtue,' commented the recording chronicler, who was also a Protestant, 'that we ought to commend it even in our enemies.'[6]

Were Tunstal and his surviving colleagues even enemies? They were opponents, certainly. And, because they were able, devout and motivated only by concerns of conscience, their continued resistance to the new religious settlement made them dangerous opponents. Too dangerous any longer to be allowed their liberty. The blow fell swiftly. In a few days, in late May and early June 1560, Mary's remaining bishops and other leading clergy were sent to the Tower. The following April, two of Mary's hard-line Catholic lay councillors were also arrested: Sir Edward Waldegrave went to the Tower, together with his wife, while Lord Hastings of Loughborough, protected by his noble status, was handed over to the custody of the Earl of Pembroke. Waldegrave died shortly afterwards, probably of the plague, and Hastings, alone of his fellows, kowtowed and conformed. The rest, however, never recovered their freedom.[7]

But the public necessity which imposed this penalty did not prevent private courtesies, and even something like friendship. Heath was quickly released from the Tower into a lightly guarded retirement on his own property at Chobham in Surrey. The arrangement was conditional on his absolute withdrawal from public life. He gave an undertaking 'not to interrupt the laws of Church or State or to intermeddle with the affairs of the realm' and he stuck by it until his death, at the age of almost eighty, in 1578. Boxall, on the other hand, remained formally in the 'free custody' of Archbishop Parker. Mostly, he lived in the Archbishop's agreeable Thames-side palace at Lambeth. But from time to time he was allowed to enjoy the rural retreats of Parker's Kentish manors at Bromley and Bekesbourne. In 1569 he wrote a touching letter to Cecil, asking permission to visit his sick and aged mother. Two years later, seriously ill himself, he was released to die in the house of a relative in London. After his death he was eulogized, like Tunstal, by his opponents as 'a person of great modesty and knowledge'.[8]

None of this is surprising. Boxall and Heath had been the members of the Marian hierarchy most amenable to Elizabeth's government and they enjoyed excellent personal relations both with the Queen's leading councillors and with the Queen herself. But even the more obviously hard-line of Mary's bishops were well-treated. They did not always repay like with like. Ex-Bishop Scot of Chester broke parole to flee to Louvain; while Watson, formerly Bishop of Lincoln, was accused of breaching his agreement to withdraw from public life by giving spiritual succour to fellow-Catholics. Eventually, along with other irreconcilables, he was banished to imprisonment in the lonely fastness of Wisbech Castle, isolated between the Wash and the Fens. Yet still there were many who mourned his loss, for Watson had been a leading light of that brightest Cambridge generation at St John's in the 1530s. This was the nursery of Elizabeth's own intellectual world and her tutor, Roger Ascham, paid warm tribute to Watson's classical learning while deploring his religious beliefs.[9]

The story was repeated scores, even hundreds, of times, as divisions of religion cut across ties of family, of friendship, of neighbourliness and of intellectual sympathy. The same, of course, had happened under Mary and Bishop Thirlby had wept bitterly when he took part in Cranmer's degradation from holy orders, which was the prelude to his burning. Cranmer was his lifelong friend and colleague and, Thirlby protested through his tears, nothing would have impelled him to act as he did, save the Queen's express commandment. Thirlby himself was more fortunate than Cranmer when, in his turn, he found himself on the losing side and he spent most of the 1560s in Archbishop Parker's gentle custody until his death in 1570.[10]

The two friends, Thirlby and Cranmer, had committed the same offence, of rejecting the religion by law established. But they suffered wildly different penalties: the one enjoyed an easy detention; the other suffered the agonies of the fire. The difference lies in one thing: the attitude of the Queen at the time. It was Elizabeth who saved Thirlby and Mary who condemned Cranmer.

Here we encounter the curious reluctance of historians to blame Mary for the Marian persecutions. Instead, some of them have sought to pin responsibility on one or other of her councillors and advisers: Gardiner, her Spanish friars, or Cardinal Pole. But in no case does the charge really stick: even Bishop Bonner, the favourite scapegoat of the

Protestants of the time, was often accused of slacking. And if you can't blame Bonner for the burnings, who can you blame? The answer, with the *mauvaise foi* too characteristic of historians, has been to put abstractions in the dock: habit, for beginning the burnings as that was the usual English way of dealing with heresy; and inertia, for continuing them as no one could think of what else to do.

Perhaps. But abstractions do not light a fire; nor, more importantly, do they sign the warrant which authorizes the lighting. Only one person could do that: the Queen. Only one person, too, bore the final responsibility for the policy that lay behind the burnings and that also was the Queen.

Contemporaries, of course, made no bones about it. Thirlby is clear that Mary is responsible for Cranmer's degradation and death and for his own role in it. But, above all, Mary condemns herself out of her own mouth. She boasts of her mission to rid England of heresy and it is clear that she willed the means as well as the end. How, in a profoundly pious and conventional woman, could it have been otherwise?

It is possible, even proper, to claim that, in the circumstances of the sixteenth century, the burnings were not a crime. It is even possible to argue, as some recent Catholic historians have done, that they were not a blunder. Elizabeth, however, thought that they were both.

She expressed her views through the careful third person of Bacon's speech at the opening of her first parliament. Their new sovereign, he said, 'is not, nor never meaneth to be, so wedded to her own will and fantasy that for the satisfaction thereof she will do anything . . . to bring any bondage or servitude to her people, or give any just occasion to them of any inward grudge whereby any tumults or stirs might arise as hath done of late days'.[11]

By the time he uttered these words, Bacon had been speaking for at least half an hour and more than one noble head was nodding. But these phrases shook into life all but the most drowsy. For the Lord Keeper was describing the '*merum motus*' – the 'mere will and motion' of the monarch that authorized all royal grants and was the mainspring of Tudor government – as, potentially, a private aberration. That was political heresy. And he was blaming rebellion, the supreme crime in the Tudor political lexicon, not on the unnatural disobedience of the subject, but on the folly and arrogance of the sovereign. This was more than heresy, it was apostasy. Yet the Queen listened approvingly. She probably went further. This was the mission statement of her government and her reign. So she nodded and smiled and lent a pretty feminine emphasis to Bacon's clumping masculine prose.

The times were indeed changing.

CHAPTER 44

Promise Fulfilled

It is easy to promise change and Bacon's speech in 1559 promised much. But it is another matter to deliver, and the listening lords and commons can be forgiven a measure of scepticism. For they had heard it all before. Mary had begun her reign by promising to seek the middle ground and to force no one's conscience. And in the Guildhall she had pledged her royal word that she would marry only with the consent of the realm. She had then, for the highest motives no doubt, jointly and severally broken these and every other assurance and promise to her people.

Like Mary, Elizabeth had begun well. But would she be any better in the long run?

At first sight the signs were not all that good. I have already talked about the doctrine of the Queen's two bodies. More important, however, from the point of view of practical government, was the distinction between the Queen's two wills: her private will and her public will. Her private will was what she actually wanted to do. Her public will was what, after taking due counsel and advice, she ought to do. Elizabeth had promised to respect this distinction, not only via Bacon's parliamentary speech but also personally in the more intimate atmosphere of her first formal council meeting when she had enjoined Cecil 'that, without respect of my private will, you will give me that counsel that you think best'.

Implicit, of course, in Elizabeth's injunction to Cecil is her promise that she would submit her private to her public will. But doing what we ought rather than what we want comes easily to none of us. It is particularly difficult for a young, high-spirited woman who happens to be Queen and happens, above all, to inherit the over-weaning self-confidence and egotism of her house.

For there is no doubt that Feria read her character correctly when he reported that 'she is a very vain and clever woman'.

What saved her and England, however, is the form that her vanity took. It was indeed colossal. But it tended to exercise itself in petty things. Elizabeth was vain about her looks, her clothes, her abnormally long thin fingers, her Latin grammar, her dancing, her keyboard playing (where, no doubt, her finger-span made octaves easy), her defiance of time and her supposedly changeless youthful appearance. All these made enormous demands on the ingenuity, the patience and, increasingly, the credulity of her courtiers and ladies-in-waiting. But on the world outside the court and the privy lodgings they made almost no impact at all.

It had been the opposite with Mary. Mary was singularly free from personal vanity and she had the honesty to worry about the effect of her unattractiveness on Philip. But, modest in little things, she was vain in big ones. Above all, she suffered from the sin of spiritual pride and when her priests and propagandists told her that she was the second Mary, Virgin and Queen, she believed them with a simple, devastating literalness. Like her namesake, she was humble yet exalted and even before her false pregnancy she swelled with a God-given certainty. God had chosen her to be His instrument to restore England to the faith; God had chosen Philip, like a latter-day Joseph, to be her spouse; God had made her pregnant, let no man think otherwise. Armed with these invincible certainties, this otherwise modest and merciful woman found the strength to brush aside her parliament, bully her council, and burn and brutalize dissenters. But, in Mary's case, out of the strong came forth, not sweetness, but the bitter harvest of repeated rebellion, political uncertainty, state bankruptcy and the final humiliation of the loss of Calais. Her only achievement was the Catholic restoration and even that, it began to appear, had been bought at too high a price.

These were awful lessons. Moreover, Elizabeth learned them, not as a detached observer but as a victim of her sister's policies and personal hatred. The Tower made a good classroom and Lady Jane Grey's scaffold an excellent text. Elizabeth never forgot these experiences nor the lessons she drew from them. And she made her memories into frequently cited maxims – of what she would not do herself and of what she would not do to others.

And, on the whole, not being Mary or behaving like her proved to be a golden rule for Elizabeth.

Elizabeth's first promise to her people was to take counsel. But counsel – advice – required council – a body of advisers. This made the choice of a council, always important, crucial to the success of Elizabeth's government. Here, as we have seen, she behaved utterly differently from Mary. Mary acquired her council as a rolling stone gathers dross. The result was a council where councillors distrusted each other and where, even worse, the Queen disliked and distrusted many of her advisers. Elizabeth, in contrast, consulted widely on the choice of councillors and then acted decisively. Within forty-eight hours of her accession she had a council, whose size, shape and personnel changed astonishingly little in the four and a half decades of her reign. The council was not homogenous, but it was reasonably harmonious and it was characterized by a profound degree of mutual trust between the Queen and at least the inner ring of her councillors.

All this made the council very powerful. Its members acquired a sense of corporate identity; they began to wear a ceremonial dress of a black silk gown laced with gold, like the modern Speaker or Lord Chancellor; and they behaved with a commensurate self-importance – like 'so many kings', as a ruffled ambassador noted. Indeed it is possible to argue that they *were* a sort of collective king, with Elizabeth, not so much as a queen regnant, as a mere consort of her council. Variations of this argument have been advanced by distinguished modern historians, who have written of the 'royal republic of Elizabethan England', and by Elizabeth's contemporaries, especially misogynist Protestants, such as Sir Thomas Smith. There is a grain of truth in it, as the boots of traditional male monarchy were too big to be quite filled even by the most forceful and ablest of females. But it is only a grain. And it is striking that the misogynist theoreticians never got very far. They might sneer at Elizabeth's feminine faults and foibles; in return, she broke or blocked their careers.

Here, however, Elizabeth broke no promise. She had said that she would rule by counsel; not that she would be ruled by her councillors. It is an important distinction and one which she enforced ruthlessly. In doing so, she used her best and her worst qualities. The latter came in particularly useful. The love-games she required of her courtiers and councillors throughout her reign, and the full-blown absurdities of the Gloriana cult in its later years, were all means of forcing a masculine élite to pay tribute to a woman. And curiously, only the fact that she was a woman made it tolerable. When a male sovereign required to be worshipped as a deity, like Richard II or, in a somewhat different fashion, Charles I, the political nation revolted. But with a woman such foibles could be tolerated since they resembled the rituals of courtship.

Courtship, of course, was usually followed by the consummation and role-reversal of marriage. Since Elizabeth remained unmarried, mankind never enjoyed this revenge on her. Instead, when grandees found their dignity or their persons injured by blows from her tongue or fist, they consoled themselves by shrugging and blaming it all on the follies of that inferior thing, a woman. At least most of them did.

Elizabeth could also deploy a more intellectual armoury. Her council was often divided on important questions; and there were other sources of counsel – in parliament, or the peerage, or, as it recovered confidence, among the bishops of the new Anglican Church. All these provided ample opportunities for divide and rule. But her greatest room for manoeuvre was given by her second promise: never to lose the love of her people and never to provoke them to rebellion. Or, in a word, to be *popular*. Her direct hold on popular affection was something that Elizabeth guarded as jealously as the sovereignty itself. Indeed, for her it was *the* key to sovereignty. She told Feria this as a boast and she repeated it, via Bacon, as a promise. But, on certain issues, the counsel of the élite and popular opinion were divergent or even directly opposed. In such circumstances, Elizabeth acted as the judge of last resort and, when it came to choosing between counsel and popularity, she invariably opted for the latter.

The tangled story of religion illustrates these points perfectly.

The Elizabethan Church, as we have seen, was a Goldilocks settlement: neither too hot nor too cold. As such, it pleased neither the orthodox Roman Catholics, for whom it went far too far, nor the hotter sort of Protestants, later known as Puritans, for whom it did not go nearly far enough. Indeed, among the élite, it probably pleased only Elizabeth.

At first sight, this looks like Elizabeth being as egotistical as her sister Mary and imposing, equally imperiously, the doctrine of *cuius regio eius religio*, which can roughly be translated as *the religion of the people is determined by that of the prince*.

In fact, the resemblance is purely formal. To begin with, Elizabeth paid much more attention to the reality of parliamentary consent. True, once she had it, she was immovable – despite the overwhelming advice of her council and clergy for further Protestant reformation. This can look like obstinacy or obscurantism and Elizabeth has been subject to heavy criticism, both at the time and subsequently, for 'failing to recognize the Protestant destiny of England'. She would have replied, had she deigned to, that her sister had recognized the Catholic destiny of England and look where that had got her. Instead Elizabeth, unlike the Protestant zealots of her own reign or the Catholic ones of her sister's, was acutely aware of the dangers of religious change. This, as well as personal preference, is why she insisted on a measure of religious continuity: in liturgy, in vestments, in the words of the consecration of the Eucharist. Elizabeth knew that there were many tens of thousands of Machyns – warmly attached to the old ways and vehement in their dislike of the new, 'Geneva ways'. These were her subjects too, and her promise not 'to give any just occasion . . . of any inward grudge' applied to them also. Her Protestant councillors might forget this; she did not.

This same caution, which was also a proper recognition of the strength of Catholicism, helps explain Elizabeth's approach to the enforcement of the settlement as well. Normally this is criticised for hesitation and inconsistency. Many areas of the country were more or less Catholic until the 1570s or later. Even then, Protestantism came only as a result of accident, when a royal progress, a political crisis or a new-broom bishop brought the unreformed state of the region to the attention of the government. No doubt, if the aim was an English New Jerusalem, this was deplorable. If, on the other hand, you wanted a quiet life, it made a lot of sense. Effectively, it meant that England was reformed by local demand: areas like London, that were enthusiastic for Protestantism, reformed quickly, but in areas like Lancashire, that were hostile, change was slow and cautious. And when circumstances forced change, the times were more propitious: the new religion was firmly established at a national level and had even put down some roots in the more stony regional soils.

The strange anomaly of the treatment of the nobility worked similarly to Elizabeth's advantage – as she had probably calculated. At first sight, it seems absurd that they, the Queen's most powerful and influential subjects, were alone exempt from swearing the oath of supremacy. Actually, the government had no choice about the concession. Only a handful of the peerage was Protestant; the rest varied from the unthinkingly conservative to the enthusiastically Catholic. Elizabeth's conversations with Feria show that she understood this. She therefore understood as well that, if the bill of supremacy had required the lords to swear the oath, the lords would have retaliated by joining the Catholic bishops in ditching the supremacy a second and decisive time. So Elizabeth, as she often did, decided to lose a battle in order to win a war. Nor did she really lose much. Her judgement was that her nobility, however much they preferred the Roman religion, remained loyal to England and their Queen. In the event, not many of them let her down. And, in any case, she lived long enough to see a new generation come along, most of whom rejected the religion of their fathers and accepted hers.

Her lenient treatment of the Roman Catholic hierarchy was based, in part, on a similar calculation about their patriotism. This proved less accurate. But Elizabeth, in defiance of a vociferous and quite under-standable Protestant demand for vengeance, persisted with her moderate policy. Only in the case of Bonner were steps taken to tender him the oath a second time and so set in motion the machinery of escalating penalties which might have led to his execution as a traitor. Londoners would have loved every slice of the knife. But they were denied their fun – and by Bonner, rather than Elizabeth. The act required that the oath be administered to Bonner by his diocesan bishop, Horne of Winchester. But Bonner, with that ferocious forensic skill, which, at least as much as his supposed blood-thirstiness, had made him so dangerous to accused Protestants, challenged the validity of Horne's consecration as Bishop. His challenge was well-founded and it needed an act of parliament to free the titles of the Elizabethan bishops from ambiguity. After he had threatened, so effectively, to pull down the pillars of the temple, it was felt safer to let Bonner die in peace. No doubt, the families and friends of his victims were disappointed but Elizabeth was relieved.

For her policy was founded on a careful combination of principle and expediency. After her own experiences under Mary, she was not, she insisted, in the business of forcing men's consciences. That alone made her reluctant to seek the death penalty. But she was also reluctant to make martyrs *per se*. In 1553, at the time of Mary's accession, Protestantism was utterly discredited – by the extremism of its clergy and by the ambition, greed and, as it turned out, the cowardice of its lay leaders such as Northumberland. Even Elizabeth had been revolted by Bishop Ridley's attack on her legitimacy. But, two years later, the steadfastness of Ridley and the rest in their martyrdom had restored the moral stature of their faith: 'We shall this day light such a candle by God's grace in England,' Latimer had told Ridley as they went to the stake together, 'as, I trust, shall never be put out.'

Elizabeth was determined not to light Roman candles to match the Protestant candles of her sister's reign. Much better to leave Bonner alive with all the resentments against him alive as well. For over a decade the policy worked remarkably well and, unfertilized by the blood of martyrs, Catholicism began its retreat into a semi-secret practice, to which a blind eye was turned if it were in private, though very public demonstration was dangerous.

The Queen's marriage was the other great issue of the first years of Elizabeth's reign. Here, once again, there is a superficial resemblance between her position and Mary's: both found themselves in a minority of one on the question. Mary had been determined to marry but her choice of Philip of Spain aroused almost universal opposition. Elizabeth, after her less than happy experience of relations with men – from Seymour's abusive gropings to the scheme for marriage with menaces to Emmanuel Philibert – seems to have been predisposed to the single life from the beginning. But everybody else wanted her to marry.

Mary had persevered, bullying her council, hoodwinking parliament and provoking a rebellion that threatened to burst into Whitehall itself less than six months after her coming to the throne. She survived and managed to effect her marriage. But at a terrible cost. For it soon became clear that Philip regarded both his wife and his English kingdom as mere pawns in the Habsburgs' great game. Mary had thought she was marrying a divinely-chosen spouse; instead, it turned out that, by pursuing 'her own will and fantasy', she had brought 'bondage and servitude to her people'. That was the charge levelled at her by Elizabeth/Bacon in the Lord Keeper's speech. The loss of Calais ensured that the verdict was 'guilty'.

Elizabeth's idiosyncratic attitude to marriage left her equally isolated and, so it appeared, just as vulnerable. But Elizabeth was saved, once again, by divided counsel. Her council, her parliament and her people all agreed that she must marry. But they could never agree on any one particular candidate: one man's meat was another's poison. Elizabeth probably realized this from the beginning. Only one other politician showed similar wisdom and that was Paget, who told Feria in November 1558 that 'there was no one she could marry outside the kingdom or within it'. Paget was not appointed to Elizabeth's council and he took his insight with him into retirement. But the situation remained objectively as he described it. And Elizabeth exploited the fact to see off with ease the multitude of attempts, public and private, to force her to marry.

It was less easy for her on the couple of occasions when her own commitment to the single life wavered. The first man she thought seriously of marrying was her childhood companion, fellow-prisoner in the Tower and Master of her Horse, Lord Robert Dudley. Dudley was strikingly similar to Seymour – in looks, physique and temperament. But whereas Seymour's seduction had involved the threat and perhaps the reality of force, Dudley's was all soft words and whirlwind charm. It was the more attractive for that. Did Elizabeth surrender and have sexual relations? She denied it absolutely – just as she had denied it with Seymour. On the other hand, powerful rumour accused her. Perhaps a Clintonesque formula will square the circle: Dudley had sex with her but she did not have it with him.

Perhaps. But the hysterical intervention of Catherine Ashley, once Elizabeth's governess and now, as chief gentlewoman of her privy chamber, her most intimate personal attendant, suggests that something serious was afoot. Catherine flung herself on her knees before Elizabeth and begged her to put a stop to the rumours about her behaviour with Dudley, which, she warned her, would occasion much bloodshed in the realm. Rather than that, Catherine protested, she wished that she had strangled her in her cradle. Strong words! Sir Nicholas Throckmorton, Elizabeth's old *éminence grise* and now her ambassador in France, wrote in scarcely more measured terms. What was said in Paris about Elizabeth's conduct, he informed Cecil, made him ashamed to be English. But it was Cecil's own intervention that was decisive. Regretfully but firmly he conveyed to Elizabeth that, should Dudley become consort, he would resign. Elizabeth was staring into the same abyss as had Mary.

But, unlike Mary, she had the wisdom to pull back. So, instead of entering into a marriage which would have torn apart the kingdom, she allowed her relationship with Dudley to evolve into a sentimental friendship that played a useful part in the elaborate dynamics of her court and council. Enough of a sexual charge remained to make the friendship interesting and, incidentally, to blight Dudley's own marital relations. But that was all.

Twenty years later, on the occasion of her second serious emotional brush with matrimony, it was that much harder. At the time of the Dudley affair, Elizabeth was still in her twenties. She was attractive and nubile and, when it was over, was able to console herself with the thought that, if not this time with this man, then at another with someone else. When, however, she tangled with the Duke of Anjou, youngest son of Catherine de Medici and Henri II of France, she was in her mid-forties. Mary had been much the same age when she had wooed Philip; now something of the same desperation entered into Elizabeth's behaviour.

It had all begun coolly and decorously as yet another diplomatic marriage proposal. But Anjou was the first of her foreign suitors to woo in person and Elizabeth found herself bowled over. She had been led to expect a deformed and disfigured degenerate; instead she found a slight, charming *joli laid*. He had none of the hearty, athletic manliness of her English 'lovers'. But he more than made up for it by a bold, yet insinuating, sexuality. Elizabeth, who had a taste for the exotic, was hooked and determined to turn her 'Frog', as she called him, into a king. But her people hated him. John Stubbs attacked the match in a savage pamphlet; oblivious for once to her popularity, Elizabeth had his right hand struck off. Worse, from her point of view, was the reaction of the council. They debated the marriage all day, and then, irretrievably split, refused to tender any opinion at all. Elizabeth raged against them for their inconsistency: for twenty years they had been begging her to marry; now, she fumed, when she presented a suitable candidate, they backed off. Once more, Elizabeth stood on the brink. In similar circumstances, Mary had pressed forward and defied both her people and her council. Instead, Elizabeth recollected herself. She wept tears, first of anger and then of self-pity, and sent her Frog back to France, loaded with gifts and fair words.

But it had been a close-run thing. Perhaps, after twenty odd years, the memory of her sister Mary was fading. If so, Elizabeth could refresh it by looking at the example of the other Mary – Mary, Queen of Scots, her cousin, prisoner and would-be successor. For Mary Stuart replicated Mary Tudor's faults, but on a larger scale and for smaller motives. After the premature death of her husband, Francis II of France, she had returned reluctantly to Scotland as a youthful widow. There she had been approached by a stream of royal suitors, most of whom had already been rejected by Elizabeth. But their wooing was never very ardent: Scotland, in comparison with England, was insignificant and Mary, as a widow, was shop-soiled goods. Not did she do herself any good by her vacillations over religion.

By 1565 she had tired of her widowed state and she married, after a whirlwind romance, her cousin Henry Darnley. Both Mary and Darnley shared a common grandmother, Margaret Tudor, eldest daughter of Henry VII. Mary descended from Margaret's first marriage to James IV of Scotland; Darnley from her second to the Earl of Arran. Darnley had been brought up in England and it is a mystery why Elizabeth allowed him to go to Scotland. Probably it was due to the malign influence of Dudley, who feared that otherwise Elizabeth would send him there himself as the unwilling husband of Mary.

Mary's marriage to Darnley forestalled any such eventuality; it also led to comprehensive disaster – for Scotland, England, Elizabeth and, above all, for Mary herself. Mary and Darnley united between them the blood-royal of three kingdoms: England, France and Scotland. Yet they behaved like a pair of ill-conditioned teenage newly-weds as the passionate love-making of their wooing degenerated, within a few months of the marriage, into the vicious bickering of two powerful but childish egos. Darnley roistered with his cronies; Mary sulked and simpered with her favourites, principal of whom was her secretary, David Rizzio. Late one night at Holyrood, as Rizzio sang and talked with the heavily pregnant Queen, Darnley's friends burst into the room, dragged the secretary from the Queen's arms and stabbed him to death in the next room. They left Darnley's dagger stuck in the dead man's belly.

Despite this terrible shock, Mary, who was tall and immensely strong, carried the child to term and on 19 June 1566 was delivered of a son, James. Elizabeth is supposed to have exclaimed bitterly: 'The Queen of Scots is lighted of a fair son but I am barren stock.'

But the envious fit cannot have lasted long. A few months later, Mary had her revenge on Darnley. Kirk o' Field, the house where he was staying, was blown up and his body, which had been strangled, was found in the orchard. The principal suspect was James Hepburn, Earl of

Bothwell, who had become close to Mary. He soon became closer still as, in quick succession, he abducted, raped and married her. Elizabeth's letter to Mary after she heard the news veers between indignation at her indecent behaviour and incredulity at her folly. The Scots reacted by forcing her to abdicate in favour of her infant son in July 1567 and imprisoning her at Lochleven Castle. The following May she escaped and rallied much of Scotland to her cause. But she was outgeneraled by her opponents; panicked; and fled to England.

Elizabeth, in one of her frequent forays into verse-writing, described Mary as 'the daughter of debate that eke [also] discord doth sow'. Scotland was now rid of her divisive presence; England and Elizabeth had to cope with it. Neither was to be the same again.

Hitherto Mary's actions had rather been to Elizabeth's advantage. In Scotland, Mary's follies had encouraged the development of a Protestant, pro-English party, which, with Mary's abdication, took over power. And, in the wider European context, they kept Spain benevolently disposed to England. This is because Mary embodied Philip of Spain's worst nightmare. Thanks to her Tudor blood, she was Elizabeth's most plausible heir. But she was also French by upbringing and inclination and, should she succeed, she would deliver England into the hands of her French cousins and in-laws. This would have created an Anglo-French alliance that would straddle the Channel and cut metropolitan Spain off from the Spanish Netherlands.

In 1553 this same calculation had led Philip to marry Mary and become King of England. In the decade following Elizabeth's accession it led Philip first to offer Elizabeth his hand, and, when that was politely refused, to continue to extend the hand of friendship. This he did despite considerable apparent provocation. In particular, the Catholic King developed a strange blindness about Elizabeth's Protestantism. He blocked any kind of European crusade against Protestant England; while in Rome his agents were successful in preventing the Pope from even issuing formal censures against the heretical Queen.

The fact that Rome remained silent had major domestic reper-cussions as well. It left English Catholics directionless and leaderless in their response to Elizabeth's new Church. This, as we have seen, was designed to be attractive to the less doctrinaire of them. In the absence of any firm doctrinal ruling against it, many of them were duly seduced. Elizabeth's policy of conversion by stealth and kindness rather than compulsion seemed well on the way to success.

But Mary's arrival in England as Elizabeth's guest-cum-prisoner threw all this into reverse. For it recreated the problem which had bedevilled English politics in the 1540s and 1550s: that of the successor. Edward VI's Protestant government had been destabilized by the fact that Mary, his designated successor, was Catholic and was expected to reverse his policies. Mary's Catholic government in turn had been undermined by Elizabeth's own Protestantism and her known determination to roll back the Catholic tide which her sister had set in motion.

For the first ten years of her reign, Elizabeth enjoyed a welcome holiday from an equivalent threat. Her father's will formally remained in force. But its rules applied less obviously to the current situation and Elizabeth was determined to keep it that way as long as possible. The official copy of the will was examined by the council and placed in the treasury of the Exchequer in a locked chest, the key of which was effectively thrown away. Elizabeth also used the amorous adventures of the surviving Grey sisters (the next claimants under Henry VIII's will) to put them under lock and key as well. And she was resolute, then and always, never to nominate a successor herself. To do so, she said, would be 'to bury herself alive' and, to drive the point home, she reminded the Lords and Commons how many of them had tried to involve her in conspiracies against her sister Mary.

But the arrival on English soil of the other Mary overwhelmed all Elizabeth's careful defences. Of course, despite Mary's wheedlings, Elizabeth never acknowledged her as heir. But that made little difference: Mary's proximity of blood was enough in itself, without any further legal endorsement. Moreover, Mary quickly turned herself into Elizabeth's political antithesis. In Scotland, she had been too busy committing the sins of flesh to worry much about her immortal soul, and events had made her more sympathetic than not to the Kirk. Once in

England, however, she quickly reinvented herself as a Roman Catholic *dévôte*. And Catholics were foolish enough to believe her. The horns of the old dilemma rose once more: Protestant Elizabeth had a Catholic successor.

The result was a chain of consequences that destroyed the careful balance of the Elizabethan polity and, at times, threatened to dethrone Elizabeth herself. Hitherto Catholic resentment had lacked a focus. Mary provided one and within a year of her arrival the north had risen in revolt in 1569. It was led by the Earls of Westmorland and Northumberland on a platform of full-scale reaction: they would overthrow Protestantism, marry Mary to the Duke of Norfolk, and proclaim the couple Elizabeth's heirs. For the moment the rebels seemed to carry all before them. Durham was occupied and, in the great cathedral, Protestant Prayer Books and Bibles were destroyed and the Mass celebrated on their burned and trampled fragments. But the tide turned and the rebellion collapsed.

The damage, however, had been done. The Northern Rebellion stirred Rome into action at last and in 1570 the Pope issued the bull known from its opening words as *Regnans in Excelsis* (reigning on high). This branded Elizabeth as a heretic, declared her deposed and absolved her subjects from their allegiance. The combination of the rebellion and the bull forced Elizabeth, unwillingly and against her better judgement, to begin to give way to her hard-line Protestant advisers. At home, she sanctioned more vigorous policies against Catholic recusants, as those who refused to submit to the Anglican Church were known. And abroad, she found herself pursuing a more aggressively Protestant policy.

As Elizabeth was, willy-nilly, captured by hard-line Protestantism, she found it more difficult to maintain the delicate *modus vivendi* with Catholic Spain. English Catholics turned to Spain to support them in their plots to overthrow Elizabeth; while Elizabeth found herself cast as the protectress of Philip's rebellious Protestant subjects in the Netherlands. Less justifiably, she was also a sleeping-partner in the piratical schemes which Englishmen such as Francis Drake launched against the Spanish empire in the New World. Neither Philip nor

Elizabeth wanted war and Elizabeth in particular postponed it as long as possible. But the logic of their respective positions increasingly drove them towards it.

The result of all this was that, by the 1580s, Elizabethan England looked uncomfortably like a mirror-image of her sister Mary's divided and distracted kingdom of thirty years before. Elizabeth herself was head of a Protestant Church and State that was increasingly extreme, intolerant and persecuting. Arrayed against her was a resurgent Catholic party. For the laws against recusants had produced martyrs, including the Jesuit Edmund Campion, who died with a prayer for the Queen on his lips. And the blood of the martyrs, as usual, was the seed of the Church. There were now Catholic exiles as well: gentlemen and merchants who supplied money and political leadership, and clergy and intellectuals who stiffened the present generation in their faith and educated the next. And at home there were plotters and conspirators. After the failure of the 1569 revolt they did not attempt full-scale rebellion; instead, more insidiously, they planned Elizabeth's own assassination. There was also the threat of foreign war. Under Mary, war with France had cost the English dependency of Calais; under Elizabeth, war with Spain, the superpower of the day, risked the loss of England itself. Finally, at the heart of it all, there was Mary, Queen of Scots. Like Elizabeth at Hatfield or Woodstock, Mary at Tutbury or Sheffield Castle had her finger in every plot against the Queen and in every project for a foreign invasion. Unlike Elizabeth, however, Mary was careless. In 1586, Sir Francis Walsingham, Elizabeth's cousin and spy-master, set a trap for her. She walked into it and sent a letter explicitly endorsing a plot to murder Elizabeth. She had signed her own death warrant.

It took much persuasion and almost as much chicanery, however, before the royal signature on Mary's death warrant could be extracted from the reluctant Elizabeth. For Elizabeth, finally, was no Mary Tudor. In similar circumstances, Mary had sought any excuse to execute Elizabeth. Elizabeth, on the contrary, clutched at any and every straw to avoid executing the much more flagrantly guilty Mary Stuart.

The difference lay partly in character. Elizabeth had the imagination, self-awareness and empathy which her sister Mary wholly lacked. She knew that she had once been in Mary Stuart's position and, to that extent, could feel sympathy for her – even though her actual conduct filled her with contempt for both its ingratitude and folly.

But, above all, it was a question of final objectives. Mary Tudor aimed for a heavenly crown; Elizabeth for an earthly one. The result was that for Elizabeth human life, in some sense, mattered. She was proud of the fact that there had been no political executions in the first decade of her reign and bitterly disappointed when the circumstances of 1569 gave the headsman work once more. This is not modern humanism, much less modern humanitarianism, as Stubbs's severed hand and some dozens of butchered Catholic bodies will testify. But nor is it the deliberate, spiritually sanctioned savagery of Mary either.

Here too, the change had been announced in Bacon's speech to her first parliament in 1559. As so much in Elizabeth's life, it went back to her father. In the wake of the Reformation, Henry had considered altering the coronation oath and a draft survives, heavily corrected in his own unmistakable hand. Its effect is to stand the traditional oath on its head. Instead of the King swearing to respect the laws and liberties of his people and Church, he promises to protect the rights and interests of the Crown. The draft was never used and is generally supposed to have been forgotten. Instead, it is clear that Elizabeth, characteristically, had her cake and ate it. The first two promises she made through Bacon, about counsel and popularity, were obligations to her people and corresponded to the traditional coronation oath. Indeed they rather went beyond it. But her third and final promise was a solemn compact between her and the Crown. And this echoed her father's draft, in gross and in detail.

Like the other two promises, the undertaking took the specific form of a critique of Mary's reign. 'Could there have happened to this imperial crown a greater loss in honour, strength and treasure than to lose that [place], Calais, I mean,' Bacon exclaimed. The implication is plain. Mary had diminished the Crown territorially by the loss of Calais, and she had reduced it jurisdictionally by surrendering the royal supremacy and putting England in 'bondage' once more to Rome. What made it the more unforgivable for Elizabeth was that Calais and the supremacy lay at the heart of their father's achievement: the supremacy was his creation, while he had refortified Calais and extended English territories in France by the conquest of Boulogne in 1544. Mary, Henry VIII's half-Spanish daughter, had lost what he had gained and degraded his achievement. Now his other daughter Elizabeth, mere English, would restore his crown and her inheritance.

To do nothing 'to the loss of any of her dominions'. That was the promise, and Elizabeth stuck by it. It was the source of the best and worst in her reign. It accounts for the terrible punishment she inflicted on the north in the wake of the rebellion of 1569 and her still more savage vengeance on the Irish rebels at the end of her reign. These men were far from her experience or her understanding. So the pity and sympathy which she could extend to the Duke of Norfolk and Mary Stuart for their treasons had no place and no mercy was shown. But equally her determination to preserve what was hers also turned her into a great war leader against Spain. She was not a general in the field nor an admiral at sea, of course, though she did wear a pretty pretend breastplate at Tilbury in 1588. Instead, and more importantly, she was a mistress of language, thinking, in her speech at Tilbury, 'full scorn that Parma or Spain or any prince of Europe should dare invade the borders of my realm'.

This same determination led her to destroy her favourite, Essex, when he dared to challenge both her crown and her popularity. And it made sure that the high command of her Church was never captured by Puritan extremists, however much they might rule the roost in the parishes and the universities. The result was that, despite four and a half decades of female regiment, she handed over to her Stuart successors something that was recognisable as the inheritance of Henry VIII. How they handled it is, of course, another matter.

Notes on Sources

The standard biographies of Elizabeth – by M. Creighton, J. E. Neale, N. Williams and, most recently, W. MacCaffery – all treat her early life fairly cursorily. By far the best treatment, despite its age, is L. Wiesener, *La jeunesse d'Élizabeth d'Angleterre* (Paris, 1878). A. Strickland's account in her *Lives of the Queens of England* (4th edition, London, 1854) remains valuable, both as a pioneering work and as a source-quarry. But the chronology is often confused.

I have tried, using the modern handbooks and *Calendars of State Papers*, to get the chronology broadly right, though I am sure many details have eluded me. In addition, for specific points, I have consulted the following printed sources. I give them roughly in the order in which they are used in the text.

The best guide to early Tudor royal ceremony is 'The Ryalle Book', which is printed in F. Grose, ed., *The Antiquarian Repertory*, 4 vols. (London, 1807–9) I, pp. 296–341. I discuss the dating and the circumstances of the composition of the document in 'Henry VI's Old Blue Gown: the English court under the Lancastrians and Yorkists' in *The Court Historian* 4 (1999), pp. 1–28. The description of the fitting-up of Greenwich Palace for the birth of Elizabeth is based on J. W. Kirby, ed., 'Building Works at Placentia, 1532–33', *Transactions of the Greenwich and Lewisham Antiquarian Society* 5 (1957), pp. 22–50.

The standard biography of Catherine of Aragon is by G. Mattingley (London, 1942). But it is hagiographic and ignores the evidence of her curious gynaecological views and her less than frank account of her first pregnancy. These are concealed in the decent obscurity of a learned tongue in G. A. Bergenroth, ed., *Supplement to Vols. I and II of the Calendar of State Papers, Spanish* (London, 1868). The best discussion of the rules of the English succession, as Henry VIII and, later, Edward VI, might have understood them, is M. J. Bennett, 'Edward III's Entail and the Succession to the Crown, 1376–1471', *English Historical Review* 113 (1998), pp. 580–609. E. W. Ives, *Anne Boleyn* (Oxford, 1986), is an exciting life and gives a useful discussion of Henry VIII's love-letters to Anne.

The principal source for Mary's youth and upbringing, which were so closely and fatally intertwined with Elizabeth's, is F. Madden, *The Privy Purse Expenses of the Princess Mary* (London, 1831), and, in particular, its superb introductory memoir. For Catherine Parr, Strickland's biography is still unsuperseded. But some useful new material is added by B. Kemeys and J. Raggatt, *The Queen Who Survived* (London, 1993). M. Dowling, *Humanism in the Age of Henry VIII* (Beckenham, 1986), is refreshingly sceptical about Catherine's role in the education of her step-children. Pending the forthcoming Chicago University Press edition of Elizabeth's complete works (letters, translations, poems and speeches), M. Perry, *The Word of a Prince* (Woodbridge, 1990), is useful; it also contains some sharp insights into her character. Henry VIII's speech to parliament in 1545, which did so much to shape Elizabeth's approach to monarchy, is printed by E. Hall, *The Union of the Two Most Noble and Illustre Families of Lancaster and York* (London, 1809). Other great influences, on both Elizabeth's thought and her language, were Cranmer's English Prayer Book and the English Bible. I have used both frequently.

Edward VI's own *Chronicle*, edited by W. K. Jordan (London, 1966), is the main source for his reign and character. The account of his coronation is drawn from C. Blair, ed., *The Crown Jewels*, 2 vols. (London, 1998), and J. Strype, *Memorials of Cranmer*, 2 vols. (Oxford, 1812). For the extraordinary career of Elizabeth's governess, Catherine Ashley, and her husband John, I have been guided by A. J. Collins, *Jewels and Plate of Queen Elizabeth I* (London, 1955), pp. 199–230. This casts some light on Elizabeth's household as princess, as does her book of household expenses for 1551–2, edited by Viscount Stranford in *The Camden Miscellany 2* (1853). But I have been unable to find any full list of her household servants. In the absence of such a list, much about Elizabeth's true part in the turbulent politics of the 1550s will remain obscure. I have pieced together the story of Elizabeth's landed endowment from *The Calendar of Patent Rolls* and an atlas. The descriptions of her main residences, Ashridge, Hatfield and Somerset House, are taken from H. M. Colvin, ed., *The History of the King's Works*, 6 vols. (London, 1963–82). The list of Elizabeth's furnishings is given in D. Starkey, ed., *The Inventory of King Henry VIII: 1. The Transcript* (London, 1998).

J. G. Nichols, ed., *The Chronicle of Queen Jane and Queen Mary*, Camden Society 48 (1850), sets out the chronology and prints the key documents for the establishment of the Marian régime. The story of Elizabeth's imprisonment at Woodstock is documented by C. R. Manning, ed., 'State Papers Relating to the Custody of the Princess Elizabeth at Woodstock', *Norfolk Archaeology* 4 (1855), pp. 133–231, and, more romantically, by 'The Miraculous Preservation of the Lady Elizabeth', printed by J. Foxe, *Acts and Monuments*, ed. J. Pratt, 8 vols. (London, 1877), VIII, pp. 600–25. My account of the Marian attempt at creating a new dynastic history and a new Catholic nationalism is based on *The Chronicle of Queen Jane and Queen Mary* and Foxe's printed text of Gardiner's sermon. W. Schenk, *Reginald Pole* (London, 1950), is a brief but sensitive life. J. A. Froude loathed Pole and Mary and everything they stood for but *The Reign of Mary Tudor* (London, 1910) remains one of the great historical narratives.

With Mary's reign, Elizabeth herself becomes a political player of the first rank. D. M. Loades, *Two Tudor Conspiracies* (Cambridge, 1965), disentangles some of the murkier events but consistently underplays Elizabeth's involvement. It is at its weakest on Ashton's role in the Dudley conspiracy. I have traced Ashton's background by trawling *Letters and Papers of the Reign of Henry VIII*; while the involvement in plotting of the members, male and female, of Elizabeth's household has been followed through C. Garrett, *The Marian Exiles* (Cambridge 1938), *The Complete Peerage* and *The History of Parliament*. Wiesener, with his mastery of the French sources, is likewise invaluable here; he also establishes a proper chronology for the proposal to marry Elizabeth to Emmanuel Philibert and was the first to denounce the *fêtes* in Wharton's *Life of Sir Thomas Pope* as forgeries.

Michieli's *relazione*, with its double character-sketch of Elizabeth and Mary, exists in several versions. I have used that in H. Ellis, ed., *Original Letters, Illustrative of English History*, 2nd series, 4 vols. (London, 1827), II, pp. 216–43. J. E. Neale, 'The Accession of Queen Elizabeth I' in *The Age of Catherine de Medici*, was the first to draw attention to Elizabeth's preparations for an armed *coup*. There is an excellent modern edition of Feria's crucial despatch of 14 November 1558 by S. Adams and M. J. R. Salgado in *Camden Miscellany* 28 (1984). My account of the formation of Elizabeth's council is based on: Throckmorton's advice to Elizabeth on her accession, printed by J. E. Neale in *English Historical Review* 65 (1950), pp. 91–8; Heath's letter to Cecil, in C. Read, *Mr Secretary Cecil and Queen Elizabeth* (1965), p. 479; Elizabeth's speeches, reported in Sir John Harington, *Nugae Antiquae*, ed. T. Park, 2 vols. (London, 1804), and, above all, on Cecil's manuscript notes in the Public Record Office: SP 12/1, fos. 3–5.

The main materials for Elizabeth's coronation are: J. M. Osborn, ed., *The Quenes Maiesties Passage* . . . (New Haven, 1960), C. G. Bayne, 'The Coronation of Queen Elizabeth', *English Historical Review* 22 (1907), pp. 650–73, and A. L. Rowse, 'The Coronation of Queen Elizabeth I', *History Today* 3 (1953), pp. 301–10. Elizabeth's costume is discussed and documented in J. Arnold, *Queen Elizabeth's Wardrobe Unlock'd* (Leeds, 1988).

N. L. Jones, *Faith by Statute* (London, 1982), supersedes all other accounts of the Elizabethan religious settlement. I have supplemented it from the verbatim texts in T. E. Hartley, ed., *Proceedings in the Parliaments of Elizabeth I*, vol. I: 1558–1581 (Leicester, 1981). The role of the Chapel Royal as the shop-window for Elizabeth's religious policy is discussed by P. E. McCullough's pioneering *Sermons at Court* (Cambridge, 1998). The description of the Preaching Place at Whitehall is based on S. Thurley, *The Royal Palaces of Tudor England* (London and New Haven, 1993), while I have followed F. Kisby, 'When the King Goeth a Procession: chapel ceremonies and services, the ritual year and religious reforms at the early-

Tudor court' (forthcoming), on the centrality of the Chapel Royal and religious observance in general in the ceremonial life of the court. The story of Elizabeth's counter-Reformation of 1560 is pieced together from J. G. Nichols, ed., *The Diary of Henry Machyn*, Camden Society 42 (1842). Machyn and his fellow-chronicler Charles Wriothesley, edited by W. D. Hamilton, *The Chronicle of England, 1485–1559*, 2 vols., Camden Society, new series 11, 20 (1875, 1877), also provide the main markers for the chronology of Elizabeth's life in the 1550s.

Abbreviations

CW   Elizabeth I, *Collected Works*, ed. L. S. Marcus, J. Mueller and M. B. Rose (Chicago and London, 2000)

DNB   *Dictionary of National Biography*

LP   *Letters and Papers, Foreign and Domestic, of the Reign of Henry VIII, 1509–47*, ed. J. S. Brewer, J. Gairdner and R. H. Brodie, 21 vols. and addenda (London, 1862–1932)

I have standardised all references to Elizabeth's speeches or writings to the version in CW. Figures after citations of LP and The Inventory of Henry VIII refer (unless otherwise stated) to numbers of documents or items; in all other cases, to pages.

References

INTRODUCTION
    CW, p. 97.

CHAPTER 1
    1. *LP* VI, 1111/1.
    2. *The Antiquarian Repertory*, ed. F. Grose, 4 vols. (London, 1807–9), I, pp. 296–341 (pp. 304–6, 333–338). I discuss the dating and the circumstances of the composition of 'The Royal Book' in 'Henry VI's Old Blue Gown: The English Court under the Lancastrians and Yorkists' in *The Court Historian* 4 (1999), pp. 1–28. See also K. Staniland, 'The Royal Entry into the World' in *England in the Fifteenth Century*, ed. D. Williams; Proceedings of the 1986 Harlaxton Symposium (Woodbridge, 1987), pp. 27–313.
    3. J. W. Kirby, 'Building Works at Placentia, 1532–33', *Transactions of the Greenwich and Lewisham Antiquarian Society* 5 (1957), pp. 22–50 (p. 49).
    4. *LP* VI, 948, 1004.

5. Renard to Charles V, 27 June 1555. Cited in J. A. Froude, *The Reign of Mary Tudor* (London, 1910), p. 214.

6. Kirby, 'Building Works', p. 50.

7. *The York Manual* (with an appendix of the Sarum Rite), ed. W. G. Henderson, Surtees Society 63 (1875), pp. 5–17; 8*–16*.

8. *LP* VI, 1069, 1070; VII, 939; *State Papers, Published under the Authority of His Majesty's Commission, King Henry VIII*, 11 vols. (London, 1830–52), I, p. 407.

CHAPTER 2

1. F. Hepburn, 'Arthur, Prince of Wales and His Training for Kingship', *The Historian* 55 (1997), pp. 4–9; *The Privy Purse Expenses of Elizabeth of York*, ed. N. H. Nicolas (London, 1830), pp. xxxi–civ. I will discuss the role of his mother in Henry VIII's upbringing fully in my forthcoming biography of the king.

2. G. Mattingley, *Catherine of Aragon* (London, 1963), p. 26.

3. *Antiquarian Repertory* II, pp. 249–331 (pp. 284–9, 302).

4. PRO, E101/414/8, fol. 27; *Privy Purse Expenses of Elizabeth of York*, p. 99; *LP* II ii, 3802. My forthcoming biography will deal fully with the nursery arrangements for Henry VIII and his siblings.

5. *Calendar of State Papers, Spanish*, ed. G. A. Bergenroth et al., 13 vols. (London, 1862–1954), II, p. 36; *Supplement to Vols. I and II*, pp. 34–44.

6. *LP* I i, 670, 698, 707.

7. *The Privy Purse Expenses of the Princess Mary*, ed. F. Madden, (London, 1831). The 'Introductory Memoir' (pp. xv–clxx) is still the best account of her upbringing.

8. The best discussion of the rules of the English succession, as Henry VIII might have understood them, is M. J. Bennett, 'Edward III's Entail and the Succession to the Crown, 1376–1471', *English Historical Review* 113 (1998), pp. 580–609.

9. The arguments are discussed by M. Levine, *Tudor Dynastic Problems* (London, 1973), pp. 50–1.

10. Mary: *LP* IV i, 1940, 2331; Richmond: *LP* IV i, 1510, 1514, 1540.

11. See J. J. Scarisbrick, *Henry VIII* (London, 1968), pp. 152–7.

12. D. Starkey, *Henry VIII: A European Court in England* (London, 1991), V. 46.

13. E. W. Ives, *Anne Boleyn* (Oxford, 1986), pp. xvii, 22–46, 49–52; *The Love Letters of Henry VIII*, ed. H. Savage (London, 1949), p. 47.

14. Ives, *Anne Boleyn*, pp. 19–20, 102–110.

15. See the facsimile reproductions appended to *The Love Letters of Henry VIII* and for Henry's comment that 'writing to me is somewhat tedious and painful', Scarisbrick, *Henry VIII*, p. 67.

16. G. Nicholson, 'The Act of Appeals and the English Reformation' in *Law and Government under the Tudors* (Cambridge, 1988), pp. 19–30.

17. Ives, *Anne Boleyn*, pp. 181–214.

CHAPTER 3

1. Kirby, 'Building Works', p. 50; *Original Letters Illustrative of English History*, ed. H. Ellis, 2nd ser., 4 vols. (London, 1827), II, pp. 78–83.

2. *State Papers Henry VIII* I, pp. 414 –15 (*LP* VI,1486), *LP* VI,1528; *CW*, p. 16.

3. *LP* VII, 296, X, 913.

4. *LP* VII, 393; *The Lisle Letters*, ed. M. St Clare Byrne, 6 vols. (Chicago and London, 1981), II, pp. 127–8; *The History of the King's Works*, ed. H. M. Colvin et al., 6 vols. (London, 1963–82), IV, p. 82 (the reference should be to 'princess' not 'prince'); *State Papers Henry VIII* I, p. 426 (*LP* IX, 568).

5. *State Papers Henry VIII* I, p. 414–15 (*LP* VI, 1486), *LP* VI, 1528; 1558; P. Friedmann, *Anne Boleyn*, 2 vols. (London 1884), I, 266–73.

6. Ives, *Anne Boleyn*, pp. 302–331; J. P. Carley, '"Her Moost Lovyng and Fryndely Brother Sendeth Gretyng": Anne Boleyn's Manuscripts and their Sources' in *Illuminating the Book: Makers and Interpreters*, ed. M. P. Brown and S. McKendrick (London, 1998), pp. 261–80.

7. *LP* VI, 1069; VIII, 1.

8. *LP* X, 141 (p. 51).

9. *LP* X, 199, 282.

10. Ives, *Anne Boleyn*, pp. 398–9.

11. Friedmann, *Anne Boleyn*, I, p. 133, n. 1; Ives, *Anne Boleyn*, pp. 401, 411; *LP* X, 909, 910.

CHAPTER 4

1. *Calendar of State Papers, Foreign: Elizabeth*, ed. J Stevenson et al., 23 vols. (London, 1863–1950), I (1558–9), pp. 524–34: letter from Alexander Alane ('Alesius') to Elizabeth; *LP* X, 199, 307, 908 (which shows that Mary and Elizabeth and their households were certainly together at the beginning of May), 913 and index *sub* 'Hunsdon'; *LP* XI 500. For the accommodation provided for Mary and (much more elaborately) for Elizabeth at Eltham for their extended stay there over the Christmas season 1535–6, see *King's Works*, ed. Colvin, IV, p. 82.

On further reflexion, I am inclined to give more credit to Alane's reminiscences: he was an eyewitness of the events of April and May 1536 but, naturally enough, as he wrote over twenty-five years after the event, his chronology is confused and compressed. I hope to discuss the matter further in my second volume.

2. *Original Letters*, ed. Ellis, 2nd ser., II, pp. 78–83 (*LP* XI, 203), *LP* XI, 312.

3. *LP* X, 968, 1110, 1136–7; *LP* XI, 132; *Privy Purse Expenses*, ed. Madden, index *sub* 'Elizabeth'.

4. *LP* X, 908, 1069; *LP* XII ii, 911; *Privy Purse Expenses*, ed. Madden, index *sub* 'Prince [Edward']; and see below.

5. *Original Letters*, ed. Ellis, 2nd ser., II, pp. 78–83 (*LP* XI, 203); *LP* X, 1187; *LP* XV, 602.

6. A. J. Collins, *Jewels and Plate of Queen Elizabeth I* (London, 1955), p. 200, n. 1; *The Itinerary of John Leland*, ed. L. T. Smith, 5 vols. (London, 1906–8), I, pp. 160, 186, 203; M. Dowling, *Humanism in the Age of Henry VIII* (London, 1986), pp. 63, 66; *CW*, p. 34; L. V. Ryan, *Roger Ascham* (Stanford and London), p. 104.

7. T. Hearne, *Syllogue Epistolarum* (Oxford, 1716), pp. 149–51, cited in N. Williams, *Elizabeth I, Queen of England* (London, 1971), p. 7.

8. *The Epistles of Erasmus*, ed. and trans. F. M. Nichols, 2 vols. (London, 1904), II, p. 201.

9. *The Chronicle and Political Papers of King Edward VI*, ed. W. K. Jordan (London, 1966), p. 3; Ryan, *Ascham*, pp. 16–22, 42; Collins, *Jewels and Plate*, pp. 201–4.

10. Ryan, *Ascham*, p. 104; *Chronicle*, ed. Jordan, p. 3; W. K. Jordan, *Edward VI: The Threshold of Power* (London, 1970), p. 21, n. 1; M. Perry, *The Word of a Prince* (Woodbridge, 1990), p. 72.

11. *King's Works*, ed. Colvin, III, pp. 254–5; IV, pp. 47–8, 149–50, 154–7; Ryan, *Ascham*, pp. 49–81, 229–30. The sylvan setting of Hunsdon is shown in the background of the portrait of Prince Edward in the color plate insert to this book.

12. *LP* XVI, 804, 1389.

13. Dowling, *Humanism*, pp. 235–6; *LP* XXI I, 1036

CHAPTER 5

1. *LP* XIX i, 780; Public Record Office, OBS 1419.

2. *LP* XVIII i, 364 (p. 214).

3. *LP* XVIII i, 740, 873; XVIII ii, 39, 501.

4. 35 Henry VIII, c. 1, cited in M. Levine, *Tudor Dynastic Problems, 1460–1571* (London, 1973), pp. 161–2; *LP* XIX i, 780.

5. O. Miller, The Tudor, Stuart and Early Georgian Pictures in the Collection of Her Majesty the Queen, 2 vols. (London, 1963), I, pp. 63–4.

6. *LP* XIX i, 864, 903, 912, 928; XIX ii, 424.

7. *CW*, p. 6; *LP* XIX ii, 688 (p. 407); Public Record Office, OBS 1419; *LP* XXI i, 1384.

CHAPTER 6

1. *CW*, pp. 5–6.

2. I have preferred the translation in Perry, *Word of a Prince*, p. 30 ('continuous whirl of human affairs'), since it seems to catch Elizabeth's meaning much better than *CW*'s rather lame '[fortune] who revolves things human'.

3. I have followed the commentary in Perry, *Word of a Prince*, p. 30. *CW*, p. 5, n. 2 agrees.

4. *LP* XIX i, 864; *King's Works*, ed. Colvin, IV, p. 136; *Chronicle*, ed. Jordan, p. 3.

5. See my forthcoming biography.

6. *LP* XIX ii, 4, 246, 688.

7. *LP* XIX ii, 688.

8. *LP* XIX i, 864, 1035/78, 88, 962–3, 967, 979, 1019, 1029 (for which also see below); *State Papers* X, pp. 28–9 (*LP* XIX ii, 39).

9. D. Starkey, 'Court, Council and Nobility in Tudor England' in *Princes, Patronage and the Nobility: the Court at the Beginning of the Modern Age*, ed. R. G. Asch and A. M. Birke (Oxford, 1991), pp. 175–203.

CHAPTER 7

1. C. A. J. Armstrong, 'The Piety of Cecil Duchess of York: a Study in Later Medieval Culture' in *For Hilaire Belloc*, ed. D. Woodruff (London, 1942); M. K. Jones and M. G. Underwood, *The King's Mother: Lady Margaret Beaufort, Countess of Richmond and Derby* (Cambridge, 1992), espec. pp. 174–83; Dowling, *Humanism*, pp. 223–231.

2. J. Foxe, *The Acts and Monuments*, ed. J. Pratt, 8 vols. (London, 1877), V, pp. 553–4.

3. All this is usefully summarized in B. Kemeys and J. Raggatt, *The Queen Who Survived: the Life of Katherine Parr* (London, 1993), pp. 53–8.

4. J. Strype, *Ecclesiastical Memorials*, 7 vols. (London, 1816), VI, pp. 312–12 (*LP* XIX i, 1029).

5. *LP* XIX ii, 58.

6. C. Lloyd and S. Thurley, *Henry VIII: Images of a Tudor King* (London, 1990), pp. 30–1, 34–5.

7. D. MacCulloch, *Thomas Cranmer* (New Haven and London, 1996), pp. 313–14, 326–7.

8. A. Strickland, *Lives of the Queens of England*, 8 vols. (London, 1854), III, pp. 259.

9. MacCulloch, *Cranmer*, pp. 326–32.

10. *CW*, p. 96.

11. *LP* XIX ii, p. 726; *CW*, pp. 6–7.

12. *CW*, pp. 6–7.

13. A. L. Prescott, 'The Pearl of the Valois and Elizabeth I' in *Silent but for the Word*, ed. M. P. Hannay (Ohio, 1987).

14. *CW*, pp. 6–7. I have followed the broad outlines of the excellent commentary in Perry, *Word of a Prince*, pp. 31–3.

CHAPTER 8

1. *LP* XVIII i, 364 (p. 214); *LP* XX ii, index *sub* 'Elizabeth'; *LP* XXI i, 969. Even the Habsburgs came round and she was (not for the first time) considered as a wife for Philip of Spain.

2. *LP* XX ii, 909/16–18.

3. *LP* XX ii, 909/50.

4. *CW*, pp. 8–10.

5. Once again, I have followed the broad outlines of the excellent commentary in Perry, *Word of a Prince*, pp. 34–9.

CHAPTER 9

1. *Handbook of British Chronology*, ed. E. B. Fryde et al., 3rd edn. (London, 1986), p. 573; Lloyd and Thurley, *Henry VIII: Images of a Tudor King*, pp. 42–3; *The Inventory of King Henry VIII: I, The Transcript*, ed. D. Starkey (London, 1998), 11243; E. Hall, *The Vnion of the Two Noble and Illustre Famelies of Lancastre and Yorke (The Chronicle)*, (London, 1809), pp. 864–6.

2. Hall, *Chronicle*, pp. 864–6. It was Erasmus, in his letter to Henry Bullock of 1516 who first told the story of the ignorant English priest. For twenty years he had said 'quod ore *mumps*imus' instead of '*sumps*imus' ('which we receive by mouth'), but, when the mistake was pointed out to him, he refused to mend his ways on the grounds that it would be to change Scripture! (*The Correspondence of Erasmus: Letters 446 to 593, 1516 to 1517, The Complete Works of Erasmus* 4 (Toronto and Buffalo, 1977), p. 46). The tale was picked up the following year by Richard Pace, Henry's secretary, in his *De Fructu qui ex Doctrina Percipitur*, which is most idiomatically to be translated as 'Glittering Prizes' (R. Pace, *De Fructu qui ex Doctrina Percipitur*, ed. F. Manley and R. S. Sylvester (New York, 1967), pp. 65 and 103. But these works were *in Latin*; it is Henry who, in this speech, first imports the words *into English* (see Oxford English Dictionary, *sub* 'mumpsimus' and 'sumpsimus').

3. *LP* XXI i, 963/37, 86, 151; *LP* XXI i, 1384; *LP* XXI ii, 475/117.

4. *Inventory*, ed. Starkey, 1922, 11798–11800; A. S. MacNalty, *Henry VIII, a Difficult Patient* (London, 1952).

5. *LP* XXI ii, 547; L. B. Smith, *Henry VIII: The Mask of Royalty* (London, 1971), pp. 255–9.

6. T. Rymer, *Foedera, Conventiones, Litterae*, 15 vols. (London, 1704–35), XV, pp. 110–17; *LP* XXI ii, 770/85; BL, Cotton MS Titus B I, espec. fols. 239r–v; 259v–261; BL, Harley MS 849, fols. 32r–v; PRO, E 36/266, fol. 76v (I owe the latter reference to Mr Christopher Coleman).

7. Foxe, *Acts and Monuments* V, p. 689.

8. D. Starkey, *The Reign of Henry VIII: Personalities and Politics* (London, 1985), pp. 159–65.

9. P. F. Tytler, *England under the Reigns of Edward VI and Mary*, 2 vols. (London, 1839), I, pp. 15–18 (*Calendar of State Papers, Domestic: Edward VI*, ed. C. S. Knighton, revised edn. (London, 1992), pp. 1–2.); *Chronicle*, ed. Jordan, pp. 3–4.

10. Strickland, *Lives of the Queens of England* IV, pp. 19–20, citing Sir John Hayward's *The Life and Raigne of King Edward the Sixt*, published posthumously in 1630, where the story seems first to appear (J. Hayward, *Annals of the First Four Years of the Reign of Queen Elizabeth*, ed. J. Bruce, Camden Society OS 7 (London, 1840), pp. xxvii and l).

CHAPTER 10

1. J. J. Scarisbrick, *Henry VIII* (London, 1968), p. 496.

2. Strickland, *Lives of the Queens of England* IV, p. 20; Strype, *Ecclesiastical Memorials* VI, pp. 506–7.

3. C. Wriothesley, *A Chronicle of England* . . . *1485–1559*, Camden Society NS 11 and 20, 2 vols. (London, 1875–7), I, p. 178; *Acts of the Privy Council of England*, ed. J. R. Dasent, 32 vols. (London, 1890–1907), II, pp. 3–7; Strype, *Ecclesiastical Memorials* VI, pp. 267–90; J. Strype, *Memorials of the Most Reverend Father in God Thomas Cranmer*, 2 vols. (Oxford, 1812), I, pp. 202–7.

4. Strype, *Ecclesiastical Memorials* VI, pp. 267–90.

5. Strype, *Cranmer* I, pp. 202–7.

6. Strype, *Ecclesiastical Memorials* VI, pp. 285.

CHAPTER 11

1. Rymer, *Foedera* XV, p. 116.

2. *Calendar of Patent Rolls: Edward VI*, 6 vols. (London, 1924–9), II, p. 20.

3. Rymer, *Foedera* XV, p. 116.

4. Wriothesley, *Chronicle* I, p. 182.

5. R. Hughey, *John Harrington* (Columbus, Ohio, 1971), p. 96: 'Of person rare, strong limbs and manly shape'.

6. BL, Additional MS 48126, fols. 6–16.

7. Rymer, *Foedera* XV, p. 116; Strickland, *Lives of the Queens of England* III, p. 259–67; *CW*, pp. 25, 29.

8. Miller, *Tudor, Stuart and Early Georgian Pictures* I, p. 65.

9. S. Haynes, *A Collection of State Papers* . . . *Left by William Cecil, Lord Burghley* (London, 1740), pp. 99–100.

10. *CW*, p. 28; Haynes, *Burghley State Papers*, pp. 99–100.

11. Perry, *Word of a Prince*, p. 54.

12. Haynes, *Burghley State Papers*, p. 96; *CW*, pp. 17–18 and n. 3.

13. *CW*, p. 20.

14. *Calendar of State Papers, Domestic: Edward VI*, ed. Knighton, p. 110; Kemeys and Raggatt, *Queen Who Survived*, p. 109; Strickland, *Lives of the Queens of England* III, p. 280–2; Haynes, *Burghley State Papers*, pp. 62, 103–4.

15. Strickland, *Lives of the Queens of England* III, pp. 283–4.

16. Tytler, *Edward VI and Mary* I, pp. 131–5 (*Calendar of State Papers, Domestic: Edward VI*, ed. Knighton, pp. 157–80); Haynes, *Burghley State Papers*, pp. 77–8.

17. Strickland, *Lives of the Queens of England* IV, pp. 27–8; Haynes, *Burghley State Papers*, p. 95.

18. Strickland, *Lives of the Queens of England* IV, pp. 28–9; Haynes, *Burghley State Papers*, p. 97.

19. Tytler, *Edward VI and Mary* I, pp. 142–6 (*Calendar of State Papers, Domestic: Edward VI*, Knighton, p. 191), Haynes, *Burghley State Papers*, p. 101.

20. Haynes, *Burghley State Papers*, p. 95; Strickland, *Lives of the Queens of England* IV, pp. 30–1.

21. Strickland, *Lives of the Queens of England* IV, pp. 32, 34; CW, p. 30; Haynes, *Burghley State Papers*, p. 96.

22. *The History of Parliament: the House of Commons, 1509–1558*, ed. S. T. Bindoff, 3 vols. (London, 1982), III, pp. 501–2.

23. Haynes, *Burghley State Papers*, pp. 70–1, 88; Strickland, *Lives of the Queens of England* IV, p. 35.

24. Strickland, *Lives of the Queens of England* IV, pp. 36–7; *CW*, p.23.

25. Haynes, *Burghley State Papers*, pp. 89, 102; *CW*, pp. 22–4.

26. *CW*, pp. 31–3.

27. Haynes, *Burghley State Papers*, pp. 107–8; Strickland, *Lives of the Queens of England* IV, pp. 40–1.

28. *CW*.

29. H. Latimer, *Sermons and Remains*, ed. G. E. Corrie, Parker Society, 2 vols. (Cambridge, 1844–5), I, pp. 161–5, pp. 228–9; Strype, *Ecclesiastical Memorials* II, pp. 204–6.

Chapter 12

1. *Calendar of State Papers, Foreign*, I (1558–9), p. xxxvi (*Calendar of State Papers, Domestic: Edward VI*, ed. Knighton, p. 181); Haynes, *Burghley State Papers* p. 89; K. T. Parker, *The Drawings of Hans Holbein . . . at Windsor Castle* (London, 1945), reproduced between p. TK and p.TK; Strickland, Lives of the Queens of England IV, p. 39.

2. Haynes, *Burghley State Papers*, p. 70; *CW*, pp. 27–8 (*Calendar of State Papers, Domestic: Edward VI*, ed. Knighton, p. 196). Elizabeth had already consulted Denny on the advisability of putting up in Seymour's townhouse (he had advised against it): see Chapter 15, n. 5 below.

3. Rymer, *Foedera* XV, p. 116.

Chapter 13

1. Collins, *Jewels and Plate*, p. 202.

2. Collins, *Jewels and Plate*, p. 203.

3. Strickland, *Lives of the Queens of England* IV, pp. 50–1.

4. *Queen Elizabeth's Englishing of Boethius*, ed. C. Pemberton, Early English Text Society (London, 1899); *The Letters of Queen Elizabeth*, ed. G. B. Harrison (London, 1968), p. xv; R. Ascham, *The Scholemaster*, ed. E. Arber (Birmingham, 1870), pp. 97, 101, 121, defends imitation and repetition and attacks originality and variation on the grounds that when something has been said well, it cannot be improved. Elizabeth evidently agreed.

5. Ascham, *Scholemaster*, ed. Arber, pp. 67–8, 96, 111–2, 154.

6. Ascham, *Scholemaster*, ed. Arber, pp. 25–31, 90.

7. Cited in Williams, *Elizabeth I*, p. 15; M. Creighton, *Queen Elizabeth* (London, 1899), p. 12; Strickland, *Lives of the Queens of England* IV, pp. 50–1.

8. Cited in Strickland, *Lives of the Queens of England* IV, pp. 54–5.

9. *King's Works*, ed. Colvin, IV, pp. 252–3.

10. 'Household Expenses of the Princess Elizabeth . . . 1551–2', ed. Viscount Strangford, *Camden Miscellany* II, Camden Society OS 55

(London, 1853), 5 note, 45 note, 46 and plate opposite, 48 note and plate opposite; see below p. 26; William Camden, *Annals or the History of . . . Elizabeth, Late Queen of England* (London, 1635), Introduction.

11. *LP* XXI i, 802.

12. Strickland, *Lives of the Queens of England* IV, pp. 47–8 (not included in *CW*); *Trevelyan Papers, Part II, 1446–1643*, ed. J. P. Collier, Camden Society OS 84 (London, 1863), pp. 19–21.

13. Collins, *Jewels and Plate*, p. 249; *Acts of Privy Council* III, p. 451; 'Household Expenses', ed. Strangford, p. 36.

14. Perry, *Word of a Prince*, pp. 73–4 (not included in *CW*).

15. Perry, *Word of a Prince*, p. 72; Strickland, *Lives of the Queens of England*, p. 249.

16. For aspects of this Edwardian ethos, see D. MacCulloch, *Tudor Church Militant: Edward VI and the Protestant Reformation* (London, 1999); M. Howard, *The Early Tudor Country House: Architecture and Politics, 1490–1550* (London, 1987); D. Starkey, 'England' in *The Renaissance in National Context*, ed. R. Porter and M. Teich (Cambridge, 1992), pp. 146–63.

CHAPTER 14

1. Tytler, *Edward VI and Mary* I, pp. 228–30 (*Calendar of State Papers, Domestic: Edward* VI, ed. Knighton, p. 396).

2. A. Hawkyard, '"The Alcibiades of England": John Dudley, Duke of Northumberland' in *Rivals in Power: Lives and Letters of the Great Tudor Dynasties*, ed. D. Starkey (New York, 1990), pp. 132–145.

3. J. Murphy, 'The Illusion of Decline: the Privy Chamber 1547–58' in *The English Court; from the Wars of the Roses to the Civil War*, ed. D. Starkey (London and New York, 1987), pp. 119–146 (pp. 127–33).

4. *Calendar of State Papers, Domestic: Edward VI*, ed. Knighton, p. 428.

5. *Calendar of State Papers, Spanish* X, p. 6.

CHAPTER 15

1. *Acts of Privy Council* II, pp. 83, 84, 92, 100, 120 and see above, Chapter 11, n. 2.

2. Haynes, *Burghley State Papers*, p. 87; Strickland, *Lives of the Queens of England* IV, pp. 28–9; *Acts of Privy Council* II, p. 393; *Calendar of Patent Rolls: Edward VI*, III, pp. 238–242.

3. Haynes, *Burghley State Papers*, p. 97; Strickland, *Lives of the Queens of England* IV, pp. 28–9.

4. *King's Works*, ed. Colvin, IV, pp. 47–8.

5. *Trevelyan Papers, Part II*, ed. Collier, p. 16; Haynes, *Burghley State Papers*, p. 102; Strickland, *Lives of the Queens of England* IV, pp. 30–1; CW, p. 26.

6. *Inventory of Henry VIII*, ed. Starkey, 15212–3 and pp. 380–382.

7. 'Household Expenses,' ed. Strangford, pp. 3, 47, 48.

8. *King's Works*, ed. Colvin, IV, pp. 149–50; *Calendar of Patent Rolls:*

Edward VI, III, pp. 71, 364–5, 415.

9. Perry, *Word of a Prince*, p. 71; see above, Chapter 3, n. 2 and 3; *Privy Purse Expenses*, ed. Madden, p. lxxviii. Edward resided at Hatfield in autumn 1546 (*LP* XXI, pp. ii, 38,86,361–4); Elizabeth joined him for at least some of this time before the two parted, probably in late November, Elizabeth going to court and Edward to Hertford (*LP* XXI, pp. ii, 502).

10. Strickland, *Lives of the Queens of England* III, p. 276.

11. *Calendar of State Papers, Spanish* X, p. 214; *Calendar of Patent Rolls: Edward VI* IV, pp. 88–93, 101–2.

CHAPTER 16

1. Strype, *Ecclesiastical Memorials* II, pp. 94–95; W. K. Jordan, *Edward VI: The Young King* (London, 1968), pp. 208–9.

2. Strype, *Ecclesiastical Memorials* II, pp. 93–4 (Privy Council letter misdated).

3. Jordan, *Edward VI: The Threshold of Power*, p. 31; *Calendar of State Papers, Spanish* X, p. 186.

4. Jordan, Edward VI: *The Threshold of Power*, pp. 258–9.

5. *Calendar of State Papers, Spanish* X, p. 215.

6. *Calendar of State Papers, Spanish* X, p. 215; F. Kisby, 'When the King Goeth a Procession: Chapel Ceremonies and Services, the Ritual Year and Religious Reform at the Early-Tudor Court, 1485–1547' (forthcoming).

7. *The Diary of Henry Machyn . . . 1550–1563*, ed. J. G. Nichols, Camden Society OS 42 (London, 1848), pp. 4–5; *Chronicle*, ed. Jordan, p. 55.

8. *Chronicle*, ed. Jordan, pp. 56–7; Jordan, *Edward VI: The Threshold of Power*, pp. 260–1 and 260, n.5.

9. Machyn, ed. Nichols, p. 16; *Calendar of State Papers, Spanish* X, p. 493.

10. *King's Works*, ed. Colvin, IV, pp. 241–3.

11. *Inventory of Henry VIII*, ed. Starkey, pp. 383–5.

12. *Machyn*, ed. Nichols, p. 16.

13. *'Household Expenses'*, ed. Strangford, pp. 3–4, 38, 42.

14. Jordan, *Edward VI: The Threshold of Power*, p. 264; *Calendar of State Papers, Domestic: Edward VI*, ed. Knighton, p. 804.

15. *CW*, pp. 37–8.

16. *CW*, pp. 38–9.

CHAPTER 17

1. *King's Works*, ed. Colvin, IV, p. 76; *The London Encyclopaedia*, ed. B. Weinreb and C. Hibbert (London, 1983), p. 244. *CW*, pp. 39–40 also suggests that Elizabeth was getting a raw deal from the Privy Council at this time.

2. Tytler, *Edward VI and Mary* II, pp. 162–3 (*Calendar of State Papers, Domestic: Edward VI*, ed. Knighton, p. 804); *King's Works*, ed. Colvin IV, pp. 252–3; *Acts of Privy Council* IV, p. 196.

3. *Machyn*, ed. Nichols, p. 16; *Acts of Privy Council* IV, pp. 244, 284, 287.

4. *Chronicle*, ed. Jordan, pp. 57–61.

5. *Chronicle*, ed. Jordan, p. 117; *CW*, pp. 36–7.

6. *The Chronicle of Queen Jane and of Two Years of Queen Mary*, ed. J. G. Nichols, Camden Society OS 48 (London, 1850), pp. 89–90 and see above Chapter 2, n. 8. The report of the Venetian ambassador (*Calendar of State Papers, Venetian*, ed. R. Brown, et. al., 38 vols. (London, 1864–1947)V, p. 536) specifically mentions Mary's lack of patriotism ('she [Mary] having clearly demonstrated how little love she bore the English nation') as one of the reasons motivating Edward to set her aside.

7. G. E. C[okayne], *The Complete Peerage*, new edn. revised by V. Gibbs and H. A. Doubleday, 13 vols. (London, 1910–49), *sub* 'Cumberland' and 'Suffolk'.

8. *Calendar of State Papers, Venetian* V, p. 536; *Calendar of State Papers, Spanish* XI, pp. 45–6; Strype, *Ecclesiastical Memorials* III, pp. 286–7; *History of Parliament*, 1509–1558, ed. Bindoff, II, pp. 61–3.

9. Jordan, *Edward VI: The Threshold of Power*, p. 514.

10. *Chronicle of Queen Jane and Queen Mary*, ed. Nichols, pp. 89–90. There is a reproduction of the beginning of the 'Device' in B. L. Beer, *Northumberland: The Political Career of John Dudley* (Kent: Ohio, 1973), p. 151.

11. Jordan, Edward VI: The Threshold of Power, pp. 514–17; Chronicle of Queen Jane and Queen Mary, ed. Nichols, pp. 91–100.

12. Jordan, Edward VI: The Threshold of Power, pp. 519, n. 4.

13. Jordan, Edward VI: The Threshold of Power, pp. 519–20.

14. Jordan, Edward VI: The Threshold of Power, pp. 522 and Calendar of State Papers, Spanish XI, pp. 79–80.

15. Chronicle of Queen Jane and Queen Mary, ed. Nichols, pp. 1–3.

16. *Chronicle of Queen Jane and Queen Mary*, ed. Nichols, pp. 3–5; 'Vita Mariae Reginae: A Short Treatise of the Deeds of Mary Queen of England by Robert Wingfield of Brantham', ed. and trans. D. MacCulloch, *Camden Miscellany* 28 (London, 1984), pp. 244–301 (pp. 252–3); Jordan, *Edward VI: The Threshold of Power*, p. 524.

17. Chronicle of Queen Jane and Queen Mary, ed. Nichols, pp. 5–8.

18. *Chronicle of Queen Jane and Queen Mary*, ed. Nichols, pp. 9–11; Wriothesley, *Chronicle* II, pp. 90–1.

19. *Machyn*, ed. Nichols, p. 37.

20. Wriothesley, *Chronicle* II, pp. 93–4.

21. Wriothesley, *Chronicle* II, p. 88. Interestingly, Latimer, Elizabeth's other *bête noire*, had already raised doubts about female succession – actually mentioning Elizabeth and Mary by name – in his Lenten sermons before Edward VI in 1549 when Elizabeth was under a cloud because of the Seymour affair (Latimer, *Sermons and Remains* I, p. 91).

22. *CW*, p. 97; G. E. C., *Complete Peerage*, *sub* 'Cumberland'.

CHAPTER 18

1. *Calendar of State Papers, Venetian*, V, p. 533.

2. *Chronicle of Queen Jane and Queen Mary*, ed. Nichols, pp. 11–12.

3. *Tudor Royal Proclamations*, ed. P. L Hughes and J. F. Larkin, 3 vols. (New Haven and London, 1969), II, pp. 5–8.

4. Strickland, *Lives of the Queens of England* IV, p. 64; Williams, *Elizabeth I*, pp. 25–6.

5. Strype, *Ecclesiastical Memorials* IV, pp. 53–7.

6. Strype, *Ecclesiastical Memorials* IV, pp. 58–62; Levine, *Tudor Dynastic Problems*, pp. 90, 171.

7. *Calendar of State Papers, Venetian* V, p. 538; Williams, Elizabeth I, pp. 26–7.

8. J. Lingard, *The History of England*, 6th edn., 10 vols. (London, 1854), V, p. 203, n. 2 citing the French ambassador Noailles.

9. Williams, *Elizabeth I*, pp. 26–7.

CHAPTER 19

1. Strype, *Ecclesiastical Memorials* IV, pp. 53–7; VII, pp. 11–13.

2. Wriothesley, *Chronicle* II, p. 95 and n. a.

3. D. M. Loades, *The Reign of Mary Tudor* (London, 1979), pp. 109–20.

4. Froude, Reign of Mary Tudor, p. 72. A soberer version is given by Weisener, *La jeunesse d'Elizabeth d'Angleterre*, (Paris, 1878), p. 141, n. 1.

5. Froude, *Reign of Mary Tudor*, pp. 73–5.

6. Loades, *Reign of Mary Tudor*, pp. 121–3; *Chronicle of Queen Jane and Queen Mary*, ed. Nichols, pp. 33–4.

7. *Chronicle of Queen Jane and Queen Mary*, ed. Nichols, p. 34.

8. *Chronicle of Queen Jane and Queen Mary*, ed. Nichols, pp. 34–5.

CHAPTER 20

1. Loades, *Reign of Mary Tudor*, pp. 124–7.

2. D. M. Loades, *Two Tudor Conspiracies* (Cambridge, 1965), pp. 23–6.

3. Loades, *Two Tudor Conspiracies*, pp. 26–46.

4. Loades, *Two Tudor Conspiracies*, pp. 47–50, 59–61; *Chronicle of Queen Jane and Queen Mary*, ed. Nichols, pp. 37–9.

5. Foxe, *Acts and Monuments* VI, pp. 414–15.

6. Wriothesley, *Chronicle* II, p. 108 (admittedly a sympathetic witness), was deeply impressed by both the content and especially the delivery of Mary's speech: '[it] was so nobly and with so good spirit declared, and with so loud a voice, that all the people might hear her Majesty, and comforting their hearts with so sweet words that made them weep for joy to hear her Majesty speak.'

7. *Machyn*, ed. Nichols, p. 53.

8. *Chronicle of Queen Jane and Queen Mary*, ed. Nichols, pp. 42–8.

9. *Chronicle of Queen Jane and Queen Mary*, ed. Nichols, pp. 48–52, corrected by Underhill's account on pp. 131–3.

10. *Chronicle of Queen Jane and Queen Mary*, ed. Nichols, p. 54.

CHAPTER 21

1. *Chronicle of Queen Jane and Queen Mary*, ed. Nichols, p. 55.

2. *Chronicle of Queen Jane and Queen Mary*, ed. Nichols, pp. 55–8.

3. Strype, *Ecclesiastical Memorials* IV, pp. 130–2.

4. Froude, *Reign of Mary Tudor*, p. 110; Tytler, *Edward VI and Mary* II, pp. 426–9.

5. Froude, *Reign of Mary Tudor*, pp. 114–15; Tytler, *Edward VI and Mary* II, pp. 310–11; *Machyn*, ed. Nichols, p. 57.

6. Loades, *Reign of Mary Tudor*, pp. 47, 91–3.

7. Tytler, *Edward VI and Mary* II, pp. 313–14; *Chronicle of Queen Jane and Queen Mary*, ed. Nichols, p. 65.

8. *Chronicle of Queen Jane and Queen Mary*, ed. Nichols, pp. 68–70.

9. Foxe, *Acts and Monuments* VIII, pp. 607–8.

10. Foxe, *Acts and Monuments* VIII, pp. 608.

11. *CW*, pp. 41–2.

12. Foxe, *Acts and Monuments* VIII, pp. 608–9.

CHAPTER 22

1. Foxe, *Acts and Monuments* VIII, pp. 608–9.

2. Foxe, *Acts and Monuments* VIII, pp. 600–25; *Chronicle of Queen Jane and Queen Mary*, ed. Nichols, pp. 70–1.

3. The rules governing Elizabeth's imprisonment in the Tower and the accommodation provided for her there are set out in the 'Articles' which preface Bedingfield's record of his custodianship of Elizabeth ('State Papers Relating to the Custody of the Princess Elizabeth at Woodstock', ed. C. R. Manning, *Norfolk Archaeology* 4 (1855), pp. 133–231 (pp. 141–4), and see Chapter 23 below). For the topography of the Tower and, in particular, the Coldharbour Gate, see *King's Works*, ed. Colvin, Plan II.

4. *Chronicle of Queen Jane and Queen Mary*, ed. Nichols, pp. 70–1; Foxe, *Acts and Monuments* VIII, pp. 610–11.

5. Foxe, *Acts and Monuments* VIII, pp. 610–11.

6. Tytler, *Edward VI and Mary* II, pp. 366–7.

7. Anticipating what was planned, the French ambassador, Noailles, who was in the confidence of the rebels, reported Elizabeth's removal to Donnington and the mustering of her forces there on 26 January (J. Gairdner, *Lollardy and the Reformation in England*, 4 vols. (London, 1908–13), IV, p. 285; Weisener, *Jeunesse d'Elizabeth*, pp. 183–7 and notes.

8. *Chronicle of Queen Jane and Queen Mary*, ed. Nichols, pp. 73–4.

9. *Machyn*, ed. Nichols, pp. 59–60.

10. *Chronicle of Queen Jane and Queen Mary*, ed. Nichols, p. 75.

11. Strickland, *Lives of the Queens of England* IV, pp. 92–3; Tytler, *Edward VI and Mary* II, pp. 371–8; Froude, *Reign of Mary Tudor*, p. 72; *Chronicle of Queen Jane and Queen Mary*, ed. Nichols, pp. 75–6.

12. 'Elizabeth at Woodstock', ed. Manning, pp. 141–4.

13. 'Elizabeth at Woodstock', ed. Manning, p. 140; Foxe, *Acts and Monuments* VIII, pp. 613–14.

CHAPTER 23

1. *Chronicle of Queen Jane and Queen Mary*, ed. Nichols, p. 75; Tytler,

Edward VI and Mary II, pp. 340–1; Wriothesley, *Chronicle* II, pp. 117–8.

2. Weisener, *Jeunesse d'Elizabeth*, p. 263; *Illustrations of British History*, ed. E. Lodge, 3 vols. (London, 1791), I, pp. 191–3.

3. Foxe, *Acts and Monuments* VIII, p. 614.

4. Tytler, *Edward VI and Mary* II, p. 396; Froude, *Reign of Mary Tudor*, p. 137, n. 1.

5. Foxe, *Acts and Monuments* VIII, pp. 614–5.

6. 'Elizabeth at Woodstock', ed. Manning, pp. 146–54.

CHAPTER 24

1. *History of Parliament, 1509–1558*, ed. Bindoff, I, pp. 408–9; *DNB* sub 'Bedingfield'; 'Elizabeth at Woodstock', ed. Manning, p. 173.

2. 'Elizabeth at Woodstock', ed. Manning, pp. 133–231.

3. 'Elizabeth at Woodstock', ed. Manning, pp. 157–9.

4. *King's Works*, ed. Colvin, IV, pp. 349–55; 'Elizabeth at Woodstock', ed. Manning, pp. 154, 202.

5. 'Elizabeth at Woodstock', ed. Manning, pp. 159–60.

6. 'Elizabeth at Woodstock', ed. Manning, pp. 160 and n. 'd', pp. 166–7, 169–71.

7. C. Garrett, *The Marian Exiles* (Cambridge, 1938), pp. 55–6, 295–6, 360; Foxe, *Acts and Monuments* VIII, p. 581.

8. 'Elizabeth at Woodstock', ed. Manning, pp. 169, 184; Garrett, *Marian Exiles*, pp. 104–9; *The History of Parliament: The House of Commons, 1558–1603*, ed. P. W. Hasler, 3 vols. (London, 1981), I, p. 538.

9. 'Elizabeth at Woodstock', ed. Manning, p. 216.

10. 'Elizabeth at Woodstock', ed. Manning, pp. 177–9.

11. 'Elizabeth at Woodstock', ed. Manning, pp. 155–6, 171.

12. 'Elizabeth at Woodstock', ed. Manning, pp. 194, 196, 211.

13. 'Elizabeth at Woodstock', ed. Manning, pp. 187–8, 220, 226.

14. 'Elizabeth at Woodstock', ed. Manning, pp. 161, 168–9, 170–1, 172–3; *History of Parliament, 1558–1603*, ed. Hasler, II, pp. 148–151.

15. 'Elizabeth at Woodstock', ed. Manning, pp. 152, 177.

16. G. E. C., *Complete Peerage*, *sub* 'Williams of Thame'; *History of Parliament, 1558–1603*, ed. Hasler, II, p. 77.

17. *Letters and papers of the Verney Family*, ed. J. Bruce, Camden Society 56 (London, 1853), pp. 55–78; 'Elizabeth at Woodstock', ed. Manning, pp. 177, 187, 194, 196, 199, 202, 211, 212, 215–17, and see Chapter 30 below.

18. 'Elizabeth at Woodstock', ed. Manning, p. 208.

CHAPTER 25

1. 'Elizabeth at Woodstock', ed. Manning, pp. 176, 179.

2. 'Elizabeth at Woodstock', ed. Manning, pp. 182–3.

3. 'Elizabeth at Woodstock', ed. Manning, pp. 185–6, 191–3.

4. 'Elizabeth at Woodstock', ed. Manning, pp. 195–6, 202–4.

5. 'Elizabeth at Woodstock', ed. Manning, pp. 161–2, 174, 203.

6. 'Elizabeth at Woodstock', ed. Manning, pp. 172, 175, 213–5.

7. 'Elizabeth at Woodstock', ed. Manning, pp. 218–19; MacCulloch, *Cranmer*, pp. 328–30; *LP* XIX, ii, 688 (p. 407).

8. 'Elizabeth at Woodstock', ed. Manning, pp. 194–6.

9. 'Elizabeth at Woodstock', ed. Manning, pp. 210–12, 215.

10. 'Elizabeth at Woodstock', ed. Manning, p. 217.

11. 'Elizabeth at Woodstock', ed. Manning, pp. 221–2, 224–5.

12. 'Elizabeth at Woodstock', ed. Manning, p. 163.

CHAPTER 26

1. 'Elizabeth at Woodstock', ed. Manning, pp. 213–14.

2. Loades, *Reign of Mary Tudor*, pp. 134–5.

3. 'Elizabeth at Woodstock', ed. Manning, p. 175.

4. *Chronicle of Queen Jane and Queen Mary*, ed. Nichols, pp. 77 and n. 'b', pp. 167–70; Wriothesley, *Chronicle* II, pp. 120–1.

5. Wriothesley, *Chronicle* II, pp. 121–2; *Chronicle of Queen Jane and Queen Mary*, ed. Nichols, pp. 144–5.

6. Foxe, *Acts and Monuments* VI, pp. 557–8; *Chronicle of Queen Jane and Queen Mary*, ed. Nichols, pp. 145–51.

7. The idea of the genealogy, with its demonstration that Philip 'was not a foreigner but a descendant of the house of Lancaster', was dreamed up by Gardiner, with the ulterior intention, so Paget claimed, of giving Philip an hereditary right to the throne: Tytler, *Edward VI and Mary* II, p. 390.

8. See note 6 above.

CHAPTER 27

1. W. Schenk, *Reginald Pole* (London, 1950), p. 128.

2. Schenk, *Reginald Pole*, p. 73.

3. Foxe, *Acts and Monuments* VI, p. 572.

4. Foxe, *Acts and Monuments* VI, pp. 568–71; *Chronicle of Queen Jane and Queen Mary*, ed. Nichols, pp. 154–9; Foxe, *Acts and Monuments* VI, pp. 568–71.

5. G. Nicholson, 'The Act of Appeals and the English Reformation' in *Law and Government under the Tudors*, ed. C. Cross et al. (Cambridge, 1988), pp. 19–30.

6. *Chronicle of Queen Jane and Queen Mary*, ed. Nichols, pp. 161–3; Foxe, *Acts and Monuments* VI, pp. 577–9.

7. *Chronicle of Queen Jane and Queen Mary*, ed. Nichols, pp. 136–70 (pp. 164–5).

CHAPTER 28

1. Froude, Reign of Mary Tudor, p. 165.

2. Tytler, *Edward VI and Mary* II, p. 455.

3. 'Elizabeth at Woodstock', ed. Manning, p. 214.

4. Foxe, *Acts and Monuments* VI, pp. 567, 584; Froude, *Reign of Mary Tudor*, p. 208; *Machyn*, ed. Nichols, pp. 84–5; Tytler, *Edward VI and Mary* II, pp. 468–9.

5. 'Elizabeth at Woodstock', ed. Manning, pp. 225–6.

6. Foxe, *Acts and Monuments* VIII, p. 620; *Machyn*, ed. Nichols, p. 85; Wriothesley, *Chronicle* II, pp. 128; Tytler, *Edward VI and Mary* II, p. 470.

7. Foxe, *Acts and Monuments* VIII, pp. 620–1.

8. Foxe, *Acts and Monuments* VIII, p. 621.

9. Weisener, *Jeunesse d'Elizabeth*, pp. 310–11 and notes.

10. Strickland, *Lives of the Queens of England* III, p. 549; Froude, *Reign of Mary Tudor*, p. 211 and n. 1; *Original Letters*, ed. Ellis, 2nd ser., II, pp. 216–243; *Privy Purse Expenses*, ed. Madden, pp. clxiv–v and notes.

11. Froude, *Reign of Mary Tudor*, p. 214; *Calendar of State Papers, Spanish* XIII, pp. 224–5.

12. *Machyn*, ed. Nichols, p. 92.

13. Froude, *Reign of Mary Tudor*, pp. 166 and n. 2, 214 and n. 1; Loades, *Reign of Mary Tudor*, pp. 217–20; *Calendar of State Papers, Spanish*, XIII, pp. 175, 223–6.

14. Loades, *Reign of Mary Tudor*, pp. 220–7.

15. *Machyn*, ed. Nichols, p. 93; Strickland, *Lives of the Queens of England* III, pp. 553–4; Weisener, *Jeunesse d'Elizabeth*, pp. 328–9 and n. 4.

16. *Machyn*, ed. Nichols, p. 94; Weisener, *Jeunesse d'Elizabeth*, pp. 329 –30; *Calendar of State Papers, Spanish*.

17. Froude, *Reign of Mary Tudor*, p. 220 and n. 2.

CHAPTER 29

1. Weisener, *Jeunesse d'Elizabeth*, p. 333.

2. Collins, *Jewels and Plate*, p. 206.

3. Loades, *Two Tudor Conspiracies*, p. 178; *History of Parliament, 1509–1558*, ed. Bindoff, III, pp. 523–5.

4. Loades, *Two Tudor Conspiracies*, pp. 180–1.

5. Loades, *Two Tudor Conspiracies*, pp. 181–2; J. E. Neale, *Elizabeth I and Her Parliaments*, 2 vols. (London, 1965), I, p. 25.

6. Loades, *Two Tudor Conspiracies*, pp. 182–3; *Verney Papers*, ed. Bruce, p. 62.

7. Garrett, *Marian Exiles*, pp. 210–13; Perry, *Word of a Prince*, p. 86 (not included in *CW*).

8. Loades, *Two Tudor Conspiracies*, pp. 183–4; Neale, *Elizabeth I and Her Parliaments I*, pp. 25–6.

9. Loades, *Two Tudor Conspiracies*, pp. 184–6; Froude, *Reign of Mary Tudor*, p. 260.

10. C. Read, *Mr Secretary Cecil and Queen Elizabeth* (London, 1965), p. 108.

11. Loades, *Two Tudor Conspiracies*, pp. 184–5.

12. Garrett, *Marian Exiles*, pp. 147–9; Loades, *Two Tudor Conspiracies*, pp. 186–8.

CHAPTER 30

1. Froude, *Reign of Mary Tudor*, p. 261.

2. *LP* XI, 1217/20; XII, ii, 157; *LP* XIII, i, 646/36, 1115/69.

3. I. Arthurson, *The Perkin Warbeck Conspiracy, 1491–1499* (Stroud, 1994), pp. 123, 195; *LP* I, i, 485/7, 563/8.

4. *LP* XIII, i, 190/25,1520 (p. 574). Incidentally, John Williams, later Lord Williams of Thame, whose attitude to Elizabeth was so curiously favourable, had the reversion of Ashton's lands in the right of his wife. For Williams's other dealings with Elizabeth, see Weisener, *Jeunesse d'Elizabeth*, p. 308 and n. 1 and *Calendar of State Papers, Spanish XIII*, pp. 145, 148.

5. *LP* X, 139; XII, i, 167, 1208; XIII, i, 245, 284; XIV, i, 223.

6. *LP* XIX, i, 273 (p. 152), 799; *Acts of the Privy Council*, ed. Dasent, II, pp. 120, 133; Garrett, *Marian Exiles*, pp. 74–5.

7. Loades, *Two Tudor Conspiracies*, pp. 188–9, 207; Froude, *Reign of Mary Tudor*, p. 260.

8. Weisener, *Jeunesse d'Elizabeth*, p. 339.

9. Loades, *Two Tudor Conspiracies*, pp. 190–8, 202–4; Froude, *Reign of Mary Tudor*, p. 268, n.3.

10. Loades, *Two Tudor Conspiracies*, pp. 221, 223–4.

11. *Verney Papers*, ed. Bruce, pp. 59, 66–7, 75.

12. Loades, *Two Tudor Conspiracies*, p. 223; Froude, *Reign of Mary Tudor*, p. 268, n. 3.

13. Loades, *Two Tudor Conspiracies*, p. 226; Collins, *Jewels and Plate*, pp. 206–7; Weisener, *Jeunesse d'Elizabeth*, p. 341; Foxe, *Acts and Monuments* VIII, p. 622; *History of Parliament, 1509–1558*, ed. Bindoff, II, pp. 443–4, III, pp. 19–20.

14. Collins, *Jewels and Plate*, pp. 206–7.

15. *DNB sub* 'Dormer'; Weisener, *Jeunesse d'Elizabeth*, pp. 342–3 and n. 3.

16. Weisener, *Jeunesse d'Elizabeth*, pp. 340, n. 3, p. 343.

17. Weisener, *Jeunesse d'Elizabeth*, pp. 343–4.

CHAPTER 31

1. Weisener, *Jeunesse d'Elizabeth*, p. 345; Perry, *Word of a Prince*, p. 110.

2. Loades, *Two Tudor Conspiracies*, pp. 265–7; *Verney Papers*, ed. Bruce, p. 73.

3. Loades, *Two Tudor Conspiracies*, p. 225.

4. Strype, *Ecclesiastical Memorials* V, p. 75.

5. *CW*, pp. 43–4.

6. *DNB sub* 'Warton'; Weisener, *Jeunesse d'Elizabeth*, pp. 300–1, n. 1.

7. Collins, *Jewels and Plate*, pp. 206–7.

8. Perry, *Word of a Prince*, pp. 108–9 (not in *CW*).

CHAPTER 32

1. *CW*, pp. 43–4.

2. Weisener, *Jeunesse d'Elizabeth*, pp. 349–50.

3. *Machyn*, ed. Nichols, p. 120; Weisener, *Jeunesse d'Elizabeth*, p. 350.

4. *Machyn*, ed. Nichols, p. 120; Weisener, *Jeunesse d'Elizabeth*, p. 351 and n. 1, p. 352.

5. Weisener, *Jeunesse d'Elizabeth*, pp. 295, 301–2. It seems to have been Paget who first thought of the Savoy marriage. His intention was to make Elizabeth as unpopular as Mary by giving her a foreign spouse, thus paving the way to exclude her from the succession: Tytler, *Edward VI and Mary* II, p. 367.

6. Weisener, *Jeunesse d'Elizabeth*, pp. 302, 303 and n. 4, p. 304 and n. 2; *Calendar of State Papers, Spanish*, XIII, p. 293.

7. Weisener, *Jeunesse d'Elizabeth*, pp. 304, n. 1, p. 351–2.

8. Weisener, *Jeunesse d'Elizabeth*, pp. 351, n. 1, p. 352.

9. G. E. C., *Complete Peerage, sub* 'Sussex'; Tytler, *Edward VI and Mary* II, pp. 108–9 (*Calendar of State Papers, Domestic: Edward VI*, ed. Knighton, p. 624.)

10. Weisener, *Jeunesse d'Elizabeth*, pp. 352–3.

11. Weisener, *Jeunesse d'Elizabeth*, pp. 357–8.

12. G. E. C., *Complete Peerage, sub* 'Northampton'.

13. Weisener, *Jeunesse d'Elizabeth*, p. 358.

14. Weisener, *Jeunesse d'Elizabeth*, pp. 354–6; Strype, *Ecclesiastical Memorials* VII, pp. 268–70; *Calendar of State Papers, Spanish*, XIII, p. 293.

CHAPTER 33

1. *Original Letters*, ed. Ellis, 2nd ser., II, pp. 218–243 (unpaginated).

CHAPTER 34

1. Weisener, *Jeunesse d'Elizabeth*, pp. 356–60; Loades, *Reign of Mary Tudor*, p. 369.

2. Loades, *Reign of Mary Tudor*, pp. 371–3.

3. Loades, *Reign of Mary Tudor*, pp. 375–9.

4. *Privy Purse Expenses*, ed. Madden, pp. clxxxv–cci.

CHAPTER 35

1. *Machyn*, ed. Nichols, pp. 166–7.

2. *Original Letters*, ed. Ellis, 2nd ser., II, 218–243 (unpaginated).

3. Loades, *Reign of Mary Tudor*, p. 390; *Calendar of State Papers, Spanish*, XIII, pp. 385, 389, 416–17, 438–9.

4. J. E. Neale, 'The Accession of Queen Elizabeth I' in *The Age of Catherine de Medici* (London, 1963), pp. 131–44 (p. 135). Markham was the brother of Isabell Markham, one of Elizabeth's ladies, who was shortly to marry John Harington the elder (R. Hughey, *John Harington of Stepney: Tudor Gentleman*, (Columbus: Ohio, 1971), pp. 243–4, notes 327–8).

5. Neale, 'The Accession of Queen Elizabeth I', p. 135; Perry, *Word of a Prince*, p. 123 (not in *CW*).

6. *Privy Purse Expenses*, ed. Madden, pp. cci–ccv.

7. Loades, *Reign of Mary Tudor*, pp. 391–2; *Calendar of State Papers, Spanish*, XIII, p.440; S. Adams and M. J. R. Salgado, 'The Count of Feria's Despatch to Philip II of 14 November 1558' in *Camden Miscellany 28* (London, 1984), pp. 328–44. For Feria's earlier visit to Elizabeth, which left

both of them 'very much pleased,' see *Calendar of State Papers, Spanish,* XIII, pp. 387, 390,395,400.

    8. *Calendar of State Papers, Spanish* XIII, pp. 437–8; Weisener, *Jeunesse d'Elizabeth,* p. 372.

    9. Lingard, *History of England* V, p. 258 and n. 4, p. 259.

    10. Strickland, *Lives of the Queens of England* IV, p. 129. Sandys's description of the message borne by the privy councillors corresponds closely to the account given by the Spanish agent: *Calendar of State Papers, Spanish* XIII, pp. 437–8.

    11. Adams and Salgado, 'Feria's Despatch', pp. 328–9.

    12. Adams and Salgado, 'Feria's Despatch', pp. 329, 338 notes 7 and 8; G. E. C., *Complete Peerage, sub* 'Lincoln' (VII, p. 693, n. 'a').

    13. Adams and Salgado, 'Feria's Despatch', pp. 329–331, 334.

    14. Adams and Salgado, 'Feria's Despatch', p. 331.

    15. Adams and Salgado, 'Feria's Despatch', p. 332.

    16. Adams and Salgado, 'Feria's Despatch', p. 332.

    17. Adams and Salgado, 'Feria's Despatch', p. 332.

    18. Adams and Salgado, 'Feria's Despatch', p. 331.

CHAPTER 36

    1. Adams and Salgado, 'Feria's Despatch', pp. 334–5; Weisener, *Jeunesse d'Elizabeth,* pp. 380–1.

    2. Schenk, *Reginald Pole,* p. 125.

    3. Foxe, *Acts and Monuments* VI, p. 704 and n. 1.

CHAPTER 37

    1. Loades, *Reign of Mary Tudor,* pp. 391–2.

    2. Schenk, *Reginald Pole,* p. 155–6.

CHAPTER 38

    1. R. Holinshed, *Chronicles of England, Scotland and Ireland,* ed. H. Ellis, 6 vols. (London, 1807–8), IV, p. 155.

    2. Strickland, *Lives of the Queens of England* IV, pp. 133–4.

    3. Read, *Cecil and Queen Elizabeth,* pp. 17–19, 63–5.

    4. Read, *Cecil and Queen Elizabeth,* pp. 25–30; Ascham, *Scholemaster,* ed. Arber, p. 17.

    5. Read, *Cecil and Queen Elizabeth,* p. 122; *Tudor Royal Proclamations,* ed. Hughes and Larkin, II, pp. 99–100.

    6. Public Record Office: SP12/1/3.

    7. *Calendar of State Papers, Spanish: Elizabeth I,* ed. M. A. S. Hume, 4 vols. (London, 1892–9), I, p. 2; *Proceedings in the Parliaments of Elizabeth I,* ed. T. E. Hartley, 3 vols. (Leicester, 1981), I, pp. 7–11 (p. 11).

    8. Sir R. Naunton, *Fragmenta Regalia or Observations on Queen Elizabeth, Her Times and Favourites,* ed. J. S. Cerovski (Washington D.C., 1985), p. 40. Naunton gives the quotation in a variant form: 'A domino factum est illud, et est mirabile in oculis meis'.

9. *CW*, pp. 51–2; Adams and Salgado, 'Feria's Despatch', p. 329.

10. *Proceedings in Parliaments*, ed. Hartley, I, p. 11. Montagu says that he heard the Queen deal with this theme 'twice' and mentions her specific injunction to counsel 'as they in conscience thought'. Nothing of the latter appears in the speech in *CW*, pp. 51–2. But there *is* such a reference in *CW*, p. 51. This other speech is supposed to have been addressed by Elizabeth to Cecil when he was sworn of the Council. But it is entirely unspecific and may well have been used to other newly sworn councillors as well.

11. *CW*, p. 52.

12. Public Record Office: SP12/1/4. Feria (*Calendar of State Papers, Spanish: Elizabeth I*, I, p. 2) gives a list of six Marian councillors who were deputed to go to Elizabeth at Hatfield. I have subtracted the names of the three who were, according to Cecil's notes, in London on the 18th.

13. *Tudor Royal Proclamations*, ed. Hughes and Larkin, II, p. 99; *Acts of Privy Council* VII, pp. 3–4; *CW*, p. 51 (but see note 10 above).

14. Calendar of State Papers, Spanish: Elizabeth I, I, p. 2.

15. *History of Parliament, 1509–1558*, ed. Bindoff, III, pp. 450–5.

16. *History of Parliament, 1509–1558*, ed. Bindoff, III, pp. 456–7; *Verney Papers*, ed. Bruce, pp. 68–9.

17. *History of Parliament, 1509–1558*, ed. Bindoff, III, pp. 458–60.

18. J. E. Neale, 'Sir Nicholas Throckmorton's Advice to Queen Elizabeth on Her Accession to the Throne', *English Historical Review* 65 (1950), pp. 91–8.

19. Read, *Cecil and Queen Elizabeth*, p. 479, n. 9.

20. Neale, 'Throckmorton's Advice', pp. 93–5, 98.

21. Loades, Reign of Mary Tudor, p. 459.

CHAPTER 39

1. Neale, 'Throckmorton's Advice', pp. 93, 98; *CW*, p. 52.

2. Public Record Office: SP12/1/4; *Acts of Privy Council* VII, p. 4.

3. *Calendar of State Papers, Spanish: Elizabeth I*, I, p. 3.

4. Wriothesley, *Chronicle* II, p. 142.

5. A. Prockter and R. Taylor, *The A to Z of Elizabethan London* (London, 1979), map sections 7 and 8; *Acts of Privy Council* VII, pp. 5–8.

6. Wriothesley, *Chronicle* II, p. 142; *Machyn*, ed. Nichols, p. 180.

7. Wriothesley, *Chronicle* II, p. 142.

8. Strickland, *Lives of the Queens of England* III, p. 581; *Machyn*, ed. Nichols, p.181.

9. *Machyn*, ed. Nichols, p. 181.

10. Strype, *Ecclesiastical Memorials* V, pp. 281–2; *Machyn*, ed. Nichols, pp. 182–4; *The Funeral Effigies of Westminster Abbey*, ed. A. Harvey and R. Mortimer (Woodbridge, 1994), pp. 55–7.

11. Strype, *Ecclesiastical Memorials* VII, pp. 397–413.

12. Strype, *Ecclesiastical Memorials* VII, pp. 408–10.

13. Strype, *Ecclesiastical Memorials* VII, p. 404.

14. *Tudor Royal Proclamations*, ed. Hughes and Larkin, II, pp. 99–100.

15. *Machyn*, ed. Nichols, p. 178; *Calendar of State Papers, Spanish: Elizabeth I*, I, p. 4.

16. Strype, *Ecclesiastical Memorials* VII, pp. 408–10.

17. E. K. Chambers, *The Elizabethan Stage*, 4 vols. (Oxford, 1923) IV, p. 77.

18. P. E. McCullough, *Sermons at Court* (Cambridge, 1998), pp. 11–27; Kisby 'When the King Goeth a Procession'; *Original Letters*, ed. Ellis, 2nd ser., II, pp. 261–3.

19. *Tudor Royal Proclamations*, ed. Hughes and Larkin, II, pp. 102–3; Wriothesley, *Chronicle* II, pp. 142–3.

CHAPTER 40

1. *Machyn*, ed. Nichols, p. 185; Strickland, *Lives of the Queens of England* IV, pp. 144–5.

2. C. H. Cooper and T. Cooper, *Athenae Cantabrigienses*, 2 vols. (Cambridge, 1861), II, p. 498; *DNB sub* 'Dee'; Tytler, *Edward VI and Mary* I, p. 479; Loades, *Two Tudor Conspiracies*, p. 229; *History of Parliament, 1558–1603*, ed. Hasler, I, p. 521.

3. *Acts of Privy Council* VII, pp. 9–10; Williams, *Elizabeth I*, pp. 53–4; J. Arnold, *Queen Elizabeth's Wardrobe Unlock'd* (Leeds, 1988), pp. 52–67, 255–6 (items 1A–6A).

4. L. G. Wickham Legg, *English Coronation Records* (London, 1901), pp. 112, 115.

5. Adams and Salgado, 'Feria's Despatch', p. 331.

6. Wriothesley, *Chronicle* II, p. 143; Wickham Legg, *English Coronation Records*, p. 113; Ives, *Anne Boleyn*, pp. 215–227.

7. *The Quenes Maiesties Passage* . . . , ed. J. M. Osborn, (New Haven, 1960), frontispiece, p. 27; Williams, *Elizabeth I*, p. 54.

8. *The Quenes Maiesties Passage* . . . , ed. Osborn, pp. 61–3.

9. See Chapter 26 above.

10. *The Quenes Maiesties Passage* . . . , ed. Osborn, pp. 31–7; *King's Works*, ed. Colvin, I, pp. 267–8, III, p. 251.

11. The Quenes Maiesties Passage . . . , ed. Osborn, pp. 37–43.

12. The Quenes Maiesties Passage . . . , ed. Osborn, pp. 44–9.

13. The Quenes Maiesties Passage . . . , ed. Osborn, pp. 53–7.

14. The Quenes Maiesties Passage . . . , ed. Osborn, p. 28.

15. *The Quenes Maiesties Passage* . . . , ed. Osborn, pp. 10–11, 60.

16. C. G. Bayne, 'The Coronation of Queen Elizabeth', *English Historical Review* 22 (1907), pp. 650–673 print the texts.

17. Bayne, 'Coronation of Queen Elizabeth', p. 671.

18. Bayne, 'Coronation of Queen Elizabeth', p. 666 and n. 67; A. L. Rowse, 'The Coronation of Queen Elizabeth I', *History Today* 3 (1953), pp. 301–310.

19. Bayne, 'Coronation of Queen Elizabeth', pp. 666–7; *Machyn*, ed. Nichols, p. 673.

20. Bayne, 'Coronation of Queen Elizabeth', pp. 667–8.

21. Bayne, 'Coronation of Queen Elizabeth', pp. 668–70.

22. Bayne, 'Coronation of Queen Elizabeth', pp. 670 and n. 89, p. 671.

23. *Tudor Royal Proclamations*, ed. Hughes and Larkin, II, pp. 104–8 (p. 107).

24. Bayne, 'Coronation of Queen Elizabeth', pp. 666, 670–1.

25. Bayne, 'Coronation of Queen Elizabeth', p. 673.

CHAPTER 41

1. *Proceedings in Parliaments*, ed. Hartley, I, p. 3.

2. *Calendar of State Papers, Venetian* VII, pp. 22–3; Garrett, *Marian Exiles*, pp. 134–6.

3. Read, *Cecil and Queen Elizabeth*, p. 479 n. 9; Neale, 'Throckmorton's Advice', p. 95; Robert Tittler, *Nicholas Bacon: The Making of a Tudor Statesman* (London, 1976), pp. 16–17, 70–1.

4. *Proceedings in Parliaments*, ed. Hartley, I, pp. 33–9.

5. *Proceedings in Parliaments*, ed. Hartley, I, pp. 34–5 and see above, Chapter 9.

6. N. L. Jones, *Faith by Statute* (London, 1982) supersedes all other accounts of the Elizabethan religious settlement.

7. N. M. Sutherland, 'The Marian Exiles and the Establishment of the Elizabethan Régime', *Archiv für Reformations Geschichte* 78 (1987), pp. 253–286 (pp. 269–71) gives an admirably clear summary of these complex events.

8. Sutherland, 'Establishment of the Elizabethan Régime', p. 271; *Proceedings in Parliaments*, ed. Hartley, I, pp. 10, 13, 16–17.

9. Neale, *Elizabeth I and Her Parliaments* I, p. 67; Jones, *Faith by Statute*, p. 74.

10. Jones, *Faith by Statute*, pp. 114–15; *Tudor Royal Proclamations*, ed. Hughes and Larkin II, pp. 109–10.

11. Jones, *Faith by Statute*, pp. 117, 120–3.

12. See Chapter 35 above; Jones, *Faith by Statute*, pp. 123–7.

13. Foxe, *Acts and Monuments* VIII, pp. 679–93 (pp. 687–92); Jones, *Faith by Statute*, pp. 20–2.

14. Jones, *Faith by Statute*, pp. 130–2, 136–7.

15. Jones, *Faith by Statute*, pp. 145–50; *Proceedings in Parliaments*, ed. Hartley, I, pp. 22–3.

16. Jones, Faith by Statute, pp. 136, 158; Neale, *Elizabeth I and Her Parliaments* I, p. 83.

17. *Tudor Royal Proclamations*, ed. Hughes and Larkin, II, pp. 117–32.

18. Neale, *Elizabeth I and Her Parliaments* I, p. 79; Jones, *Faith by Statute*, p. 151.

19. *Tudor Royal Proclamations*, ed. Hughes and Larkin, II, pp. 119, 122.

20. Neale, *Elizabeth I and Her Parliaments* I, p. 83. R. Bowers, 'The Chapel Royal, the First Edwardian Prayer Book, and Elizabeth's Settlement of Religion, 1559', *Historical Journal* 43 (2000), pp. 317–44, which appeared

after the British publication of this book, confirms most of my account of Elizabeth's conservative religious inclinations and their role in the Settlement.

21. *Proceedings in Parliaments*, ed. Hartley, I, pp. 46–51 (pp. 47, 51); Jones, *Faith by Statute*, p. 150.

22. Jones, *Faith by Statute*, pp. 141–44; C. Cross, *The Royal Supremacy in the Elizabethan Church* (London, 1969), pp. 130–1, 133–4.

CHAPTER 42

1. *Machyn*, ed. Nichols, p. 200; Wriothesley, *Chronicle* II, p. 145; P. Collinson, *Archbishop Grindal, 1519–1583: The Struggle for a Reformed Church* (London, 1979), p. 91.

2. Wriothesley, *Chronicle* II, p. 145.

3. Wriothesley, *Chronicle* I, p. xiv, II, p. 146; *Machyn*, ed, Nichols, pp.207–8.

4. McCullough, *Sermons at Court*, pp. 42–9.

5. *Machyn*, ed. Nichols, pp. vii–xiii.

6. Prockter and Taylor, *A to Z of Elizabethan London*, map section 21.

7. *Machyn*, ed. Nichols, pp. 226–8.

8. *Machyn*, ed. Nichols, pp. 226, 229; Garrett, *Marian Exiles*, p. 80; Read, *Cecil and Queen Elizabeth*, p. 134.

9. Collins, *Jewels and Plate*, pp. 5–6, 24–6, 160–3, 307–316.

10. *Machyn*, ed. Nichols, p. 230.

11. *Machyn*, ed. Nichols, p. 232.

12. See Chapter 39 above; *Machyn*, ed. Nichols, pp. 228, 247, 272–3.

13. Collinson, *Archbishop Grindal*, pp. 233–65.

14. Kisby, 'When the King Goeth a Procession.'

CHAPTER 43

1. C. Cross, *The Royal Supremacy in the Elizabethan Church* (London, 1969), pp. 130–1, 133–4.

2. *Machyn*, ed. Nichols, pp. 200–1; *DNB sub* 'Bonner'.

3. *Machyn*, ed. Nichols, p. 203; *DNB sub* 'Heath' and 'Thirlby'.

4. C. Sturge, *Cuthbert Tunstal: Churchman, Scholar, Statesman, Administrator* (London, 1938), pp. 316–17, 320.

5. R. McEntegart, 'England and the League of Schmalkalden: Faction, Foreign Policy and the English Reformation' (unpublished Ph.D. dissertation, Cambridge, 1992), pp. 258–261; Sturge, *Cuthbert Tunstal*, pp. 322–7.

6. Sturge, *Cuthbert Tunstal*, pp. 329.

7. *Machyn*, ed. Nichols, pp. 235, 237–8, 256, 266.

8. *DNB sub* 'Heath' and 'Boxall'.

9. *DNB sub* 'Scot' and 'Watson'; Cooper and Cooper, *Athenae Cantabrigienses*, I, p. 494.

10. *DNB sub* 'Thirlby'.

11. *Proceedings in Parliaments*, ed. Hartley, I, p. 36.

Index

About the author

About the book

Read on

Insights,
Interviews
& More . . .

Profile of a "Resolute Executioner"

by Daniel Snowman

"Being told you have disappointed David Starkey must be akin to the feeling early Christians experienced when they found themselves being eyeballed by a pride of lions. This is a man for whom the phrase 'doesn't suffer fools gladly' could have been coined. The 'rudest man in Britain,' as the Daily Mail *called him, is famed for the viciousness of his putdowns, such as when he said of the former Archdeacon of York on* The Moral Maze: *'Doesn't he genuinely make you want to vomit—his fatness, his smugness, his absurdity?' "* —The Independent *(London)*

"Openly gay, stupendously cross (his early radio performances had him splashed across the tabloids), unrepentantly right-wing, he represents the coming together of several highly respectable archetypes not often found in the same personality. We have had plenty of waspish Tory academics before, but rarely those who turned incandescent on the airwaves or lived with their male partners in north London bliss. Or, for that matter, turned in viewing figures ahead of Ali G." —Independent on Sunday *(London)*

"FORGIVE ME . . ." Words that can strike terror into the hearts of witnesses on the BBC Radio Four's *Moral Maze* as Dr. David Starkey stalks his next mouse in preparation for the kill. Like Henry VIII or Queen Elizabeth, Starkey can be a resolute executioner when he thinks fit. Like them, too, he strives to clothe a combative personality in the correct outward forms. Immaculately garbed and groomed, formidably and forcefully articulate, Starkey can appear armed cap-a-pie to those who fear him. Yet the apparel probably conceals as much of the man as it proclaims. Despite his public prickliness, Starkey can be a kind and conscientious colleague and teacher sensitive to the person beneath the posture, deeply conscious of the historical impact of human nature with all its frailties, of the all-important role of individual quirk and nuance, preference and prejudice. To Starkey, the pageantry of the past is prelude to the real drama—the intersection of power and personality.

Like Thomas Cromwell and Cardinal Wolsey, Starkey rose from the ranks. Born in Kendal, the only child of working-class Quakers, Starkey learned—especially from his mother—the value of education and self-improvement. Overcoming the childhood disability of a double club foot, he developed into a bookish lad who soon came to love history and enjoyed nothing more than to curl up with Arthur Mee's *Children's Encyclopaedia* or the Quennells' *History of Everyday Things in England* or to gaze at books depicting the gods of Ancient Greece. From Kendal to Cambridge may have been a long jump socially and geographically. But intellectually, says Starkey, he had been well prepared by excellent teachers at the local grammar school. When he took up his scholarship at Fitzwilliam in 1964, he already knew as much history as some of his supervisors.

Starkey remembers himself as clever, bumptious, and doubtless difficult to manage, and he pays tribute to the mature understanding of his director of studies, the kindly Leslie Wayper, who patiently taught his young charge how to organize material, marshal arguments, and write with clarity. Starkey also discovered starrier figures in the Cambridge ▶

> 66 Overcoming the childhood disability of a double club foot, [Starkey] developed into a bookish lad who soon came to love history. 99

galaxy such as Denis Brogan, Moses Finley, Jack Plumb. It was inevitable that at some stage the young comet would come blazing into the orbit of Geoffrey Elton.

Elton had not only mastered the minutiae of Tudor history and refashioned the subject—an astonishing achievement for a refugee from Hitler's *Mitteleuropa*—but, by sheer industry and force of character, had all but obliterated those who tried to gainsay him. Admired and feared in equal measure, Elton would enthrall an undergraduate audience in the morning, enjoy a genial pub lunch—and then hasten back to Clare to mow down the latest wrong-headed hack with a rebarbative review. Once David Starkey opted for the Tudor Special Subject in his third year, the immovable Elton and the irresistible Starkey soon found themselves playing a deadly game of father and son.

For his postgraduate research, Starkey staked out core Elton territory: the Tudor bureaucracy. To Elton, with his continental background and influenced perhaps by the theories of Max Weber, Tudor England was best understood—celebrated indeed—as the fountainhead of the emerging liberal bureaucracy from which Britain had benefited ever since. Elton wrote and lectured with unconcealed enthusiasm about what he saw as the administrative institutions of Tudor government (and its great innovator Thomas Cromwell) and tended to pass over with disdain such vulgarities as the religious or personal passions of the time or the shenanigans of court life. The revolution of which Elton wrote was essentially a conceptual one: the creation of an efficient and essentially benign bureaucracy. To Starkey, by contrast, the institutions of Tudor government were constantly shifting in response to the individuals and circumstances around them. Political change in general, says Starkey, tends

66 Political change in general, says Starkey, tends to be erratic, episodic, unpatterned. Real history means real people interacting with each other at specific times and places. 99

to be erratic, episodic, unpatterned. Real history means real people interacting with each other at specific times and places. For a while, as Starkey went on to do postgraduate work under Elton, the two men with their radically different views of the way the world wags lived out a kind of heightened intellectual symbiosis, elks whose entangling antlers presumably sharpened each.

The subject of Starkey's thesis, the King's Privy Chamber, was not one of the great offices or institutions of state but the informal body of advisers around the monarch (a bit like today's Downing Street staff). What emerged was a dissertation that, far from distinguishing between court and government as Elton had done, clamped them firmly together. Courtiers were councilors under Henry VII and Henry VIII, argued Starkey.

The relationship went from sour to bitter. Starkey got his doctorate but was denied a job at Cambridge. Soon, however, he obtained a post at the London School of Economics (where he was to stay until 1998) and, as both men grew in prominence, they came to criticize each other in print with the kind of vehemence that, a generation earlier, had characterized Trevor-Roper's attacks on Tawney. There are indications of what was to come in Starkey's first solo book, *The Reign of Henry VIII* (1985), in which he writes of the "great game of politics" over which Henry presided. It is an engaging volume, well illustrated, with chapters on the main political events and personalities of the age, all expressed with Starkey's characteristic combination of the lapidary and the colloquial. Cognoscenti doubtless noted the emphasis on the "small group of intimates and personal attendants" who "acted as the land-line between the King and the formal machinery of government," men who "were supremely well-placed to rig politics and patronage to their own benefit." And they ▶

66 Starkey's first solo book, *The Reign of Henry VIII* (1985), . . . is an engaging volume, well illlustrated, with chapters on the main political events and personalities of the age, all expressed with Starkey's characteristic combination of the lapidary and the colloquial. 99

would certainly have spotted Starkey's references to, and avowed distancing from, the work of Elton in a brief bibliographical appendix.

But the real assault came in the next two books, both of them edited and introduced by Starkey: *Revolution Reassessed: Revisions in the History of Tudor Government and Administration* (1986, coedited with Christopher Coleman) and *The English Court from the Wars of the Roses to the Civil War* (1987). Not the sort of titles to appeal to the general reader, perhaps. But scarlet rags to the Elton bull. Starkey's young toreros, among them star performers such as John Guy (on the Privy Council) and Kevin Sharpe (on the court and household of Charles I), set out to question and reassess the Elton legacy. Elton, provoked, duly charged.

In particular, he charged Starkey with errors of fact. Starkey, instead of turning aside on an elegant heel, held his ground. Elton's criticisms of *The English Court*, he showed in a point by point rebuttal of his former mentor's review, were themselves misinformed. In a prominent footnote that dropped the academic courtesies, Starkey accused Elton of bullying some of the younger contributors "for no worse crime than agreeing with me."

Starkey looks back on his protracted battles with Elton with profound and mixed feelings. Fully aware at the time of the psychological implications of what he was doing, he found as he tore into Elton that he was reliving some aspects of his relationship with his mother, the other defining mentor whom he had outgrown. Soon afterwards, Elton's health deteriorated and there is no doubt that he was deeply wounded by the broadsides he received from Starkey and others towards the end of his life. Yet Starkey's sadness is mitigated by a sense of intellectual satisfaction. He and his colleagues had argued persuasively for a fluid, flexible interpretation

> 66 Starkey looks back on his protracted battles with [Geoffrey] Elton with profound and mixed feelings. . . . He found as he tore into Elton that he was reliving some aspects of his relationship with his mother, the other defining mentor whom he had outgrown. 99

of the various Tudor and early Stuart courts, showing how a subtly different pattern of government had emerged under each. Thus (for example), Henry VII, Elizabeth, or Charles I retained a certain distance from their councilors, while Henry VIII and James I were more "participatory" monarchs. Elton's somewhat static, institutional interpretation of the Tudor and Stuart administrations was no longer adequate.

Not only are personalities important, says Starkey. So are places and objects. You can't understand Henry VIII, for example, unless you know where and how he lived, whom he controlled, what he owned. As historical consultant to the Greenwich Exhibition to mark the 500th anniversary of Henry's birth (1991), Starkey clustered the display around a series of specific dates and events, many of them linked to Greenwich Palace: a particular joust or reception, the christening of Elizabeth, the marriage to Anne of Cleves. The exhibition contained a cornucopia of authentic and evocative objects—clocks and scientific instruments, books and manuscripts, gold and silver plate, armor and weaponry, musical instruments, tapestries, maps, paintings, and sculpture. In a richly illustrated catalogue, Starkey emphasized the importance of these objects—the real things, not smoke-and-light reproductions—to a true understanding of King Henry, the man who "built and accumulated more than any other English king." It was the work on the exhibition that kick-started Starkey into what was to become one of his most important contributions to scholarship: the annotating, indexing, and publication of the massive *Inventory of Henry VIII*, a vast project of which Starkey was and remains overall editor.

Starkey was working on a major ▶

> ❝ [It has] become one of [Starkey's] most important contributions to scholarship: the annotating, indexing, and publication of the massive *Inventory of Henry VIII*, a vast project of which Starkey was and remains overall editor. ❞

biography of Henry VIII that would reflect the crucial links between the private aspects of Henry's life—his marital woes, his insatiable acquisitiveness, the membership of his Privy Chamber—and the history of England and its neighbors during his reign. In the course of his researches in the Royal Archives, Starkey extended his interests to encompass the history of the monarchy as such. Indeed, for a while he was considering writing a history of the English monarchy. Neither project materialized. But the preliminary work had important consequences.

Britain at the time was much exercised with its monarchy and, in particular, with the evidently deteriorating relationship between its heir to the throne, Prince Charles, and his wife Princess Diana. Was a royal divorce possible? Could a divorced Prince Charles remarry? If so, who would eventually become queen, his first or second wife? Should Charles perhaps renounce (or be denied) his right to the throne? These were juicy tabloid issues. But they were also major constitutional ones, and everybody wanted to know the precedents. Enter David Starkey, whose knowledge of the Tudors, augmented by a little prepping on Queen Caroline and Mrs. Simpson, led to his being taken up by the media and re-branded as one of Britain's leading constitutional experts. Always ready with a pithy yet highly informed answer to a complex question, Starkey clearly enjoyed the frisson he could create with the forcefulness and clarity of his views. As he strayed ever further from his academic center of gravity, some found him opinionated and objectionable. But that only increased the media currency of a man who wrote easily for the popular press, had been taken up as one of the regular inquisitors on Radio Four's *Moral Maze* (from 1992), and was soon to have his own commercial radio talk show (on Talk Radio UK in 1995–98). In the aftermath of the death of the Princess of Wales in 1997, Starkey achieved international celebrity as he explained the arcane implications with exemplary and indefatigable lucidity to American television audiences.

Starkey continues to pursue what is, in effect, a double career: academic specialist and media celebrity. He is not the first to attempt this balancing act, nor the first academic to be criticized for trying. The two careers came together in 1998, the year in which, having finally parted company with the LSE, he accepted a visiting fellowship at his old Cambridge college, Fitzwilliam. That year, in addition to publishing *The Inventory*, Starkey wrote and presented a three-part television series on Henry VIII for Channel 4. Then, two years later, his four-part series about Elizabeth for C4 achieved national headlines (and a *Times* editorial) for having achieved higher ratings than the popular comedian Ali G.

Now in his mid-fifties, Starkey radiates a sense of immense energy, ambition, and intellectual power. He is negotiating to buy a house in France, with a lot of library space, to which he can repair from time to time to concentrate on his writing. Meanwhile, he enjoys his periodic bouts of teaching and (especially) working with colleagues and graduate students on shared projects. He sits on the Blue Plaque Committee, helping to decide which illuminati from the past should be commemorated on the buildings they once inhabited, and is president of the recently founded Society for Court Studies. Anyone inclined to scoff from within the ivory tower would do well to read Starkey's article on "Henry VI's Old Blue Gown" from the Society's excellent journal *The Court Historian* (April 1999)—or to await publication of the volumes of commentary on the Henry VIII inventory, ▶

> ❝ In the aftermath of the death of the Princess of Wales in 1997, Starkey achieved international celebrity as he explained the arcane implications with exemplary and indefatigable lucidity to American television audiences. ❞

a genuinely collegiate enterprise which Starkey speaks about with great enthusiasm. There is another Channel 4 series in discussion, this time about the wives of Henry VIII. Familiar territory, perhaps. But Starkey, with his outspoken views about the court and constitution, will undoubtedly illuminate the past while throwing irresistible sidelong flickers of light upon the present. He also has an ambition to write about the twentieth-century monarchy. Few serious historians nowadays write about the modern monarchy, Starkey says. Those who do tend to miss the point, either building up its supposed constitutional significance or, *par contre*, seeing it as a mere plaything of the nation's powerbrokers. To Starkey, drawing on his intellectual roots in Tudor history, the interesting thing about the monarchy in recent times is the way it has intersected with a quasi-moral debate about the nature of our national identity. "Who are we?" people ask—and are immediately forced, willy-nilly, to address the role of the monarchy.

Is Starkey still a historian? Of course. You don't have to have an academic position or institution, he says, to qualify (and many who do, don't!). He says he is cutting back on the more "tinselly" aspect of his work, preferring to concentrate on a smaller number of large projects. These include not only his television series and the Elizabeth exhibition, but also the publication of the commentaries on the Inventory and, above all, the eventual completion of his biographies of Henry VIII and Elizabeth—the works on which, I suspect, his ultimate reputation as a historian may depend. For these, he will need to give himself time, space, and tranquility. Let's hope his proposed foray to France proves furnished with cloth of gold.

[Interview, January 2001]

66 Few serious historians nowadays write about the modern monarchy, Starkey says. Those who do tend to miss the point, either building up its supposed constitutional significance or, *par contre*, seeing it as a mere plaything of the nation's powerbrokers. 99

10

Epilogue

David Starkey writes:

"You have a monkey mind," my mother (who was single-mindedness itself) told me sternly over half a century ago. Daniel Snowman says much the same, though at greater length and more kindly. It is true, of course. And it has been the bane of my life. But also the blessing, since I can't stand being bored.

That is why, a year after I had spoken with Daniel at the turn of 2000–01, I left *The Moral Maze*. We were debating the same topics; the same phrases echoed in my head; I had the same, stage-managed confrontations with chairman Michael Buerk. So I went.

But equally there is a time to return, and I am now presenting the rather similar *Last Word* on More 4. For this, however, Dr. Rude of *The Moral Maze* has been joined by my other persona of Mr. Nice. It is the classic "good cop, bad cop" routine that panelists tell me it is hard to get the measure of.

That makes it sound as though the media half of my career has overwhelmed the historical. It hasn't. Instead, I just manage to keep the two in balance, with the one fertilizing the other. *Six Wives*, on which I had begun work in 2001, turned into another Channel 4 series, which got even larger audiences than *Elizabeth* and was likewise BAFTA-nominated. It also became a big book—big enough, I hope, to escape the charge of being hastily written which Snowman levels at *Elizabeth*.

But one man's hasty is another man's (or woman's, more likely) lightness of touch. At all events, *Elizabeth* was a number-one bestseller in both hardback and paperback and has sold hundreds of thousands here and abroad. This matters because history belongs not only in the academy, but in the marketplace and on the public stage. That is also why I have become ▶

> ❝ 'You have a monkey mind,' my mother (who was single-mindedness itself) told me sternly over half a century ago. . . . It is true, of course. And it has been the bane of my life. But also the blessing, since I can't stand being bored. ❞

Profile of a "Resolute Executioner" *(continued)*

involved in the Prince of Wales Summer Schools, which are trying to rescue history teaching in schools by reintroducing narrative and content.

The quadricentennial Elizabeth exhibition duly took place at the National Maritime Museum, Greenwich, in 2003, and was both a popular and critical success. I shall also be guest-curating the Antiquaries exhibition at the Royal Academy in 2007 to mark the tercentenary of the Society of Antiquaries. Otherwise 2007 will be devoted to finishing the writing of *Henry VIII* in time for the 500th anniversary of his accession in 2009.

Nor have other projects mentioned by Snowman been forgotten. The edition of *The Inventory of Henry VIII*, after lying fallow for some years, has sprung back to life, and the first volume of the *Commentary* is due for publication next year.

Similarly, the history of the monarchy has become a fifteen-part TV series, broadcast over three years, 2004–06. Underlying it is a broad reinterpretation of the last five centuries of our history, which I see as driven not by the birth pangs of the British state (as yesterday's academic fashion had it), but by Henry VIII's assumption of the Royal Supremacy, the resulting tensions between religion and politics and their eventual resolution (at least before 9/11).

No, I haven't got bored!

Daniel Snowman was born in London and educated at Cambridge and Cornell. He became a lecturer in American Studies at Sussex University at age twenty-four and was later the BBC's chief producer (features, radio). He is the author of many books on social and cultural history, including The Hitler Emigres *(2002). He is currently writing a book about the social history of opera.*

"Profile of a 'Resolute Executioner' " appears under a different title in Historians, *by Daniel Snowman, and is reprinted by permission of Palgrave Macmillan Publishers.*

> 66 2007 will be devoted to finishing the writing of *Henry VIII* in time for the 500th anniversary of his accession in 2009. 99

David Starkey
In His Own Words

On his parents
"My father, Robert, a washing machine factory foreman, was very emotional but it was all bottled up. I had no relationship with him on an emotional level. He was a man in internal exile. My mother, Elsie, was immensely possessive of me and there was no room for anyone else. He adored her but she was a very difficult woman. She was a cut above him socially and once told me that she'd married him out of pity." —*The Times* (London)

"My mother was both monstrous and wonderful. For all her sexual repression, she was profoundly sensual. My father revealed that they had slept together before their marriage, and that she couldn't get enough until I was born." —*The Times* (London)

On growing up with club feet
"The worst thing about it wasn't the pain, which could be appalling at times, but the fact that I was sentenced to being different from everyone else. Flimsy-soled shoes were fashionable while I had to wear hugely heavy boots. And it limited me physically, not that I was ever interested in sport. In fact, I later got out of a whole term of cross-country running because of a bad thumbnail, which I thought pretty good."
—*Daily Mail* (London)

On love
"I don't rate physical fidelity highly, and could hardly take the moral high ground, but it is important that whatever the other person does, you remain the center of their life. I am with the Prince of Wales on love—'whatever that may be.' The difficulty with love is that it has been contaminated with the soupiness of romance." —*The Times* (London)

> ❝ I later got out of a whole term of cross-country running because of a bad thumbnail, which I thought pretty good. ❞

David Starkey: In His Own Words *(continued)*

On his intelligence

"I am very intelligent. . . . Although, having said that, it isn't a Mensa intelligence. Or a Mastermind intelligence. What I have is a very strong analytical intelligence plus a very powerful verbal intelligence. My mother always used to say to me: 'David, that tongue will be the ruin of you.'" —*Daily Mail* (London)

On being called "the rudest man in Britain"

"I'm not the rudest man in Britain. I'm the plainest-speaking."

On being a historian

"What I have is high malice. . . . It is something almost inherent in the profession of historian. Most academics are people who have sacrificed the chance of action. We have an intellectual power but an impotence in practical affairs. It's this frustration which is the real source of the bitchiness, the sharpness and the malice." —*Daily Mail* (London)

"What has been a profound influence on me and makes me a social historian is the relation between things and ideas, how buildings form power structures like Whitehall. I like things: buildings and furniture, but what has given me most pleasure recently is a silver fish slice from the 1790s, it's pure sculpture." —*The Guardian* (London)

❝ My mother always used to say to me: 'David, that tongue will be the ruin of you.' **❞**

A Conversation with David Starkey

The following interview, conducted by Philippa Sheppard, appeared in the Globe and Mail *(Canada), February 17, 2001.*

DAVID STARKEY arrives for tea looking the perfect English gentleman, even to the starched handkerchief in his jacket pocket. The tortoiseshell glasses suggest the scholar, but time in front of the camera, presenting Channel 4's *This Land of England* and *Henry VIII*, is evident in his polished grooming—atypical of a Cambridge don. He is passionate when talking about his fifth book on Tudor history, *Elizabeth: The Struggle for the Throne.*

PHILIPPA SHEPPARD: You focus on Elizabeth before she became queen, claiming that she appeals to you most as a young woman. But were you also driven to find an original academic niche?

DAVID STARKEY: No, because Elizabeth is not in any sense an academic book; in fact, it was written to accompany a television series. This book is unashamedly popular. Popular history has got it right and academic history has got it wrong. Academic history loves abstract nouns, isms and ologies; people love color, stories. The royal story starts before the subject gets to the throne; youthful experience is the key thing. With Elizabeth, it's even better because her formative years were so bizarre.

P.S.: You often cite Continental sources. Were you consciously seeking an outsider's perspective?

D.S.: No. England at this period was a secondary power caught like a bone between the dogs of ▶

About the book

❝ This book is unashamedly popular. Popular history has got it right and academic history has got it wrong. Academic history loves abstract nouns, isms and ologies; people love color, stories. ❞

15

A Conversation with David Starkey *(continued)*

France and Spain, great powers. So the French and Spanish ambassadors quoted are major players. My concern is to get authentic voices, eyewitness accounts of people who knew and interacted with Elizabeth. Whenever I used a witness, I tried to give a sense of his life as well, so you have interlocking lives. I also make heavy use of two English diarists, the Tower official, who is anonymous, and Henry Machyn, who was a man in the street. We can literally watch how the Queen's policies are responded to, and how she changes people's opinions of her. Machyn starts off hostile, and ends up a fan.

P.S.: Have you gotten into trouble for suggesting that sixteen-year-old Elizabeth enjoyed the sexual abuse of Thomas, Lord Seymour?

D.S.: There's been some sucking in of lips, but when people read the book, it becomes obvious that was the case. One of the keys to my Elizabeth is that the virginity business doesn't come easily. It's pretty clear she had a vigorous sexual appetite.

P.S.: Perhaps she inherited that from her father, Henry VIII, whom you reveal as bigamous, married to both Catherine of Aragon and Elizabeth's mother, Anne Boleyn, for several months.

D.S.: I'm just now doing a short television series and a book on Henry's marriages in England. What we haven't fully grasped is that all Henry's marriages were secret. It's rather unusual, but the circumstances of the marriages are pretty damned awkward. Also, Henry seems to have regarded marriage as fundamentally private— this business of being happy. The key to understanding why Henry marries so often is why we have so many divorces today—because

> 66 One of the keys to my Elizabeth is that the virginity business doesn't come easily. It's pretty clear she had a vigorous sexual appetite. 99

we expect an awful lot; when we don't get it, we behave childishly. With Anne Boleyn, because of the secrecy, we don't really know when it took place. He may even have married her twice.

P.S.: Religion is central to Tudor histories. Your book is unbiased, but where do you stand personally?

D.S.: I was brought up in a poor Quaker family—extreme Protestant. But I very early decided that religion was not for me. On the radio show that I do, *The Moral Maze*, my stance is as radical atheist. But when I'm a historian, that's silly. I try to be reasonably fair, and enter imaginatively into what their feelings are. While I make no bones about Mary [Tudor] being responsible for all those horrible burnings, equally, I try to make clear that she thinks she's saving England.

P.S.: You include a teaser in the book suggesting you might finish Elizabeth's story later. Do you have plans for a sequel?

D.S.: Two, actually. I haven't done enough research to produce serious evidence, but I think my argument will be that Elizabeth gradually withdraws from policy, so that you get an element of the quasi-constitutional during her reign. I also want to challenge the traditional, one-dimensional view that she was invariably cautious. As a young woman, she's an adventuress, a kind of Becky Sharp, on the make. There's a wonderful vulgarity about Elizabeth. As far as men were concerned, she liked rough trade, she liked rascals.

"A Conversation with David Starkey" has been reprinted by permission of Philippa Sheppard. ❧

> ❝ There's a wonderful vulgarity about Elizabeth. ❞

Read on

Have You Read?
More by David Starkey

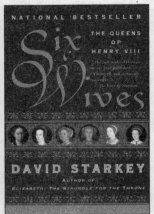

SIX WIVES: THE QUEENS OF HENRY VIII

No one in history had a more eventful career in matrimony than Henry VIII. His marriages were daring and tumultuous, and made instant legends of six very different women. In this remarkable study, David Starkey argues that the king was not a depraved philanderer but someone seeking happiness—and a son. Knowingly or not, he elevated a group of women to extraordinary heights and changed the way a nation was governed.

Six Wives is a masterful work of history that intimately examines the rituals of diplomacy, marriage, pregnancy, and religion that were part of daily life for women at the Tudor Court. Weaving new facts and fresh interpretations into a spellbinding account of the emotional drama surrounding Henry's six marriages, David Starkey reveals the central role that the queens played in determining policy. With an equally keen eye for romantic and political intrigue, he brilliantly recaptures the story of Henry's wives and the England they ruled.

"Eminently interesting. . . . A boon to fans of English royal history, full of murder and mayhem, but also of solid analysis of a maddeningly complicated era."
—*Kirkus Reviews* (starred review)

Excerpt: *Six Wives: The Queens of Henry VIII*

ROYAL WEDDINGS in the early sixteenth century, like royal weddings now, were an opportunity for lavish public ceremony. And none was more magnificent than the first of the century: the marriage, on 14 November 1501, of the Prince and Princess of Wales in St. Paul's Cathedral.

The preparations had been going on for weeks. A great elevated walkway had been erected from the west doors to the steps of the chancel, nearly six hundred feet away. The walkway was built of wood covered in red cloth and trimmed with gilt nails and it stood at head height. In the middle of the nave the walkway broadened out into a stage, several steps high. The wedding itself took place on the upper part of the stage, with the rest of the officiating clergy standing on the lower steps, so as not to obscure the view. After the wedding the young royal couple, both dressed in white, walked hand in hand (the gesture had been arranged beforehand) along the remaining section of the walkway towards the high altar for the nuptial mass. The musicians, placed high up in the vaults for maximum effect, struck up again and the cheers resounded. They became even louder, when, just before entering the gates of the sanctuary, the couple turned to face the crowds, "that the present multitude of people might see and behold their persons." Was there a royal wave? Did they actually kiss?

This was the sort of occasion that was tailor-made for Henry. He was handsome, tall for his age and already with an indefinable star quality (two years earlier Erasmus, on first meeting him, had noticed "a certain royal demeanor" in his bearing). It was a show for stealing, and steal it Henry duly did. All eyes were on him when he escorted the bride along the walkway. All eyes were on him again ▶

> **This was the sort of occasion that was tailor-made for Henry. He was handsome, tall for his age and already with an indefinable star quality. . . . It was a show for stealing, and steal it Henry duly did.**

Read on

Excerpt: *Six Wives:*
The Queens of Henry VIII (continued)

when at the wedding ball he cast his cloth of
gold gown—and his dignity—to the winds
and "danced in his jacket." His father and
mother applauded indulgently.

This was Henry's first experience of a
wedding. But it was not his own. It was the
wedding of his elder brother, Prince Arthur.
Eight years later, it was his own turn. ⌒

Don't miss the next
book by your favorite
author. Sign up now for
AuthorTracker by visiting
www.AuthorTracker.com.